From The Women's Press Ltd
124 Shoreditch High Street, London E1

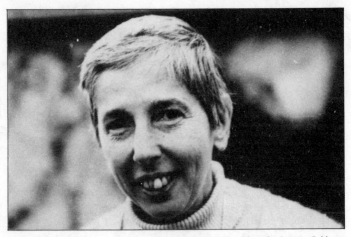
Dr Rosalie Bertell   *Photo by Leonie Caldecott*

Dr Rosalie Bertell was born in Buffalo, New York in 1929; she received her PhD in mathematics from the Catholic University of America, Washington in 1966. Between 1974 and 1978 she worked as Assistant Research Professor at the Graduate School, State University of New York at Buffalo, and between 1970 and 1978 as Senior Cancer Research Scientist at Roswell Park Memorial Institute, Buffalo. She has acted as consultant for the National Council of Churches Energy Task Force and for the Citizen's Advice Committee of the President's Commission on the Accident at Three Mile Island.

She has published over eighty academic papers, addresses and articles in an international range of environmental, peace and health journals and books; she has been called as an expert witness before the United States Congress, and in licensing hearings for nuclear power plants before the United States Nuclear Regulatory Commission. In the international arena Dr Bertell has testified before the Select Committee on Uranium Resources in Australia in 1980, and at the Sizewell Enquiry in Britain in 1984. A member of the Order of Grey Nuns, she now researches low-level radiation as Director of Research of the International Institute of Concern for Public Health in Toronto, Canada, and campaigns internationally against the dangers of nuclear technology.

DR ROSALIE BERTELL

No Immediate Danger

Prognosis for a Radioactive Earth

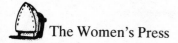 The Women's Press

First published by The Women's Press Limited 1985
A member of the Namara Group
124 Shoreditch High Street, London E1 6JE

British Library Cataloguing in Publication Data

Bertell, Rosalie
  No immediate danger: prognosis for a radioactive earth.
  1. Ionizing radiation.—Physiological effect
  I. Title
  363.1'79   RA569

  ISBN 0–7043–2846–1
  ISBN 0–7043–3934–X Pbk

Typeset by MC Typeset, Chatham, Kent
Printed and bound in Great Britain
by Nene Litho and Woolnough Bookbinding
both of Wellingborough, Northants

To my mother

Acknowledgement

The author is grateful for permission to quote from a large number of books and periodicals, which are listed in the bibliography.

In addition, acknowledgement to W.W. Norton and Company Inc., New York, and Angus & Robertson (UK), London, for the extracts from *Nuclear Disaster in the Urals* by Zhores Medvedev.

Contents

Peace – An Accident at Chalk River, Canada – The International
Commission on Radiological Protection – A Soviet Nuclear Accident – A
US Mistake in Early Nuclear Waste Management – The Commercial
Nuclear Industry – The Arms Race – Bilderbergers – Beginnings of
Dissent

A Full-Blown Commercial Nuclear Technology – The US Federal
Radiation Council – Windscale (Sellafield) – The SL1 Reactor, Idaho
Falls – Big Rock Nuclear Plant – Hearings and Licensings – Formation of
a 'Generic' Argument – National Security and Control of Workers – The
French Experience – The Trilateral Commission – The OPEC Crisis – The
Canadian Experience – Dissenting Voices – Nuclear Proliferation – Living
with Titan II Since 1963 – Rocky Flats, a Nuclear Trigger Factory –
Insuring the Uninsurable – Dr Alice Stewart – Is Peace Desirable? –
Breaking through the Secrecy – Kalkar: A Lesson in Upright Walking –
Creys-Malville – The Alternatives – Media Coverage – West Valley, New
York – Nuclear Workers' Strike – Consensus within the Nuclear
Establishment: The Australian Experience – Opinion Moulders –
Spaceship Earth

Food – The Oceans – Women's Consciousness – The First Stage:
Overcoming Denial of Species Death – The Second Stage: Dealing With
Rage – The Third Stage: Barter – The Fourth Stage: Acceptance of
Reality – On Making Decisions – The Role of Women –
Internationalisation – The Energy Crisis – Human Health

Preface

Prognosis for a Radioactive Generation

This book grew out of a request to me from Patricia and Gerald Mische, authors of *Towards a Human World Order*, to write a global analysis of the nuclear energy era. As I undertook the task, I realised that nuclear energy, both because of its origin and through its implementation, is rooted in war, oppression, secrecy and manipulation: what is sometimes referred to as 'the rule of the fist'. Because of this, I have chosen to highlight the development of nuclear energy policy as a convenience for superpower military strategists, and consider nuclear technology, whether for war or energy, as a single human endeavour.

No Immediate Danger was begun in 1976 and has grown and deepened with my own evolving social analysis. It was influenced by the many people whose lives have touched my own, to whom I have listened and with whom I have stayed in my many travels. I trust I have faithfully recounted their stories and represented their positions. I am grateful for their enrichment of my analysis and their nourishment of my sense of hope.

The call from without for a comprehensive analysis of the dilemmas of the nuclear, post-industrial world was matched by an equally strong call from within, to share with others my understanding of the biological effects of exposure to ionising radiation, an integral part of the technology now being used to produce both nuclear power and nuclear weapons, and the political secrecy which has surrounded each. My most compelling personal reason for undertaking this task is that – at this stage – only scientists are fully aware of the subtle cumulative nature of damage from low-level radiation and of the prolonged waiting time before such damage becomes obvious in an individual, in his or her children, or – as the

American Indians say – in our great-grandchildren's great-grandchildren. I have been grieving for the 16 million casualties already produced by our nuclear industries and weapon-testing and I believe their tragedies must be made visible and be clearly known by everyone.

For the individual exposed to low-level radiation, the damage suffered is of the same kind as and not easily distinguished from that caused by natural background radiation, which none of us can escape. Natural background radiation causes us to age gradually, and increasing that background exposure will accelerate the ageing process. The wear and tear caused by radiation results in the gradual accumulation of mistakes in the body's homeostatic mechanisms; for example, we may no longer be able to produce an antibody to counteract some environmental irritant, so we become 'allergic'; or, we cannot produce usable natural insulin, so we become diabetic; or, our cells multiply without having sense enough to rest, and we get a tumour. The reduced quality of life may not always be life-shortening; and usually only tumours – if they are malignant and cause death – are counted by authorities as being 'of concern' to the victim. This is, however, a value-judgement masquerading as scientific objectivity.

Nor is radiation damage limited to the subtle harm done to the person exposed. It can also injure the ovum and sperm cells from which all future generations derive. This damage may range from severe to mild. Severe damage usually results in infant or childhood death, sterility or institutionalisation of the victim. In these instances, the tragedy is seldom perpetuated since these victims seldom reproduce. With milder damage – such as asthma or allergies, juvenile diabetes, congenital heart defects and sense organ or motor dysfunctions – the individual can live a semi-normal life and perpetuate the damage in succeeding generations. During my research in Buffalo, New York, working on the analysis of the Tri-State Leukemia Survey, I discovered that such slightly damaged individuals are roughly twelve times as susceptible to radiation-related leukaemia as are those of robust health. Radiation can both increase the proportion of slightly damaged people in the population and also make their survival more difficult. What then are the long-term effects of this dual destructive action on the human species?

My frustration with the mindless assurance which automatically follows every nuclear accident or radiation spill, namely, that there is 'no immediate danger', can be quickly grasped. A greater effort is required, however, to learn the unfamiliar jargon, to grasp in detail the human health implications of radiation exposure and to understand nuclear technology. But this is necessary if we are – together – to give visible form and expression to a global consensus now birthing against nuclear options. It is only the full realisation of our shared self-destructive behaviour, whether of Eastern or Western bloc, northern or southern hemisphere, which can adequately move us to change. I have called this change a time to bloom.

Beyond the mainstream of global social and economic disintegration, this change appears to be already happening. A vibrant minority is engaged in birthing a new social order, one which will be sustainable into the foreseeable future and one which will more equitably distribute wealth, power, knowledge and services within the human community.

This book deals with the death-throes of the constricting nation-state society and provides motivation for allowing the new and more fruitful human phase to unfold and come to birth. The present national constraints to life and growth in the developing world – hunger, poverty and repression – and those in the developed world – unemployment, cancer and nuclear threat – are clearly interrelated.

The present dangers to the foetal stage of the new social structure are extreme. Yet we find hope and joy in the growing global consensus among ordinary people that war is as anachronistic as cannibalism, slavery and colonialism. New international relationships based on justice and law must be forged. A new technology in harmony with life and earth, and a new social order which assures a secure future for the world's children must be developed.

Although I can point the way and identify the promising directions, I cannot spell out the nature of this new social order. A mother cannot sketch a picture of her child-to-be-born. We do not plan a flower. It is our part to nourish the good growth and to provide a welcoming environment. We must neither give birth in fear nor abort through cowardice. It is necessary to trust the creativity and vitality of the life process itself.

I would like to express my gratitude to Audrey Mang, who believed in me before anyone else did and encouraged my growth; and to Patricia Mische, who constantly prodded me to write; and to Sister Lois Anne Bordowitz, who took care of numerous requests, giving me time to write. I am appreciative of the School Sisters of

Notre Dame with whom I have lived and my religious congregation, the Grey Nuns of the Sacred Heart, for their moral support during the years of writing. Carol Walker, a blind woman who typed the first draft and became involved in the early stages of the book, deserves special mention. She kept me in touch with the non-scientists who are eager for an intellectual glimpse of the mysterious nuclear world. Other invaluable typing assistance was given by Kathy Brouwer and Joyce Troy. Numerous people read the manuscript – or sections which pertained to them or to knowledge they have – and added to the accuracy or removed what was wrong or misleading. I owe them a great debt and hope they find the book honest and well-balanced.

If you find the book helpful, please share it with a friend. If you find parts unfair or erroneous, please let me know and see that I have access to the truthful story you have to tell. I want this book to inform and to help, rather than to be merely contentious, in the resolving of this tragic dilemma our generation faces.

Rosalie Bertell
Toronto, 1985

Introduction:
'Days Are Not Important
Unless They Are Good Ones'

In the dim night light of a hospital room, seven-year-old Jimmy was remembering the day on which he was told he had leukaemia. He remembered his mother's tears, his father's bewildered anger, the alien feeling of the hospital environment. Then his mind replayed the nausea and diarrhoea caused by radiation therapy and chemotherapy, his hair falling out and kids laughing at him, all the highs and lows over the last eight months' battle with a disease which was now demanding his total attention. Then he knew his answer, and, mentally relieved, fell into a peaceful, refreshing sleep.

Later that morning, when all the hospital ablutions were concluded, Jimmy's mother and father arrived with Dr K. whom Jimmy had learned to love and trust. After the usual greetings and kidding around which had come to be a ritual, helping them all to cope with the tragic situation, Jimmy broke his news with unusual conviction and seriousness. 'I don't want to try the new medicine. It will only give me more days, and I'll die anyway. Days aren't important unless they're good ones.'

The doctor quietly prepared for Jimmy to go home, counselling his parents on supportive medical care and assuring them he would be available for all possible emergencies.

Jimmy died at home, surrounded by familiar objects, loving parents and a younger brother who couldn't understand what was happening. Jimmy died gently, utterly exhausted by having lost so much blood. His tissue had broken down completely, and he was bleeding from every body opening. His bed looked like a battlefield.[1]

1

This story about Jimmy is related to the subjects of national defence, economic development and energy policies. Leukaemia is related to exposure to benzene[2] (a petroleum derivative), micro-wave radiation,[3] X-ray and nuclear fission products (radioactive chemicals emitted from nuclear-related industries).[4] These in turn are part of strategies for national growth and development, as well as advances in the art of war. Energy mix and a weapon strategy inseparably involve human consequences in terms of increased incidence of leukaemia, other cancers, neonatal and infant mortal-ity, mental retardation, congenital malformations, genetic diseases and general health problems.

Omnicide: The Stark Reality of Species Death

The acceptance of the fact of one's personal death is mitigated by the experienced continuity with both the past and the future. For adults this continuity is most obviously linked with biological parenting, but it also occurs because of human memory, culture, literature and scientific endeavours. One can continue to affect history even after one's death. A child survives through the cherishing memories of its parents. Personal death is natural, although it may be premature as Jimmy's was, or violent as happens in war.

The concept of species annihilation, on the other hand, means a relatively swift (on the scale of civilization), deliberately induced end to history, culture, science, biological reproduction and memory. It is the ultimate human rejection of the gift of life, an act which requires a new word to describe it, namely omnicide.* It is more akin to suicide or murder than to a natural death process. It is very difficult to comprehend omnicide, but it may be possible to discern the preparations for omnicide and prevent its happening.

The closest analogous human act which we can find in history on which to ground our thoughts about omnicide would be genocide, the deliberate ending of family lines. Hitler deliberately set out to annihilate all Jewish men, women and children, so that they and their offspring would disappear from history. Hitler also tried so to decimate Poland that it would be lost as a nation and culture, and its surviving people reduced to slave labour. Hitler did not declare war, however, against the earth: plants and animals, air, water and

* The term omnicide was first used by John Somerville in 'Human Rights and Nuclear War', *The Churchman*, 196: 10–12, January 1982.

food. It was the Second Indochina War in Vietnam which first witnessed the extensive wartime use of technological power to devastate the living environment of earth. So even the genocidal plans of Hitler fell far short of the omnicidal plans for a nuclear holocaust, of which Vietnam was only a 'clip'.

The Jewish and Polish people condemned to death by authorities within the Third Reich were carefully 'managed' and deceived, so that they would co-operate with the death plans, at least until it was too late to save themselves. The process is not unlike that of a cancer which lives off the body of its unaware victim in its early stages. It is important to examine the deliberate isolation of the Jewish people and the misrepresentation of 'outside' reality as the precursor of genocide. It will give us clues for understanding omnicide in its early stages.

The Jews were first forced to live in ghettos. These were euphemistically called 'Jewish quarters' or later 'epidemic zones' in order to prevent panic among the Jews themselves or international attention from without. The Jews were told that the special quarters were necessary if the local police were to protect them from 'public prejudice'. The ghettos were later walled or surrounded with barbed-wire fences. Jews were bound by curfews, 'for their own protection', and gradually eliminated from many occupations. Their ability to resist, both biologically and psychologically, was systematically reduced by food rationing and deterioration of living conditions. Eventually TB rates soared and other infectious diseases reached epidemic proportions among the overcrowded, undernourished Jews, making them a threat to even the sympathetic non-Jewish population, and providing further 'reasons' for confining them to the ghettos.

The ghettos were only an intermediate stage in the extermination process. In January 1942, the Nazis met in the office of the International Criminal Police Commission, *Am Grossen Wannsee*, to decide the 'final solution of the Jewish Question', '*Endlösung der Judenfrage*'. After this conference the extermination camps were established, and millions of Jewish men, women and children were sent to their death by 'resettlement orders'. The Jews within the camp were unaware of these 'outside' decisions which sealed their fate.

In his book *Hunger and Disease*,[5] Emil Apfelbaum wrote about the motivation for the Nazi escalation of violence from isolation and slow death from unemployment, disease, malnutrition and starvation, to death camps:

In the view of their inventor, the Warsaw ghetto walls, peppered with broken glass at the top, were meant to serve one and only one aim. The aim was mass murder which was committed by means of mass hunger. That was the sense and the essence of the modest brick-and-glass composition. But the wall-maker must have been slighty disappointed when he learned that his scheme had gone awry. Nurtured by the pathological soil of the ghetto enclosure, it had grown to become one of the pathological paradoxes of life: smuggling, an essentially negative phe-nomenon, was our salvation. That force was in constant motion, around the clock. Smuggling put a brake on mass starvation, slowed down its tempo and made it less all-embracing . . . It was then that hunger was replaced by deportations.

The Warsaw Ghetto, on 16 November 1940, had been sur-rounded by a nine-foot-high wall topped with an extra three feet of barbed wire. Exit points were manned by German Schutzpolizei troops, and internal 'order' was kept by the Ordnungsdienst. The length of the ghetto wall was policed by car and motorcycle patrols. Jews, close to 500,000, were crowded inside. If discovered outside the ghetto, Jews could be shot. Any Pole discovered helping a Jew would also be shot.

The ghetto became a 'self-governing' enclave, and Jewish capitalists were enlisted to establish factories and harness Jewish production potential. Small Jewish-managed manufacturing enter-prises were set up and Jewish firms were given licences to produce consumer goods. The profits went to the Nazi entrepreneurs. Jewish leaders were charged with political control of the ghetto. The Nazis, however, kept control of the health services. They also kept for themselves the power to allocate living accommodation, control food supplies and education, and they managed com-munications within the ghetto and between the ghetto and the rest of the Polish population outside.

In my view, it will be very helpful to allow the Jews themselves, inside the Jewish quarters, as much leeway as possible in running their own affairs: the absolutely correct analysis of the actual state of affairs has type cast the chairman of the Jewish Council as a really loyal worker. Should any shortcomings occur, the Jews will vent their displeasure on the Jewish self-government body rather than grumbling against the German authorities. [Heinz Auerswald, Commissar's Files, Vol. 1, letter dated 24 November 1940.]

What little money was earned in the ghetto was taken away by 'capitation tax' or spent on food. Jewish energy within the ghetto was largely spent in competition for jobs and housing, or on political action against their own leaders. As food and jobs became more scarce, they concentrated on survival within the oppressive system.

Everyone, whether wall-builders, Nazi police, Jewish manufacturers, corporate financiers or community leaders, played a role in the death system. Although their actions ultimately bred death, they also bought a few more days of life, however uncomfortable, for each individual in the ghetto. The vast majority of people quietly resigned themselves to co-operate, and to pretend things were 'normal'. It seemed to be important for the time being to wait out the reign of terror and try to survive. Anyone who tried to sound an alarm was met with disbelief.

19 July 1942: Reichsführer SS Heinrich Himmler ordered the deportations and liquidation of Warsaw Jews to be begun on 31 December 1942.
9 January 1943: Himmler inspected the Warsaw ghetto in person and judged the 'success' of the resettlement operation.
18 January 1943: about 1,200 Jews were shot to death within the ghetto and 6,000 transported to concentration camps. The Jews became aware of the liquidation plans. They *saw* death.
18 to 22 January 1943: the Jewish Combat Organization fought back for the first time.
19 April 1943: SS units, attempting to deport more Jews from the Warsaw ghetto, were met with armed Jewish resistance.

The Jewish choice was similar to the choice of Jimmy, the young boy dying of leukaemia, whose story began our questioning: 'I don't want more days unless they are good ones.' The choice was too late, the death process too far advanced. No one will ever know whether or not it could have been stopped had action come earlier.

The resistance was brutally crushed.

In crushing the Warsaw uprising the Nazis burned buildings, flooded or blew up underground sewers; and machine-gunned women, children and old people. The estimated slaughter in that incident alone was numbered at about 65,000 Jews. Himmler ordered a concentration camp set up in Warsaw to recover 'usable bricks, scrap iron and other debris' from the ghetto ruins. The Nazis eventually converted the levelled Jewish ghetto into a park area. The earth, the trees, the grass, the flowers, were all allowed to live. It was genocide, not omnicide.

Out of approximately 6 million European Jews killed between 1939 and 1945, about 3 to 4 million were put to death in concentration camps, 0.7 million in the ghettos and 1.4 to 2 million were slaughtered in other places. In addition to this number, about 2.7 Polish Jews and an unknown number of Russian Jews were killed. They were killed only because they were Jewish.

Another group of people targeted for death were the Poles. On 22 August 1939, Hitler called for a solution to the 'Eastern question', namely the destruction of Poland and the annihilation of her 'living resources'.[6] During the 1939–45 German occupation of Poland about 6 million Polish citizens were killed, almost 3,000 each day. An additional 2.5 million Poles were driven from their homes and farms, and German families on the western fighting front were moved into the furnished Polish apartments and homes. An older German woman told me how badly she had felt, moving into a Polish home, using all the linen, dishes, silverware, etc. She didn't know who the owners were, where they had gone, or what she could do to help. German families co-operated with their evacuation from the western war zone, hardly comprehending the overall Nazi plan.

A handful of people wanting to depopulate Europe to make room for the race and nationality of their choice caused six years of terror and about 40 million casualties. Countless others were disabled for life because of injuries suffered in combat, in extermination camps, or in cruel pseudo-scientific medical experiments. Millions of victims and victimisers co-operated in silent passivity, ignorant or wilfully blind to the overall violence.

It is difficult to tell what the human species learned from the genocides of the 1940s. We find even today widespread attempts to 'manage public opinion', a euphemism for lying, even as it was managed in the ghetto. Decisions are being made which affect the future of nations and peoples, without their knowledge or consent. Ghetto-like isolation and national liberation struggles exist within an injust international arena which sets the boundaries of their action.

Managing Modern Ghettos

Professor Howard R. Raiffa of the Harvard Business School calls the new public-relations style 'strategic misrepresentation'. Budding negotiators and business managers are taught deceptive tactics designed to enhance 'competitive decision-making'. A massive public relations industry is designed to 'sell' government and industry decisions to the public, spawning general public distrust but also political paralysis or ineffectiveness. Unemployment is blamed on 'cheap foreign competition', while businesses deliberately move jobs to ghetto-like cheap labour pools in the third world. Labour is pitted against environmentalists, men against women, the middle class against the poor, the 'guerillas' against the 'paramilitary', 'right' against 'left', the first world against the third. People expend energy to gain political independence within an internationally controlled ghetto-like environment where national leaders play a difficult if not impossible role.

Peru is a case in point. Peru, after the 1968 liberation struggle, made some attempts at land reform and nationalisation of industries owned by foreign transnationals. These efforts have been effectively weakened, and even reversed, through outside economic pressures and decisions made internationally without Peruvian participation. Income from Peru's agricultural exports increased by 19 percent between 1963 and 1973. However, her agricultural import costs rose by 26 percent and manufactured imports by 42 percent for the same period. With 1973 came the OPEC oil price increase, and billions of dollars from oil-exporting Middle East countries were deposited in European and US banks. This prompted the first-world banks to pursue an aggressive policy of selling loans to third-world countries, and in 1974, with copper and sugar exports selling at high prices and predicted oil finds for the Peruvian jungle territory bordering on Ecuador, Peru began to borrow money. The largest single loan was one billion (US) dollars for a pipeline to be built for the yet-to-be-discovered oil. In 1975, world copper and sugar prices fell, and not enough oil was found to meet even Peru's domestic needs. Peru's expenditure on imported food rose again from \$92 million in 1973 to \$200 million in 1975.[7]

By 1976, Peru's debt had risen to \$3.7 billion, requiring interest payments as high as \$30 million per year. Her trade deficit had grown to \$1.2 billion and her balance of payment deficit to \$1.6 billion.

The economic situation had worsened to such a point in 1978 that Peru, unwilling to accept International Monetary Fund (IMF) assistance because of its harsh conditions, made an agreement with eight private banks. Peru became a 'hostage' of a consortium of US banks[8] which claimed the right to monitor Peru's internal policies for four to five years. The lending conditions required:

devaluation of currency and increase in the price of food;
cut in government expenditures for health, education and other social services, while maintaining military and police expenditures (for internal control);
re-privatisation of some nationalised industries, depreciation and tax concessions for US-owned industries;
abandonment of Peruvian legislation which granted workers a 50 percent ownership in enterprises;
halt in Peruvian land reform and redistribution of farmland, leaving a million families in need of food;
repression of organised labour unions, anti-strike legislation and stiff reprisals against dissent.

For 90 percent of the Peruvian people the cost of such 'development' was exploitation, and for many it came to mean death by malnutrition and starvation. Even the head of the Peruvian central bank admitted that the lives of 500,000 children was Peru's price for the US loans. Today most Peruvians are spending 85 percent of their income on food, while food consumption has halved and malnutrition is widespread. Consumption of milk and protein is going down drastically and is now about 46 percent of the minimum level required for normal health. TB and other infectious diseases are reaching epidemic proportions. Only 40 percent of the Peruvian workforce has full employment, and 30 percent are unemployed. Those who lend money determine how it will be spent and who receives any profit from the enterprises. Much of the profit goes to foreign entrepreneurs. This is precisely the same economic organisation which was designed by the Nazis for 'managing' the Warsaw and other ghettos.

General Juan Alvarado Valasco, who had initiated the Peruvian economic reform after the 1968 coup, was replaced as President by his former Minister of the Economy, General Morales Bermudez, in a 'bloodless coup' in August 1975. Bermudez tried to enforce the 'austerity conditions', imposed on Peru by the banks because of its economic bankruptcy, through harsh military repression of popular resistance. Finally, unable to meet debt payments without further loans, Bermudez agreed to IMF conditions, even harsher than those of the private banks, in October 1977. There were sharp increases in the price of petrol, food and transport, sparking increasingly militant labour union protests against the Peruvian government. By 1980, Peru's total foreign debt had reached 9.3 billion, and inflation was running at 80 percent. Less than 30 percent of the workforce had full-time employment above the minimum wage level. The average hourly wage for industrial workers dropped from $0.62 in 1973 to $0.45 in 1978, making Peru more 'attractive' for foreign investors. The profits from the cheap labour went to transnational corporations abroad. These same transnationals closed non-profitable first-world factories, blaming the resulting unemployment on 'unfair foreign competition'.

The IMF policy of demanding that the Peruvian government reprivatise mining, anchovy fishing and manufacturing has benefitted only the foreign investors, who are now receiving profits at about the 1973 level. It has not helped Peru. If Peruvian workers unite to gain better wages, their plants are shut down and relocated elsewhere, the high unemployment rate assuring replacement with another cheap workforce. Sometimes management lodges legal complaints against workers so that the government, not the company, acts to repress the desperate workers. Well over half of Lima's four million inhabitants live without public water services, sewage systems or electricity. People in the *barrios* (squatter slums) must buy water in 20- to 30-gallon drums from private vendors, and in trying to make a drum last as long as possible they often unwittingly breed typhoid fever and death. Companies like Coca Cola, on the other hand, pay low prices for unlimited supplies of the country's scarce water.

Under IMF conditions, Peru's tariffs on imports were lifted and a substantial number of Peruvian companies went bankrupt, unable to compete with cheaper imports and the local subsidiaries of transnational corporations. This, of course, increased Peru's unemployment and foreign dependency. It seals the ghetto-like helplessness of this once proud nation-state. It also speaks clearly of the integral nature of the global community. Species survival requires national health *and* international health. National liberation struggles must be linked to international liberation. An organ transplant into a sick body can be futile.

Major trade union strikes by workers in the Federation of Peruvian Fishermen in 1976, the general strikes called by Commando Unitario de Lucha in 1977 and 1978, and the strike of the 40,000 copper miners in August of 1978 were all harshly put down by the Peruvian military. Thousands of workers were fired and blacklisted so that they could not get employment again. The Communist Party of Peru heads the most important Trade Union Confederation, hence the strikes could be labelled 'communist infiltration'. Eventually the strikes spread to public sector workers, hospital employees and teachers. The government resorted to beatings and gassing to break up the strikes. The most successful strike was the teachers' strike in 1978, which was backed by parents, students, the labour movement and the Church. American news services tended to present Peru's problems as due to 'unreasonable demands of workers'. The uninformed American public generally saw Peru as ungrateful for and undeserving of further US aid. Communication between the American people and the Peruvian people was non-existent or controlled.

The reader can easily discern that the ghetto system established in Peru without, of course, a deliberate genocidal intention to relocate Peruvians into death camps, has many similarities with the Nazi system. Certainly the bleeding of the people is wilful, and the deaths probably rationalised as helpful because the earth is 'overpopulated'. As the national political leaders find themselves caught between the demands of the external financial establishment and the desperation of the poor of their own country, they resort to internal violence, thinking each financial loan will 'save' them from death. Arrest of labour leaders prior to strikes, military tanks lining highways to make sure workers go to the factories, partial wage concessions for 'key' sectors of the economy and a time-staggered schedule for raising food prices so as to diffuse public opposition, were adopted as policies in Peru and will be elsewhere. Striking workers have been killed and imprisoned; schools occupied by parents and children have been attacked and, in some cases, children have been killed when hit by tear-gas bombs or stray bullets from police bent on 'keeping order'.

The re-election of Fernando Belaunde Terry as President of Peru, 18 May 1980, put an ironic touch to this tragic story. Terry was the Peruvian President who had been ousted by the 1968 military coup. It remains to be seen whether or not the global village will recognise and eliminate the ghetto situation of Peru and so many other third-world nations before the violence against the people there causes even more widespread death from malnutrition and reduces survivors to slave labourers. Thus dehumanised, Peru, just as the Warsaw ghetto, could be used to continue to provide cheap labour for consumer products and for the megaproducts and war-oriented economies of North America and Europe. At least it can be so exploited for a short time, until the mega-ambitions result in mega-death and catastrophic global horror.

There are obvious analogies in the East, as for example the escalation of food costs in Poland after the government's attempt to break the Solidarity union movement, and the use of Polish government officials to discipline the Polish people.

Even more compelling than the ghetto analogy is the analogy between the growth of violence and a human cancer, which takes over the life-processes of the human body, feeding its disordered self at the cost of killing living tissue. Once a cancer destroys a vital organ, the human victim as well as the parasite-cancer dies. So too the violent money/power interests are feeding off the people of the world. The compartmentalisation caused by national sovereignty prevents global mobilisation from counteracting this violence. It delays or destroys the formation of a global infrastructure which would include international labour unions, churches, and peasant organisations which might be able to check the malignant growth and restore local control over the essentials of survival. Only international human solidarity can make possible a new period of human fruitfulness.

Yet there is even more to be learned from Peru. It exemplifies not only the desperate human needs not being provided for in the global village, but also the false purchases which offer a dream of remnant survival to the nation's rulers.

Far from effectively extracting the nation from the death process, Peruvian leaders appear to be clutching at the false hope of joining the international 'nuclear club'. On a 125-hectare site at Huarangal, about 28 kilometres north-east of Lima, the Peruvian government is erecting a 10-megawatt nuclear reactor, purchased from the US for more than \$80 million. So far Peru has borrowed about \$67.7 million from Argentina alone for this research project and training centre. The reactor will not produce electricity; it is merely to be a training site for nuclear engineers and technicians. A meteorological station, radioisotope production laboratory and eleven other facilities are to be added to this project.

The site of the nuclear reactor is at known risk from major earthquakes. A major quake would mean death for hundreds of thousands of people and uninhabitability of the land for more than 100 years. The nuclear project compounds Peru's international debt problems, and does nothing to alleviate the critical internal problems of the people.

From the point of view of global investors, this Peruvian venture will provide new 'skilled' workers for the nuclear age. It will be a stimulus for Peru to develop its unmined uranium for export. It will also put pressure on Peru to meet US demands for cobalt, needed for nuclear weapons. Peru has a potentially rich cobalt vein in the Sur Chico area in Ica, which would be a source of foreign capital to repay foreign investors for the purchased foreign technology. Although Peruvian legislation forbids the export of radioactive concentrates to a foreign nation for weapons production, these laws can be circumvented. The uranium can be sold for the 'peaceful atom programme', regardless of the fact that after being so used the plutonium produced will be extracted for weapons. The cobalt is not radioactive, so not covered by the letter of the law.

A holistic understanding of this national death process must include both the struggle of the Peruvian people for the basic necessities of life, and the struggle of the Peruvian nation, represented by its government, for a dignified place in the family of nations. Any other nation, even the US or USSR, can be substituted for Peru. No people are exempted from the effects of international lawlessness and internal oppression.

Competition for Survival

One key to understanding the priority which a country places on the health consequences of its strategy for national survival, either military or commercial, is the precision of its predictions of such health effects when planning, together with the honest communication of the trade-offs proposed. Its on-going measurement of the accuracy of its predictions once plans are implemented serves as an audit of benefits and losses, whether in health, standard of living, jobs or personal security.

Another key to understanding the implications of national choices between military and technological options is the examination of decision-making within its historical context, noting the immediate national pressures affecting the decisions.

There are two basic fears which can compete for attention in rational choices of national defence systems, national energy strategies and national development plans: the fear of national extinction and the fear of partial or total human species extinction. Unlike past eras when survival of the human species was generally taken for granted, in the present nuclear age both realities are painfully comprehended as possible and imminent. The usual avoidance mechanisms which suppress one or both possibilities from conscious thought seem to be operative at international, national and personal levels in both developed and developing nations. Whether the cry is 'Everyone will freeze in the dark', 'The Russians are coming', or 'Better active today than radioactive tomorrow', the perception of a crisis caused by competing self-interests is apparent in the first-world. Developing countries recognise food, jobs and control over decisions affecting their survival as out of their control. Death from starvation appears more imminent than death from nuclear war. But death is death.

Neither denial nor scare tactics can change the reality of the present crisis of choice and its consequences. Nor can choice be avoided through pretending the current crises are indefinitely sustainable. Internal national disputes between governing and governed, and international manipulation of scarce resources for the economic supremacy of a privileged class, serve only to increase tension and may prevent unified action, making drifting into catastrophic war or ecological disaster inevitable.[9] There must be some agreement on concerted international action in today's interdependent world, to resolve these two basic questions of national and human survival. With the right choices one can hope for a new human consensus and a period of flowering beyond anything ever seen on earth before.

The National Strait-Jacket[10]

This book examines national energy choices in the light of national perceptions of tactics needed for survival economically and militarily in a lawless international arena. A national security mentality has so profoundly affected decision-making with respect to energy policy that it has actually become necessary in many countries for armed guards to enforce national energy policy, sometimes violently, against the protests of unarmed citizens. The growing citizen protest movement against nuclear power and nuclear weapons is a manifestation of the primitive human drive for survival. Nuclear technology has become the symbolic centre of the survival crisis. This movement is drawing attention to the inadequate inclusion of the supra-national dimension of society in national political and social planning, even in the most sophisticated 'developed' countries.

Looking at the energy crisis as a structural crisis in the development of a global community – a crisis which can catalyse a new era of human flowering and unprecedented fruitfulness or, equally, a catastrophic destruction and disintegration of civilisation – can shed new light on creative steps which can be taken now to choose life for future generations. The energy crisis is symptomatic of a deeper crisis. The human ability to expand the psyche from an awareness of personal and national identity and self-interest to identification with the planetary community and concern for its survival and well-being is the basis of hope for a fruitful future. People making diametrically opposite energy choices can share this basic vision of one world free of the age-old plague of war. Perhaps unmasking the historical choices which led to the polarisation between and within nations will help to resolve the divisive trends now being experienced.

The fragile unifying shoots which can grow into a realised human world order are already visible within the human community. This book has been written in the hope of more clearly elucidating these globally unifying trends and of suggesting more responsive new organisational structures for the present which will incorporate the commonalities sought by peoples and further their growth. Once freed from underlying militarism, intertwined with national vested and myopic self-interests, global energy/health questions stand a better chance of being resolved in a way compatible with human survival. This experience can spread to other areas of human life and work. Only then can the people of the world experience, without undue fear of species destruction or national obliteration, a time of unprecedented sharing of goods and services made possible through technological growth. Ours is a time when human values can flourish and justice can become a way of life. What was formerly a utopian dream is now a viable option. It has become viable because in the nuclear age the ways of war and force are no longer tolerable.

PART ONE

The Problem:

Nuclear Radiation and its
Biological Effects

The Seed

The future of humankind is present today within the bodies of living
people, animals and plants – the whole seedbearing biosphere. This
living biosystem which we take so much for granted has evolved
slowly into a relatively stable dynamic equilibrium, with predictable
interactions between plants and animals, between microscopic and
macroscopic life, between environmental pollutants and human
health. Changes in the environment disturb this balance in two
ways: first, by altering the carefully evolved seed by randomly
damaging it, and second, by altering the habitat, i.e. food, climate
or environment, to which the seed and/or organism has been
adapted, making life for future generations more difficult or even
impossible.

Although examples of maladaptation in nature and resulting
species extinction abound, our focus here is on human seed, the
sperm and ovum, and the effect on it and on the human habitat
resulting from increasing ionising radiation in the environment.

The increased use of radioactive materials, which is a direct
outgrowth of the current military and energy policies of the
developed world, provides an opportunity for gauging what priority
these countries give to the health and well-being of individual
citizens, and for gauging governments' understanding of the tension
between individual and national survival. The first indicator of
underlying national priorities is the precision or lack of precision
with which health effects are predicted, and the thoroughness with
which an audit is taken and the predictions checked against reality.
The audit findings should be reported to the person or people
affected, and their participation sought in formulating changes in
policy to remedy any unanticipated problems. The individual's

Glossary

1. **ABCC** Atomic Bomb Casualty Commission.
2. **Alpha particle** an electrically charged (+) particle emitted from the nucleus of some radioactive chemicals, cf. plutonium. It contains 2 protons and 2 neutrons, and is the largest of the atomic particles emitted by radioactive chemicals. It can cause ionisation.
3. **Beta particle** an electrically charged (−) particle emitted from some radioactive chemicals. It has the mass of an electron. Krypton 85, emitted from nuclear power plants, is a strong beta emitter. Beta particles can cause ionisation.
4. **Curie** a measure of radioactivity. One curie equals 3.7×10^{10} nuclear transformations per second. Ci is the symbol used.
   (a) Microcurie: one-millionth of a curie.
      (3.7×10^4 disintegrations per second) mCi is the symbol used.
   (b) Picocurie: one-millionth of a microcurie.
      (3.7×10^{-2} disintegration per second) pCi is the symbol used.
5. **Dose** energy imparted to matter by nuclear transformations (radioactivity).
   (a) **Rad** = 100 ergs per gram.
      I GRAY = 100 rad = 10,000 ergs per gram.
   (b) **Rem** = rads × Q where Q is a quality factor which attempts to convert rads from different types of radioactivity into a common scale of biological damage.
      I SIEVERT = 100 rad.
6. **Gamma ray** short wave-length electromagnetic radiation released by some nuclear transformations. It is similar to X-ray

sense of self-preservation and personal benefit, in such an ideal system, would give realistic feedback to governments on the acceptability of national policy. The combined experiences of governing and governed would forge a national consensus on future directions.

The Fissioning Process and its Consequences

In order to understand nuclear technology and its impact on human health, three atomic-level events must be understood: fissioning, activation and ionisation. Fissioning, i.e. the splitting of the and will penetrate through the human body. Iodine 131 emits gamma rays. Both gamma and X-rays cause ionisation.

7. **Half-life, biological** time required for the body to eliminate one-half of an administered dose of a radioactive chemical.

8. **Half-life, physical** time required for half of a quantity of radioactive material to undergo a nuclear transformation. The chemical resulting from the transformation may be either radioactive or non-radioactive.

9. **Ionisation** sufficient energy is deposited in a neutral molecule to displace an electron, thus replacing the neutral molecule with positive and negative ions.

10. **Radiation** the emission and propagation of energy through space or tissue in the form of waves. It usually refers to electromagnetic radiation, classified by its frequency: radio, infrared, visible, ultraviolet, X-ray, gamma ray and cosmic rays.

   (a) **Natural background radiation** – emissions from radioactive chemicals which are not man-made. These chemicals include uranium, radon, potassium and other trace elements. They are made more hazardous through human activities such as mining and milling, since this makes them more available for uptake in food, air and water.

   (b) **Background radiation** – includes emissions from radioactive chemicals which occur naturally and those which result from the nuclear fission process. The meaning of this term is vague. In a licensing process it includes radiation from all sources other than the particular nuclear facility being licensed, even if the source includes a second nuclear facility located on the same site (US regulations). Radioactive chemicals released from a nuclear power plant are called 'background' after one year.

uranium or plutonium atom, is responsible for producing radio-active fission fragments and activation products. These in turn cause the ionisation of normal atoms, leading to a chain of microscopic events we may eventually observe as a cancer death or a deformed child.

Radioactive fission products are produced in nuclear reactors. They are variant forms of the ordinary chemicals which are the building blocks of all material and living things. The radioactive forms of these chemicals were, prior to 1943, present in only trace quantities in isolated places in the environment as, for example, in South Africa where it appears that a small nuclear fission reaction occurred spontaneously hundreds of thousands of years ago.

Nuclear Fission

Nucleus of an uranium 235 atom:
143 neutrons
 92 protons

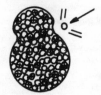

Neutron bombardment causing
uranium 235 to split into
krypton 90, xenon 137 and
8 neutrons.

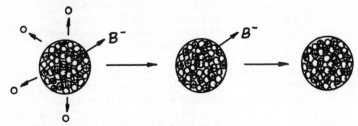

Krypton 90:
54 neutrons
36 protons

Rubidium 90:
53 neutrons
37 protons

Strontium 90:
52 neutrons
38 protons

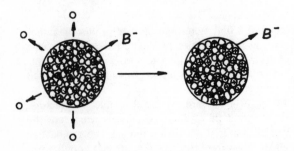

Xenon 137:
83 neutrons
54 protons

Cesium 137:
82 neutrons
55 protons

When a uranium atom is split or fissioned, it does not always split in the same place. The two pieces, called fragments, are chemicals of lower atomic weight than uranium. Each fragment receives part of the nucleus and part of the electrons of the original large uranium atom. The uranium atoms, of course, cease to exist after they are split. Instead, more than 80 different possible fission products are formed, each having the chemical properties usually associated with their structure, but having the added capability of releasing ionising radiation. X-rays, alpha particles, beta particles, gamma rays (like X-rays) or neutrons can be released by these 'created' chemicals. All these can cause 'ionisation', i.e. by knocking an electron out of its normal orbit around the nucleus of an atom they produce two 'ions', the negatively charged electron and the rest of the atom which now has a net positive electrical charge.

The atomic structure of fission fragments is unstable; for example, a nucleus may be proportionately heavier than usual due to inclusion of an extra neutron. The atom will at some time release the 'extra' particle and return to a natural, low-energy, more stable form. Every such release of energy is an explosion on the microscopic level.

The violence of the fissioning process is such that it can also yield what are called activation products, i.e. it can cause already existing chemicals in air, water or other nearby materials to absorb energy, change their structure slightly and become radioactive. As these high-energy forms of natural materials eventually return to their normal stable state, they can also release ionising radiation. About 300 different radioactive chemicals are created with each fissioning.[1] It takes hundreds of thousands of years for all the newly formed radioactive chemicals to return to a stable state.

In a nuclear power plant the fissioning takes place inside the zirconium or magnesium alloy cladding which encloses the fuel rods. Most of the fission fragments are trapped within the rods. However, the activation products can be formed in the surrounding air, water, pipes and containment building. The nuclear plant itself becomes unusable with time and must eventually be dismantled and isolated as radioactive waste.

After fissioning, the fuel rods are said to be 'spent'. They contain the greatest concentration of radioactivity of any material on the planet earth – many hundreds of thousands of times the concentration in granite or even in uranium mill tailings (waste). The spent fuel rods contain gamma radiation emitters (which are similar to X-ray emitters) so they must not only be isolated from the

Activation

Nucleus of stable cobalt 59:
32 neutrons
27 protons

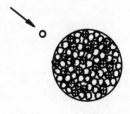

Bombardment of the stable cobalt with a neutron, released in fissioning. The neutron is 'captured' by the cobalt 59.

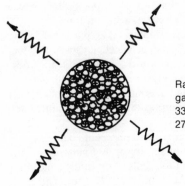

Radioactive cobalt 60 emitting gamma rays:
33 neutrons
27 protons

biosphere, but they must also be shielded with water and thick lead walls. Direct human exposure to spent fuel rods means certain death.

In reprocessing, spent fuel rods are broken open and the outer cladding is dissolved in nitric acid. The plutonium is separated out for use in nuclear weapons or for fuel in a breeder or mixed oxide nuclear reactor. The remaining highly radioactive debris is stored as liquid in large carbon or stainless steel drums, awaiting some kind of solidification and burial in a permanent repository. Waste of lower radioactivity is buried in dirt trenches or – as in Windscale (Sellafield) in England – piped out to sea. The spent nuclear fuel rods and liquid reprocessing waste are called 'high level radioactive waste'. It must be kept secure for hundreds of thousands of years – essentially forever. Lower level waste may be equally long-lived, but it is less concentrated.

In above-ground nuclear weapon testing, there is no attempt to contain any of the fission or activation products. Everything is released into the air and on to the land. Some underground tests are also designed to release most of the radioactive particles; these are called crater shots or shots with unstemmed holes. Even when below-ground shots are designed to be contained, they normally lose the radioactive gases and some particulates. The radionuclides trapped in the ground can also migrate downwards in the earth to water reservoirs which provide irrigation and drinking water for human purposes, although this process is slow. Radioactive debris piped out to sea can be washed back on shore or can contaminate fish.

In all nuclear reactions, some radioactive material – namely the chemically inert or so-called 'noble' gases, other gases, radioactive carbon, water, iodine, and small particulates of plutonium and other transuranics (i.e. chemicals of higher atomic number than uranium) – is immediately added to the air, water and land of the biosphere. In the far-distant future, all the long-lived radioactive material, even that now stored and trapped, will mix with the biosphere unless each generation repackages it. Our planet earth is designed to recycle everything.

The radioactive chemicals which escape to the biosphere can combine with one another or with stable chemicals to form molecules which may be soluble or insoluble in water; which may be solids, liquids or gases at ordinary temperature and pressure; which may be able to enter into biochemical reactions or be biologically inert. The radioactive materials may be external to the body and still give off destructive penetrating radiation. They may also be taken into the body with air, food and water or through an open wound, becoming even more dangerous as they release their energy in close proximity to living cells and delicate body organs. They may remain near the place of entry into the body or travel in the bloodstream or lymph fluid. They can be incorporated into the tissue or bone. They may remain in the body for minutes or hours or a lifetime. In nuclear medicine, for example, radioactive tracer chemicals are deliberately chosen among those quickly excreted by the body. Most of the radioactive particles decay into other radioactive 'daughter' products which may have very different physical, chemical and radiological properties from the parent radioactive chemical. The average number of such radioactive daughters of fission products produced before a stable chemical form is reached, is four.

Besides their ability to give off ionising radiation, many of the radioactive particles are biologically toxic for other reasons. Radioactive lead, a daughter product of the radon gas released by uranium mining, is a cause of lead poisoning and brain damage, just as any lead. Plutonium is biologically and chemically attracted to bone as is the naturally occurring radioactive chemical radium. However, plutonium clumps on the surface of bone, delivering a concentrated dose of alpha radiation to surrounding cells, whereas radium diffuses homogeneously in bone and thus has a lesser, localised cell damage effect. This makes plutonium, because of its concentration, much more biologically toxic than a comparable amount of radium. Some allowance for this physiological difference has been made in setting plutonium standards, but there is evidence that there is more than twenty times more damage caused than was suspected at the time of standard setting.[2]

The cellular damage caused by internally deposited radioactive particles becomes manifest as a health effect related to the particular organ damaged. For example, radionuclides lodged in the bones can damage bone marrow and cause bone cancers or leukaemia, while radionuclides lodged in the lungs can cause respiratory diseases. Generalised whole body exposure to radiation can be expressed as a stress related to a person's hereditary medical weakness. Individual breakdown usually occurs at our weakest point. In this way, man-made radiation mimics natural radiation and causes the ageing or breakdown process to be accelerated.

Penetrating Power of Different Types of Radiation

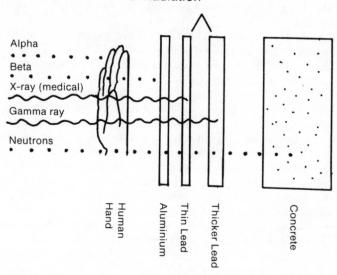

Alpha

Beta

X-ray (medical)

Gamma ray

Neutrons

Human Hand

Aluminium

Thin Lead

Thicker Lead

Concrete

Radioactive Particles and Living Cells: Penetration Power

Radioactive fission products, whether they are biochemically inert or biochemically active, can do biological damage when either outside the body or within.

X-rays and gamma rays are photons, i.e. high-energy light-waves. When emitted by a source, for example, radium or cobalt, located outside the body, they easily pass through the body, hence they are usually called penetrating radiation. The familiar lead apron provided for patients in some medical procedures stops X-rays from reaching reproductive organs. A thick lead barrier or wall is used to protect the X-ray technician. Because X-rays are penetrating, they can be used in diagnostic medicine to image human bones or human organs made opaque by a dye. These internal body parts are differentially penetrable. Where bones absorb the energy, no X-rays hit the sensitive X-ray film, giving a contrast to form the picture of the bones on the radiation-sensitive X-ray plate. High-energy gamma rays, which easily penetrate bone, would be unsuitable for such medical usage because the film would be uniformly exposed. In photography jargon, the picture would be a 'white out' with no contrasts. No radiation remains in the body after an X-ray picture is taken. It is like light passing through a window. The damage it may have caused on the way through, however, remains.

Some radioactive substances give off beta particles, or electrons, as they release energy and seek a stable atomic state. These are small negatively charged particles which can penetrate skin but cannot penetrate through the whole body as do X-rays and gamma rays.

Microscopic nuclear explosions of some radioactive chemicals release high-energy alpha particles. An alpha particle, the nucleus of a helium atom, is a positively charged particle. It is larger in size than a beta particle, like a cannon-ball relative to a bullet, having correspondingly less penetrating power but more impact. Alpha particles can be stopped by human skin, but they may damage the skin in the process. Both alpha and beta particles penetrate cell membranes more easily than they penetrate skin. Hence ingesting, inhaling or absorbing radioactive chemicals capable of emitting alpha or beta particles and thereby placing them inside delicate body parts such as the lungs, heart, brain or kidneys, always poses serious threats to human health.[3] Plutonium is an alpha emitter, and no quantity has been found to be too small to induce lung cancer in animals.

The skin, of course, can stop alpha or beta radiation inside the body tissue from escaping outwards and damaging, for example, a baby one is holding or another person sitting nearby. Also, it is impossible to detect these particles with most whole body 'counters' such as are used in hospitals and nuclear installations. These counters can only detect X-rays and gamma rays emitted from within the body.

Splitting a uranium atom also releases neutrons, which act like microscopically small bullets. Neutrons are about one-fourth the size of alpha particles and have almost 2,000 times the mass of an electron. If there are other fissionable atoms nearby (uranium 235 or plutonium 239, for example) these neutron projectiles may strike them, causing them to split and to release more neutrons. This process is called a chain reaction. It takes place spontaneously when fissionable material is sufficiently concentrated, i.e. forms a critical mass. In a typical atomic bomb the fissioning is very rapid. In a nuclear reactor, water, gas or the control rods function to slow down or to absorb neutrons and control the chain reaction.

Neutrons escaping from the fission reaction can penetrate the

Alpha Radiation

Radium 226 nucleus:
138 neutrons
 88 protons

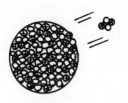

alpha particle

radon 222

Radium 226 loses an alpha particle, 2 protons and 2 neutrons, and is reduced to a radon gas nucleus:
136 neutrons
 86 protons

human body. They are among the most biologically destructive of the fission products. They have a short range, however, and in the absence of fissionable material they will quickly be absorbed by non-radioactive materials. Some of these latter become radioactive in the process, as was noted earlier, and are called activation products.

Standard-setting Preliminaries

The complexity of setting health standards for exposure to the mixture of radioactive chemicals and ionising particles released in fissioning should be apparent. As a first move towards a reasonable subdivision of the hazard itself, separate standard setting was done for external radiation exposure, i.e. when the radioactive source was outside the body, and internal radiation exposure, i.e. when the radioactive source was inside the body.

Both these categories can then be subdivided into exposures to particular parts of the body or particular internal organs. The biological effect of an X-ray of the pelvic area differs from the biological effect of a dental X-ray, even if the radiation dose to the skin is the same. Plutonium lodged in the lungs has a different biological consequence from plutonium lodged in the reproductive organs. One can also consider exposures to X-rays, gamma rays, alpha or beta particles and neutrons separately, taking each as internal or external to the body.

There are further differences in health effects based on differences between people receiving the radiation. Special consideration needs to be given to those who, because of heredity or previous experience, are more susceptible to further damage than the norm or average. Special consideration should be given to an embryo or foetus, a young child, the elderly or those chronically ill.

The severity of health effects caused by internal exposures will depend on the biological characteristic of the radioactive chemical and the length of time it may be expected to reside in the body. Radioactive cesium, for example, lodges in muscles and is probably completely eliminated from the body in two years. Radioactive strontium lodges in bone and remains there for a lifetime, constantly irradiating the surrounding cells. The usual time required by the body to rid itself of half the radioactive chemical is called the 'biological half-life' of that chemical.

Some radiation health effects are observable in the persons exposed; some effects are only seen in their children or grandchildren because the damage was to sperm or ovum.

X-rays, gamma rays and neutrons are able to inflict harm on humans even when the radioactive chemical emitting them is outside the body. Beta particles outside the body can cause serious burns and other skin anomalies, including skin cancer. Ionising radiations emitted from within the body by radioactive chemicals taken in by inhalation, ingestion or absorption are even more damaging because they are so close to delicate cell structures. The body is not able to distinguish between radioactive and non-radioactive chemicals and will as readily incorporate the one as the other into tissue, bone, muscle or organs, identifying them as ordinary nutrients. The radioactive chemicals remain in the body until biologically eliminated in urine or faeces, or until they decay into other chemical forms (which may or may not be radioactive). These daughter products and their chemical and radiological properties may be quite different from those of the parent radioactive chemical, for example, radioactive carbon decays into nitrogen. Radiochemical analysis of urine or faeces is the preferred test for most types of internal contamination with alpha or beta particles.

Within the Living Cell

The chaotic state induced within a living cell when it is exposed to ionising radiation has been graphically described by Dr Karl Z. Morgan as a 'madman loose in a library'.[4] The result of cell exposure to these microscopic explosions with the resultant sudden influx of random energy and ionisation may be either cell death or cell alteration. The change or alteration can be temporary or permanent. It can leave the cell unable to reproduce (or replace) itself. Radiation damage can cause the cell to produce a slightly different hormone or enzyme than it was originally designed to produce, still leaving it able to reproduce other cells capable of generating this same altered hormone or enzyme. In time there may be millions of such altered cells. This latter mechanism, called biological magnification, can cause some of the chronic diseases and changes we usually associate with old age. One very specific mutation which can occur within the cell is the destruction of the cell's mechanism for resting which normally causes it to cease reproductive activities after cell division. This inability to rest results in a runaway proliferation of cells in one place, which, if not destroyed, will form a tumour, either benign or malignant. The abnormal proliferation of white blood cells is characteristic of leukaemia; red blood cell proliferation results in what is called polycythemia vera.

If the radiation damage occurs in germ cells, the sperm or ovum, it can cause defective offspring. The defective offspring will in turn produce defective sperm or ova, and the genetic 'mistake' will be passed on to all succeeding generations, reducing their quality of life until the family line terminates in sterilisation and/or death.[5] A blighted or abnormal embryonic growth can result in what is called a hydatidiform mole instead of a baby.

Exposure to radiation is also known to reduce fertility, i.e. women become unable to conceive or give birth.

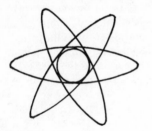

Ionization

Normal atom with the positively charged particles in the nucleus equal in number to the negatively charged electrons in ornit around the nucleus.

neutral atom

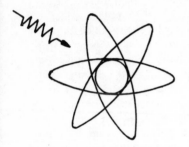

Gamma ray directly affecting an orbiting electron, giving it an escape velocity.

gamma radiation
exposure

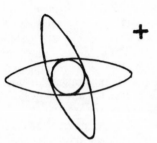

The electron 'escapes' orbit becoming a negatively charged ion. The rest of the atom, now positively charged, is also called an ion.

positive and negative
ions

Radiation can also damage an embryo or foetus while it is developing within the mother's womb. This is called teratogenic damage, or the child is said to have a congenital malformation rather than genetic damage. This means the damage is not automatically transmitted. For example, a deaf person, made so by a pre-birth injury, may have children with normal hearing.

The damage done within cells by random releases of the energy of photons, alpha, beta or neutron particles can occur indirectly through an effect called ionisation. As the energised photons or particles speed through the cells, they give energy to the electrons of chemicals already within the cells, enabling some electrons to break free from the rest of the atom or molecule to which they are attached. On the macro-level this would be comparable to an atomic explosion of a magnitude great enough to drive the earth or another planet out of its orbit around the sun. What was an electrically neutral atom or molecule is split into two particles – a larger positively charged atom or molecule missing one of its electrons, and a small negatively charged electron expelled from its orbit around the nucleus of the atom. Both are called ions and the process is called ionisation.

The complex molecules making up living organisms are composed of long strands of atoms forming proteins, carbohydrates and fats. They are held together by chemical bonds involving shared electrons. If the ionising radiation displaces one of the electrons in a chemical bond, it can cause the chain of atoms to break apart, splitting the long molecule into fragments, or changing its shape by elongation. This is an 'ungluing' of the complex chemical bonds so carefully structured to support and perpetuate life. The gradual breakdown of these molecular bonds destroys the templates used by the body to make DNA and RNA (the information-carrying molecules in the cell) or causes abnormal cell division. The gradual natural breakdown of DNA and RNA is probably the cellular phenomenon associated with what we know as 'ageing'. It occurs gradually over the years with exposure to natural background radiation from the radioactive substances which have been a part of the earth for all known ages. There is evidence that exposure to medical X-rays accelerates this breakdown process.[6] There is ample reason to think that fission products lodged within the body will cause the same kind of acceleration of ageing. However, unlike medical X-rays, these radioactive chemicals damage cells by their chemical toxicity as well as their radiological properties.

The gradual breakdown of human bio-regulatory integrity

through ionising and breakage of the DNA and RNA molecules gradually makes a person less able to tolerate environmental changes, less able to recover from diseases or illness, and generally less able to cope physically with habitat variations.

When the DNA of germ plasm is affected by radiation it can result in chromosomal diseases, such as trisomy 21, more commonly known as Down's Syndrome. Mentally retarded children, victims of Down's Syndrome, have been reported in Kerala, India, an area of high natural radioactivity.[7] Recently, cases of Down's Syndrome have been tentatively linked to women exposed to radioactive releases from the large plutonium fire at Sellafield (Windscale) in 1957.[8] While Down's Syndrome babies have long been associated with births to older women (those with higher accumulated exposure to natural background radiation),[9] the Sellafield-related cases involve women with an average age of 25 years.

So far we have considered the types of ionising radiation, the location of the source outside or within the body, and the difference between exposures to different parts of the body or to different people of various ages and states of health. These will all be important considerations underlying standard setting. Next, we need to be able to measure radiation, i.e. to quantify exposure.

Measuring Radiation

One way to approach the measurement of radiation is to count the number of nuclear transformations or explosions which occur in a given unit of radioactive substance per second. This measure is usually standardised to radium, the first radioactive substance to be discovered and widely used. One gram of radium undergoes 3.7×10^{10} nuclear transformations or disintegrations per second. The activity of 1 gram of radium is called 1 curie (Ci), named for Madame Marie Curie, a Polish-born French chemist (1867–1934). Marie Curie discovered the radioactivity of thorium, polonium and radium by isolating radium from pitchblende. She and her daughter Irene were among the earliest known radiation victims, both dying of aplastic anaemia.

In recent radiation protection guides, the curie is being replaced by the becquerel, which indicates one atomic event per second. One gram of radium would equal 1 curie of radium or 3.7×10^{10} becquerels of radium.

The energy released in nuclear disintegrations has the ability to do work, i.e. to move matter. In physics, the erg is a very small unit of work done. Lifting 1 gram of radium 1 centimetre requires 980 ergs of work. Any material exposed to the force from nuclear disintegrations at a rate of 100 ergs/gm is said to absorb one rad, i.e. *r*adiation *a*bsorbed *d*ose. There is no direct conversion from curies, which is related to the number of atomic events, to the rad dose, which is energy absorbed in tissue. The curie gives one an estimate of the number of microscopic transformations or explosions per second and the rad is an estimate of the energy release, absorbed by the surrounding tissue. On the macro-level, the word 'explosion' tells us only of an event in time. A dynamite explosion or hydrogen bomb explosion adds information about the energy released.

Sometimes radioactivity is measured in counts per minute on a Geiger counter. A nuclear transformation within an energy range measured by the instrument and close enough to the instrument causes a noise or 'count'. Most Geiger counters cannot detect alpha particle emitters like plutonium.

The radioactivity of elements which experience nuclear disintegrations is measured relative to radium. For example, it would take more than 1 million grams of uranium to be equivalent in radioactivity, i.e. to have the same number of nuclear events per second as 1 gram of radium has per second. Both 1 million grams of uranium and 1 gram of radium would be measured as 1 Ci. It has been the custom in the past to limit human exposure to uranium more for its toxic chemical properties (it is a heavy metal) than for its radioactivity. This practice may have underestimated damage caused by the biological storing of uranium in the liver.

When uranium decays, it passes through about 12 radioactive forms, called daughter products, before reaching a stable chemical form of lead. One of the radioactive daughter products of uranium is radium. Uranium released into drinking water or incorporated into food and human tissue today will eventually plague the world as radium and its other disintegration products: radon gas and the radioactive forms of polonium, lead and bismuth. The environmental and biochemical forces which may tend to reconcentrate these toxic materials in living cells are not well known. Although uranium occurs naturally, it has become much more available for entering into water, food, living cells and tissue since the mining boom which began shortly after the Second World War.

The activity which takes place in the nucleus of the uranium or radium atom is a 'haphazard' event obeying the laws of random probabilities. An atom is characterised by its atomic number, that is, the positively charged particles in its nucleus, and by its atomic mass, expressed in atomic mass units (similar to the concept of weight), which includes both the number of protons (the atomic number) and the number of neutrons in the nucleus. Carbon, the most frequently occurring chemical in living material, is taken as having exactly 12 atomic mass units and other atoms are measured in relation to this. Carbon 14, which is radioactive, has two extra neutrons in its nucleus.

Hydrogen, another example, has an atomic number of 1 and an atomic mass of 1. Isotopes of hydrogen have the same atomic number (that is, the same number of positively charged particles in the nucleus and electrons in orbit around the nucleus) but a higher atomic mass. Deuterium or hydrogen 2, an isotope of hydrogen, has an atomic number of 1 and an atomic mass of 2. It is not radioactive. The increased atomic mass is due to an added neutron in the nucleus. Deuterium is in the 'heavy water' used in the Canadian CANDU nuclear reactor. Hydrogen 3, called tritium, is radio-active, with two neutrons and a proton in the nucleus. It is produced in a nuclear reaction.

When radium 226 decays, it loses a positively charged alpha particle from its nucleus. An alpha particle has two protons (positive electrical charges) and a mass of 4 atomic units. This means a reduction in both radium's atomic number *and* atomic mass. Loss of the alpha particle changes radium 226 (transmutes it) into another element, radon 222. While radium 226 is a radioactive solid under normal conditions, radon 222 is a radioactive gas. Loss of one or more protons changes the chemical element into a different chemical. Absorption or loss of a neutron gives an isotope of the same chemical since chemical properties are determined by the number of protons and electrons in an atom.

The time required for half of any amount of radium 226 to transmute to radon 222 by these small explosions which emit alpha particles is 1,622 years. This is called the physical half-life of radium. Half of the radium literally disappears in that length of time, but radon gas is produced to replace it. Radon gas is radioactive and more mobile in air and water (it dissolves) than the solid radium. The half-life of radon is 3.82 days, after which half the gas will have disintegrated, again releasing alpha particles and transmuting into radioactive polonium 218, which is a solid. With a wind of 10 mph (or kph), the radon gas could travel 1,000 miles (or kilometres) from the point of origin before half of it would have decayed into its solid daughter products and been deposited on soil,

Hydrogen Isotopes

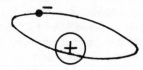

Stable hydrogen atom:
0 neutron
1 proton
1 electron

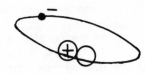

Deuterium, a stable heavy
hydrogen atom:
1 neutron
1 proton
1 electron

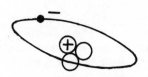

Tritium, radioactive hydrogen:
2 neutrons
1 proton
1 electron

leafy vegetables, tobacco, groundwater, human skin, lung tissue, etc. If the material receiving the radioactive daughter product is living, then it can carry the particles into its cells. Such contamination cannot be washed off.

When a negatively charged beta particle is released, there is a transmutation in which a neutron in the nucleus of the atom splits into a proton and an electron, the proton remaining in the nucleus and the electron given off as a fast-moving microscopic bullet. Beta particles are extremely small. The mass of an alpha particle is about 7,400 times that of a beta particle. Thorium 234 decays to uranium 234 (with a short-lived radioactive intermediary) by losing beta particles. Uranium and thorium are different elements, but have the same mass (atomic weight) since a neutron and proton have about the same mass. The thorium neutron becomes the uranium proton. The half-life of thorium 234 is 24.1 days, while the half-life of uranium 234 is 2.50×10^5, or 250,000 years. As was pointed out earlier, uranium nuclear events are not as frequent as those in radium, although they are destructive when they occur.

Given 12 grams of thorium 234, we would have 6 grams after 24.1 days, 3 grams after 48.2 days, 1.5 grams after 72.3 days, 0.75 grams after 96.4 days, etc. At the same time, the stock of uranium 234 would be increasing as the thorium decays into the new radioactive chemical.

There is no simple physical or chemical process such as temperature change or chemical bonding which can prevent these radioactive elements from decaying. Their nucleus is unstable and because all elements seek a stable low-energy state, they must at some time release particles in an effort to reach a resting state. The decay takes place in the nucleus of the atom regardless of whether the atom exists singly or is part of a molecule; is in the solid, liquid or gaseous state; is within the body or outside, and so on. The decay product after a radioactive disintegration may itself be radioactive, so disintegration does not put an end to the biological problems generated by these small explosions. This decay process must be taken into account when estimating the biological effects of internal exposure to radioactive material. Inhaled radon gas quickly becomes radioactive lead, bismuth or polonium in the blood-stream.

One should not confuse physical half-life with biological half-life, i.e. the time required to eliminate half of the material from the body through exhalation, urine or faeces. Cesium 137 and strontium 90 both have physical half-lives of almost thirty years, but cesium 137 is

Beta Radiation

Thorium 234 nucleus:
144 neutrons
 90 protons

Thorium 234 with two neutrons
about to eject their negative
charge, becoming protons.

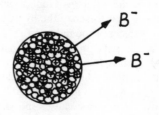

Uranium 234 nucleus:
142 neutrons
 92 protons

normally excreted from the body within two years while strontium 90 can be incorporated in bone for a lifetime.

One more measure needs to be introduced before radiation protection guides can be understood. Since the various kinds of radiation exposures need to be evaluated for biological impact and not just for the amount of energy absorbed by the tissue, the term rem, *r*oentgen *e*quivalent *m*an (or woman), was introduced. The rem dose is the rad dose times a quality factor Q. For external radiation Q is usually taken as 1, and rads and rems are used interchangeably. However, to reflect the greater biological damage done by alpha particles when inside the body, the rad dose may be multiplied by 20 to give the rem dose. This is another way of saying that the alpha particle does damage of an order of magnitude (20 times) greater when lodged within a tissue, bone or organ. For example, alpha particles giving a 2 rem (or rad) dose to skin would give a 40 rem dose to sensitive lung tissue when inhaled.

Theoretically, the rem dose measures equivalent biological effect, so that damage from X-rays, for example, would be the same as damage from alpha particles, when the dose in rem was the same. Unfortunately, living systems are too complex for such an approach to provide anything more than a good guess.

Sometimes references are made to a 'fifty-year effective dose equivalent'. This is the full dose that would be received from an internal radionuclide if the dose were given at one time instead of being spread over two to fifty years.

Linear Energy Transfer (LET)

Measurement of the number of ionisations which radiation causes per unit distance as it traverses the living cell or tissue is called the linear energy transfer of the radiation. The concept involves lateral damage along the path, in contrast to path length or penetration capability. Medical X-rays and most natural background radiation are low LET radiation, while alpha particles have high LET. On the average, fission fragments have high LET.

The density of ionisation causes special problems in sperm and ova because the damage (protein breakage) is concentrated within a few cells. The two-year sterility of Japanese fishermen exposed to fallout from the 1954 hydrogen bomb test is probably an example of this effect. Sperm and the cells which produce sperm were damaged beyond their capability of prompt repair.

As a young girl in St George, Utah, USA, Elizabeth Catalan used to stand outdoors and watch the mushroom clouds raised by the Nevada nuclear tests float overhead. She has never been able to have children. She, like some other women in St George, is unable to carry a foetus to birth. Elizabeth's father, president of a local college, died prematurely of leukaemia. He used to go horse-riding with three friends and was frequently outside when the grey clouds laden with radioactive chemicals went over. Three of the four men are now dead from cancer.

Elizabeth's sister died in her late twenties of a thyroid disease which may have been caused by the radioactive iodine released in the atomic blasts. Elizabeth and her mother attribute many of their abnormal health problems, and those of family and friends, to the atomic fallout. No government studies have been undertaken to confirm or deny these claims. However, the situation was so widely recognised as abnormal by the local population that the Governor of Utah has filed a court claim against the US Federal Government for wrongful deaths of the people of Utah. About a thousand individual damage claims have entered the courts in the USA, and as part of the trial preparations Dr Carl Johnson undertook a detailed study of the Mormon population of Utah exposed to the fallout. It is reasonable to conclude that the health problems reported by the people of Utah are typical of what could be expected on the basis of theoretical radiobiology.[10]

On 10 May 1984, US District Court Judge Bruce S. Jenkins ruled on the first twenty-four claims of US government negligence in its conduct of nuclear testing. He has awarded $2.6 million in damages to ten claimants. This landmark, 489-page, carefully worded decision is expected to be appealed against by the US Federal Government.

In order to have a quantitative sense of the frequency of the different cell effects caused by radiation exposure, imagine a colony of 1,000 living cells exposed to a 1 rad X-ray (about the dose for one X-ray spinal examination). There would be two or three cell deaths, two or three mutations or irreparable changes in cell DNA and about 100,000 ionisations in the whole colony of cells – ranging from 11 to 460 ionisations per cell.[11] While cells can repair some damage, no one claims that there is perfect repair even after only one such X-ray.

A comparable 1 rad exposure to neutrons which have higher linear energy would be expected to cause more cell deaths and more mutations. The ionisations caused would range from 145 to 1,100 per cell.

Alpha particles which occur naturally would cause roughly 10 times as many cell deaths and mutations, and 3,700 to 4,500 ionisations per cell. Alpha particles have high linear energy transfer.

The average number of cell deaths and mutations caused by fresh fission particles (i.e. those present soon after detonation of a nuclear bomb) would be even greater, with the ionisations as frequent as 130,000 per cell.[12] In nuclear reactors, most of these extremely high-energy early fission fragments are enclosed within the fuel rod. In a nuclear bomb blast, they are all released but they decay very quickly and do not persist long in the environment.

If instead of thinking of a colony of living cells, we think of a person exposed to 1 rad (again about the skin dose from one spinal X-ray) of 1 MeV (million electron volts) energy, this corresponds to 2.2 billion (US) photons per cm^2 acting on the body. In the words of Karl Morgan, 'It is inconceivable that all the billions of irradiated and damaged cells would be completely repaired.'[13] This unrepaired damage accumulates, eventually causing a reduction in the level of health that is normal for a particular age.

Stated very simply, ionising radiation seriously disrupts the chemistry of the cell. It can also kill or permanently change the cell. Every exposure to ionising radiation has this effect, and it is not possible for the body to perfectly repair all of the damage. Whether or not the residual unrepaired damage is of concern to the individual exposed is a personal value judgment. It is not at all clear that ordinary people find the damage 'acceptable' unless it initiates a fatal cancer, and yet this is the basis on which radiological safety standards are set in all nations of the world.

R. M. Sievert, the famous radiologist, who had supervised radiation therapy since 1926 at the Karolinska Institute in Stockholm, pointed out at an international meeting in 1950 that 'there is no known tolerance level for radiation'.[14] A tolerance level is a level below which there is no damage (sometimes called a threshold). A safety level is ordinarily a fraction (one-tenth) of the tolerance level.[14]

Cell Damage Expressed as a Health Problem

An example to show the connection between cell damage and observable illness in the person exposed might help in understanding the problems posed by radionuclide (radioactive chemical) uptake, i.e. their ingestion, inhalation or absorption with food, air and water, into human bodies, with subsequent cell damage. The thyroid gland contains cells which produce thyroid hormone, which when released into the bloodstream causes the body functions such as breathing, digesting and reacting to stress to proceed at a certain rate. If the thyroid is 'overactive', one might notice in the person increased pulse rate, nervousness, excitability, loss of body weight and, in females, more frequent menstruation. Such a person is often called 'hyperactive' (hyper-thyroidism). A normal amount of thyroid hormone in the blood produces a normally active individual. An 'underactive' or 'hypoactive' thyroid can result in sluggishness, listlessness, weight gain and irregular and/or infrequent menstrual flow in women (hypothyroidism).

If radioactive iodine (I 131 or I 129) is ingested with food it will enter the blood and tend to accumulate in the thyroid. Radioactive iodine emits high-energy gamma radiation which can destroy thyroid cells, thus reducing total thyroid hormone production in the individual so affected.

A small amount of I 131 would probably kill only a few cells and have little or no noticeable effect on health. However, if many cells are destroyed or altered, the hormone level would noticeably drop or the hormone itself would be slightly changed. The individual would become lethargic and gain weight. If properly diagnosed and severe enough to require medical intervention, this hypoactive thyroid condition can be controlled with artificially ingested thyroid hormone. A mild exposure experienced by a large population could cause a decrease in average thyroid hormone levels and an increase in average body weight, such as is occurring now in the North American population. The USA has been polluted with nuclear industries since 1943 and with radioactive iodine from weapon testing since 1951. Radioactive iodine is routinely released in small quantities by nuclear power plants and in large quantities by nuclear reprocessing plants. It is not part of the natural human environment. The connection between this pollution and the overweight problem has, unfortunately, never been seriously researched. There is no evidence to confirm or deny the hypothesis, but weight increase is a well-known biological response to radioactive iodine. The hypothesis is certainly plausible under the circumstances.

It is possible for thyroid cells to be altered but not killed by the radiation. The cellular growth mechanism may be damaged, allowing a runaway proliferation of cells. This results in a thyroid tumour, either cancerous (malignant), or non-cancerous (benign). Other possible radiation damage includes changes in the chemical composition of the individual's thyroid hormone, altering its action in the body and causing clinically observable symptoms not easily diagnosed or corrected.

There is an extremely remote possibility that these changes will be desirable, but the overall experience of randomly damaging a complex organism like the human body is that it is destructive of health.

An atomic veteran who participated in the nuclear tests which were conducted by the USA in the Bikini atoll in the late 1940s reported that he gained 75 lbs in the four years following his participation. The doctor diagnosed his problem as hypothyroidism. He also suffered from high blood pressure, chronic asthma and frequent bouts of bronchitis and pneumonia. He has had six tumours diagnosed since 1949, when he returned home from military service. Four have been surgically removed.

Damage to the thyroid of a developing foetus can cause mental retardation and other severe developmental anomalies.[15]

Other radionuclides will lodge in other parts of the body. If the trachea, bronchus or lung are exposed, the damage eventually causes speech or respiratory problems. If radioactive particles lodge in the stomach or digestive tract, the heart, liver, pancreas or other internal organs or tissues, the health problems will be correspondingly different and characteristic of the organ damaged. Radionuclides which lodge in the bone marrow can cause leukaemia, depression of the immune system (i.e. the body's ability to combat infectious diseases) or blood diseases of various kinds.

If the radiation dose is high, there is extensive cell damage and health effects are seen immediately. Penetrating radiation doses at 1,000 rad or more cause 'frying of the brain' with immediate brain death and paralysis of the central nervous system. This is why no one dared to enter the crippled Three Mile Island nuclear reactor building during the 1979 accident. An average of 30,000 roentgens (or rads) per hour were being reported by instruments within the containment building. This would convert to a 1,000 rad exposure for two minutes spent inside the building. Such a dose to the whole body is invariably fatal.

The radiation dose at which half the exposed group of people would be expected to die, i.e. the 50 percent lethal dose, is 250 rad. The estimate is somewhat higher if only young men in excellent health (e.g. soldiers) are exposed. Between 250 and 1,000 rad, death is usually due to gross damage to the stomach and gut. Below 250 rad death is principally due to gross damage to the bone marrow and blood vessels. A dose of about 200 rad to a foetus in the womb is almost invariably fatal.

Penetrating radiation in doses above 100 rad inflicts severe skin burns. Lower doses produce burns in some people. Vomiting and diarrhoea are caused by doses above about 50 rad. There are some individuals who are more sensitive to radiation, however, showing typical vomiting and diarrhoea radiation sickness patterns with doses as low as 5 rad. An individual may react differently at different times of life or under different circumstances. Below 30 rad, for most individuals, the effects from external penetrating radiation are not immediately felt. The mechanism of cell damage is similar to that described for minute quantities of radioactive chemicals which lodge within the body itself, and our bodies are incapable of 'feeling' damage to or death of cells. Only when enough cells are damaged to interfere with the function of an organ or a body system does the individual become conscious of the problem.

By sharpening our perceptions more subtle radiation effects can often become observable where once they went unnoticed. For example, a series of X-rays received by a young child may cause temporary depression of the white blood cells, and ten days to two weeks after the exposure the child will get influenza or some other infectious disease. Ordinarily the parent views the two events as unconnected.

Sometimes one can observe a mutation in a person who has experienced loss of hair after radiation therapy to kill tumour cells: hair that was formerly very straight can be curly when it grows again.

A plant whose flowers are normally white with red tips but which begins to form uniformly red flowers has mutated. Such an event has been observed by persons living in the vicinity of Sellafield in the United Kingdom.

The use of radiation therapy to destroy malignant cells also has observable results. It is rather like surgery in that it is deliberately used to kill the unwanted tumour cells.

Radiation and Heredity

In 1943, Hermann Müller received a Nobel Prize for his work on the

Probable Health Effects resulting from Exposure to Ionising Radiation

Dose in rems (whole body)	Health effects Immediate	Delayed
1,000 or more	Immediate death. 'Frying of the brain'.	None
600–1,000	Weakness, nausea, vomiting and diarrhoea followed by apparent improvement. After several days: fever, diarrhoea, blood discharge from the bowels, haemorrhage of the larynx, trachea, bronchi or lungs, vomiting of blood and blood in the urine.	Death in about 10 days. Autopsy shows destruction of hematopoietic tissues, including bone marrow, lymph nodes and spleen; swelling and degeneration of epithelial cells of the intestines, genital organs and endocrine glands.
250–600	Nausea, vomiting, diarrhoea, epilation (loss of hair), weakness, malaise, vomiting of blood, bloody discharge from the bowels or kidneys, nose bleeding, bleeding from gums and genitals, subcutaneous bleeding, fever, inflammation of the pharynx and stomach, and menstrual abnormalities. Marked destruction of bone marrow, lymph nodes and spleen causes decrease in blood cells especially granulocytes and thrombocytes.	Radiation-induced atrophy of the endocrine glands including the pituitary, thyroid and adrenal glands. From the third to fifth week after exposure, death is closely correlated with degree of leukocytopenia. More than 50% die in this time period. Survivors experience keloids, ophthalmological disorders, blood dyscrasis, malignant tumours, and psychoneurological disturbances.
150–250	Nausea and vomiting on the first day. Diarrhoea and	Symptoms of malaise as indicated above. Persons in poor

	probable skin burns. Apparent improvement for about two weeks thereafter. Foetal or embryonic death if pregnant.	health prior to exposure, or those who develop a serious infection, may not survive. The healthy adult recovers to somewhat normal health in about three months. He or she may have permanent health damage, may develop cancer or benign tumours, and will probably have a shortened lifespan. Genetic and teratogenic effects.
50–150	Acute radiation sickness and burns are less severe than at the higher exposure dose. Spontaneous abortion or stillbirth.	Tissue damage effects are less severe. Reduction in lymphocytes and neutrophils leaves the individual temporarily very vulnerable to infection. There may be genetic damage to offspring, benign or malignant tumours, premature ageing and shortened lifespan. Genetic and teratogenic effects.
10–50	Most persons experience little or no immediate reaction. Sensitive individuals may experience radiation sickness.	Transient effects in lymphocytes and neutrophils. Premature ageing, genetic effects and some risk of tumours.
0–10	None	Premature ageing, mild mutations in offspring, some risk of excess tumours. Genetic and teratogenic effects.

genetic effects of radiation and was a dominant figure in developing early radiation exposure recommendations made by the International Commission on Radiological Protection (ICRP).[16] He showed through his work with Drosophila, a fruit fly, that ionising radiation affects not only the biological organism which is exposed but also the seed within the body from which the future generations are formed.

In 1964 Hermann Müller published a paper, 'Radiation and Heredity', spelling out clearly the implications of his research for genetic effects (damage to offspring) of ionising radiation on the human species.[17] The paper, though accepted in medical/biological circles, appears not to have affected policy makers in the political or military circles who normally undertake their own critiques of published research. Müller predicted the gradual reduction of the survival ability of the human species as several generations were damaged through exposure to ionising radiation. This problem of genetic damage continues to be mentioned in official radiation-health documents under the heading 'mild mutations'[18] but these mutations are not 'counted' as health effects when standards are set or predictions of health effects of exposure to radiation are made. There is a difficulty in distinguishing mutations caused artificially by radiation from nuclear activities from those which occur naturally from earth or cosmic radiation. A mild mutation may express itself in humans as an allergy, asthma, juvenile diabetes, hypertension, arthritis, high blood cholesterol level, slight muscular or bone defects, or other genetic 'mistakes'. These defects in genetic make-up leave the individual slightly less able to cope with ordinary stresses and hazards in the environment. Increasing the number of such genetic 'mistakes' in a family line, each passed on to the next generation, while at the same time increasing the stresses and hazards in the environment, leads to termination of the family line through eventual infertility and/or death prior to reproductive age. On a large scale, such a process leads to selective genocide of families or species suicide.[19]

It soon became obvious that the usual method determining a tolerance level for human exposure to toxic substances was inappropriate for ionising radiation. The health effects were similar to normally occurring health problems and were quite varied, ranging from mild to severe in a number of different human organ systems, and their appearance could be delayed for years or even generations.

Permissible Levels of Exposure

The US National Council on Radiation Protection and Measurement gave expression to the theoretical resolution of this human dilemma by articulating the implicit reasoning behind subsequent radiation protection standards development:[20]

1. A value judgment which reflects, as it were, a measure of psychological acceptability to an individual of bearing slightly more than a normal share of radiation-induced defective genes.
2. A value judgment representing society's acceptance of incremental damage to the population gene pool, when weighted by the total of occupationally exposed persons, or rather those of reproductive capacity as involved in Genetically Significant Dose calculation.
3. A value judgment derived from past experience of the somatic effects of occupational exposure, supplemented by such biomedical and biological experimentation and theory as has relevance.

This is now an internationally accepted approach to setting standards for toxic substances when no safe level of the substance exists.

In short, this elaborate philosophy recognises the fact that *there is no safe level of exposure to ionising radiation*, and the search for quantifying such a safe level is in vain. A *permissible* level, based on a series of value judgments, must then be set. This is essentially a trade-off of health for some 'benefit' – the worker receives a livelihood, society receives the military 'protection' and electrical power is generated. Efforts to implement these permissible standards would then logically include convincing the individual and society that the 'permissible' health effects are acceptable. This has come to mean that the most undesirable health effects will be infrequent and in line with health effects caused by other socially acceptable industries. Frequently, however, the worker and/or public is given the impression that these 'worst' health effects are the only individual health effects. A second implication of the standards-based-on-value-judgments approach is that unwanted scientific research resulting in public scrutiny of these value judgments must be avoided.

The genetic effect considered by standard setters as most

From a column in the *Yomiuri Shinbun* (19 January 1965;
evening edition)

A nineteen-year-old girl in Hiroshima committed suicide after leaving
a note: 'I caused you too much trouble, so I will die as I planned
before.' She had been exposed to the atomic bomb while yet in her
mother's womb nineteen years ago. Her mother died three years
after the bombing. The daughter suffered from radiation illness; her
liver and eyes were affected from infancy. Moreover, her father left
home after the mother died. At present there remain a grandmother,
age seventy-five; an elder sister, age twenty-two; and a younger
sister, age sixteen. The four women had eked out a living with their
own hands. The three sisters were all forced to go to work when they
completed junior high school. This girl had no time to get adequate
treatment, although she had an A-bomb victim's health book.

As a certified A-bomb victim, she was eligible for certain medical
allowances; but the [A-bomb victims' medical care] system provided
no assistance with living expenses so that she could seek adequate
care without excessive worry about making ends meet. This is a
blind spot in present policies for aiding A-bomb victims. Burdened
with pain and poverty, her young life had become too exhausted for
her to go on . . . .

There is something beyond human expression in her words 'I will
die as I planned before.'

Quoted in Kenzaburō Ōe, *Hiroshima Notes*, YMCA Press, Tokyo (English
translator Toshi Yonezawa; English editor David L. Swain).

unacceptable is serious transmittable genetic disease in live-born
offspring. These severely damaged children are usually a source of
suffering for the family and an expense for society which must
provide special institutions for the mentally and physically disabled.
Severely handicapped people rarely have offspring; many die, are
sterile or are institutionalised before they are able to bear children.
Workers and the public are told that the probability of having such
severely damaged offspring after radiation exposure within permissible levels is slight. By omission, a mildly damaged child or a
miscarriage is implied to be 'acceptable'.

Standard setters judge that the most severe damage done directly
to the person exposed is a fatal radiation-induced cancer, and again,
this is a rare occurrence when exposure is within permissible levels.
All other direct damage is by omission considered 'acceptable'.

In its 1959 report recommending occupational standards for internal radiation doses (i.e. radioactive chemicals which are permitted to enter the body through air, water, food or an open wound), the International Commission on Radiological Protection (ICRP) formed the following definition:

A *permissible genetic dose* [to sperm and ovum], is that dose [of ionising radiation], which if it were received yearly by each person from conception to the average age of childbearing [taken as 30 years], would result in an *acceptable* burden *to the whole population.*[16] [Emphasis added.]

This might be paraphrased to say that the general public (governments) may be willing to accept the number of blind, deaf, congenitally deformed, mentally retarded and severely diseased children resulting from the permissible exposure level. Defined this way, the problem becomes primarily an economic one, since society needs to estimate the cost of providing services for the severely disabled. Once reduced to an economic problem, some nations may choose to promote early detection of foetal damage during pregnancy and induced abortion when serious handicap is suspected. When a foetus is aborted prior to sixteen weeks' gestation the event may not need to be reported and included in vital statistics. It becomes a non-happening, and the nation appears to be in 'good health', having reduced the number of defective births.

Mild mutations, such as asthma and allergies, are ordinarily not even counted as a 'cost' of pollution. The economic burdens, 'health costs', fall more on the individual and family than on the government. Their pain and grief do not appear in the risk/benefit equation. Parents and children are unaware of the 'acceptable burden' philosophy.

The prediction of the magnitude of the burden of severe genetic ills on an exposed population is essential to this philosophy. However, the data accumulated at Hiroshima and Nagasaki did not give the desired answers. Either through ineptitude or loss of survivors of the bombing, who died before their story was told, the researchers failed to find any severe genetic ills clearly attributed to the parental exposure to radiation at low doses.[21] Probably the more fragile individuals in the population died from the blast, fire and trauma of the bombs, the women not surviving long enough to become pregnant.[22]

Governments could not use the research on genetic damage in

children of medical radiologists,[23] although this damage was measurable, because, in the early days, radiation exposure to physicians was not measured. No quantitative dose/response estimates could be derived.

Animal studies of radiation-related genetic damage abounded, and the recommending body, ICRP, used (and still uses) mouse studies as a basis of its official predictions of the severe genetic effects of ionising radiation in humans.

As late as 1980, a US National Academy of Science publication from its committee on the Biological Effects of Ionising Radiation[24] stated:

New data on induced, transmissiblé genetic damage expressed in first generation progeny of irradiated male mice now allow direct estimation of first generation consequences of gene mutations on humans . . . As with BEIR I, a major obstacle continues to be the almost complete absence of information on radiation-induced genetic effects in humans. Hence, we still rely almost exclusively on experimental data, to the extent possible from studies involving mammalian species [i.e. mice].

These mouse studies are used as the basis of prediction, and permissible doses are set so that the expected number of severe transmittable genetic effects in children of those exposed could be presumed to be an *acceptable* burden for governments choosing a nuclear strategy.

The introductory section of ICRP Publication 2, 1959, states:

The permissible dose for an individual is that dose, accumulàted over a long period of time or resulting from a single exposure, which, in the light of present knowledge carries a negligible probability of *severe* somatic [damage to the individual] or genetic [damage to the offspring] injuries, furthermore, it is such a dose that any effects that *ensue more frequently* are limited to those of a minor nature that would not be considered *unacceptable* by the exposed individual and by competent medical authorities. Section 30.[16] [Emphasis added.]

Mild mutations are notably happenings of a minor nature, normally neither reported nor monitored in the population. They are likely to be statistically hidden by normal biological variations and unconnected in the mind of the individual or his/her physician with the exposure. The publication continues:

The permissible doses *can therefore be expected to produce effects* [illnesses] that could be detectable only by statistical methods applied to large groups. Section 31.[16] [Emphasis added.]

In spite of this clarity, no such statistical audit of all health effects including chronic diseases in exposed people and mild mutations in their offspring has ever been done. More than 25 years have expired since this document was published and the world is more than 35 years into the nuclear age.

As late as 1965, ICRP Publication 9[25] stated:

The commission believes that this level [5 rems radiation exposure per 30 years for the general public] provides *reasonable latitude* for the expansion of atomic energy programs in the foreseeable future. It should be emphasised that the limit may not in fact represent a proper balance between possible harm and probable benefit because of the uncertainty in assessing the risks and benefits that would justify the exposure. [Emphasis added.]

The committee protected itself against accusations of wrongdoing but failed to protect the public from its possible error. It defines its role as recommending, with the responsibility of action to protect worker and public health resting with individual national governments. Governments in turn tend to rely on ICRP recommendations as the best thought of internationally respected experts.

In spite of this uncertainty about responsibility and safety levels for exposure of the public, 5 rem per *year*, rather than per 30 years, was permitted for workers in the nuclear industry. The 5 rem per 30 years was set as the *average* dose to a population, with a maximum of 0.5 rem per year (15 rem per 30 years) for any individual member of the public.

For twenty years, between 1945 and 1965, health research on the effects of ionising radiation exposure has focused on *estimating* (not measuring) the number of *excess* radiation-induced fatal cancers and *excess severe* genetic diseases to be expected in a population (i.e. a whole country) given the *average estimated* exposure to radiation for the country. Disputes among scientists usually have to do with the magnitude of these numbers. Omitted from this research are other radiation-related human tragedies such as earlier occurrence of cancers which should have been deferred to old age or even might not have occurred at all because the individual would have died naturally before the tumour became life-threatening.

These are not *excess* cancers, they are accelerated cancers. This approach also omits other physiological disorders such as malfunctioning thyroid glands, cardio-vascular diseases, rashes and allergies, inability to fight off contagious diseases, chronic respiratory diseases and mildly damaged or diseased offspring. The implications of such 'mild' health effects on species survival seem to have either escaped the planners of military and energy technology, or to have been deliberately not articulated. Other obvious limitations of this national averaging approach include the failure to deal with global distribution of air and water with the result that deaths and the cumulative damage to future generations are not limited to one country.

The usual procedure for setting the standard for a toxic substance or environmental hazard is to decide the relevant medical symptoms of toxicity and determine a dose level below which these symptoms do not occur in a normal healthy adult. This cut-off point is sometimes called the tolerance level and it represents a sort of guide to the human ability to compensate for the presence of the toxic substance and maintain normal health. The tolerance level for a substance, if one can be determined, is then divided by a factor (usually 10) to give a safe level. This allows for human variability with respect to the tolerance level and also for biological damage which may occur below the level at which there are visible signs of toxicity, i.e. sub-clinical toxicity.

Human experience with ionising radiation had been recorded for more than fifty years prior to the nuclear age, the early history of handling radioactive material having been fraught with tragedy. The discoverer of the X-ray, W. K. Roentgen, died of bone cancer in 1923, and the two pioneers in its medical use, Madame Marie Curie and her daughter, Irene, both died of aplastic anaemia at ages 67 and 59 respectively. At that time, bone marrow studies were rarely done, and it was difficult, using blood alone, to distinguish aplastic anaemia from leukaemia. Both diseases are known to be radiation-related. Stories of early radiologists who had to have fingers or arms amputated abound. There were major epidemics among radiation workers, such as that among the women who painted the radium dials of watches to make them glow in the dark. Finally, there were the horrifying nuclear blasts in Hiroshima and Nagasaki.

The painful period of growth in understanding the harmful effects of ionising radiation on the human body was marked by periodic lowering of the level of radiation exposures permitted to workers in

radiation-related occupations. For example, permissible occupational exposure to ionising radiation in the United States was set at
52 roentgen (X-ray) per year in 1925,[26] 36 roentgen per year in
1934,[27] 15 rem per year in 1949[28] and 5 to 12 rem per year from 1959
(depending on average per year over age 18) to the present.[29]
Recently there has been an effort to increase permissible doses of
ionising radiation to certain organs such as thyroid and bone
marrow[30] in spite of research showing the radiosensitivity of these
tissues. This newer trend probably reflects economic rather than
physiological pressures, especially given the lack of an acceptable
audit of physiological cost.

Radiation Protection Standards

In 1952 the International Commission on Radiological Protection
(ICRP) issued its recommendations for limiting human exposure to
external sources of radiation. The newly formed organisation
accepted the standard agreed upon by nuclear physicists from the
USA, Canada and the UK after the Second World War.[31] In 1959 it
issued its recommendations for limiting human exposure to internal
sources of radiation. The early ICRP dose limits per year were: 5
rem to the whole body, gonads or active bone marrow; 30 rem to
bone, skin or thyroid; 75 rem to hands, arms, feet or legs; and 15
rem to all other body parts. These standards applied only to
'man-made' sources, other than medical exposures for diagnostic or
therapeutic purposes of benefit to the patient exposed.

ICRP Publication 2, in 1959, recommended no more than 5 rem
per year external or internal exposure to the whole body due to
inhalation, ingestion or absorption of radioactive chemicals into the
body. Sometimes this was misinterpreted and workers were
permitted to receive up to 5 rem internal and 5 rem external
radiation exposure during one year. Another clause allowing
averaging doses over years beyond age 18, gave excuse for still
higher doses.

In terms of the amount of whole body dose received in a chest
X-ray (about 0.03 rem at the present time), this recommendation
for workers allowed the equivalent of 400 chest X-rays in some
years with a 170 (present-day) chest X-ray average (external and
internal) dose a year. Prior to 1970 some X-ray machines used in
mass chest X-ray programmes gave as high as 3 rem per chest X-ray.

When one looks at dose to bone marrow, the permissible levels are even more troubling. By 1970 the average bone marrow dose for a chest X-ray was 0.001 to 0.006 rem averaging about 0.005 rem. In terms of dose to bone marrow, the ICRP radiation recommendation for workers permits up to the equivalent bone marrow dose of 1,000 chest X-rays per year.

ICRP recommended that members of the general public should receive no more than one-tenth of the occupational exposure or 0.5 rem per year, the equivalent bone marrow dose of about 100 present-day chest X-rays per year. The bone marrow dose is important for estimating the likelihood of causing bone cancer, leukaemia, aplastic anaemia or other blood disorders. Medical X-rays are less penetrating of bone than of soft tissue, making them valuable for 'picturing' the bones. For this reason comparisons between radiation exposures of nuclear workers and medical X-ray exposures are more appropriately based on the bone marrow dose of each than on dose to soft tissue.

These radiation exposure recommendations stayed essentially the same until 1978, when in ICRP Publication 26 a recommendation was made to *raise* the levels of radiation permitted to humans from man-made sources of radiation (excluding that for medical purposes). For 'internal consistency' of the recommendations there was some valid argument for scaling the standards for particular organ exposure in proportion to whole body exposure recommendations – but scaling down as well as up would have accomplished this. For example, the ICRP reasoned that if the whole body could receive 5 rem per year, the active bone marrow should not be limited to 5 rem per year. This was used as a reason for increasing the permitted bone marrow dose from 5 rem to 42 rem with apparently little regard for the increased damage to bones and blood-producing organs.

ICRP Publication 26 also reiterates the need to allow human exposure in order to enjoy the 'economic and social benefits' of the nuclear industries. It is difficult to understand how this conclusion was reached when so much new research is available documenting human illness associated with the present permissible exposure levels.[32] Perhaps, in view of contemporary scientific concern for *lowering* radiation exposures, ICRP Publication 26 recommendations are a political move to hold the line at present regulatory levels. At any rate, it appears to be a document with a political rather than a scientific purpose.

Some national regulatory agencies, such as the Atomic Energy Control Board in Canada, promptly implemented ICRP Publication

26 by increasing allowable radium levels in drinking water, thus reducing the clean-up cost for the uranium mining companies. Since some members of the national radiation protection community in Canada and elsewhere hold seats on ICRP, responsibility for what they recommend nationally cannot credibly be attributed to an international recommending body.

Failure to Audit Health

ICRP Publication 2 (1959) is one of special interest since it clearly states that radiation-induced severe genetic defects and cancer deaths resulting from the recommended standards would be expected to be rare and hardly distinguishable from 'natural' variations due to non-radiation causes. The document goes on to point out that mild mutations in offspring and general ill health in those exposed would be the most frequent health effects of exposure, but these could not be 'detected' except by epidemiological surveys. ICRP Publication 2 made no recommendation that this more subtle widespread degradation of public health *be* measured, although they mentioned that it *could* be measured.[33] At no time has there been an effort on the part of governments to document fully the more subtle health effects.

Workers, military service personnel and the general public have been given the impression that exposure to radiation involves a slight risk of dying of cancer and that one's chances of escaping this are better than the chances of escaping an automobile accident. The probabilities of early occurrence of heart disease, diabetes mellitus, arthritis, asthma or severe allergies – all resulting in a prolonged state of ill health – are never mentioned. Most people are unaware of the fact that ionising radiation can cause spontaneous abortions, stillbirths, infant deaths, asthmas, severe allergies, depressed immune systems (with greater risk of bacterial and viral infections), leukaemia, solid tumours, birth defects, or mental and physical retardation in children. Most of the above-mentioned tragedies affect the individual or family unit directly and society only indirectly. Dr R. Mole, a member of ICRP and the British NRPB, stated: 'The most important consideration is the generally accepted value judgment that early embryonic losses are of little personal or social concern.'[34] There are similar value judgments made with respect to other health effects. The health problems are externalised, i.e. placed beyond the responsibility of government, and they are borne by individuals and their families.

The risk/benefit decision making which arose from balancing 'health effects' against 'economic and social benefit' is based on risk and benefit to *society*, i.e. governments, rather than cost to the individual or family unit. Value judgments have been made as to the level of health effects and deaths 'acceptable' to the public. Because of military control of A-bomb studies and military need for personnel to handle radioactive materials, many of these value judgments were cloaked in secrecy for the sake of 'national security'. The subject was made to seem complicated to outsiders; the decisions were reserved for the experts. The now famous words of President Dwight D. Eisenhower, 'Keep the public confused'[35] about nuclear fission so that the government could gain public acceptance of above-ground weapon testing in Nevada, have certainly been accomplished. A growing number of people in the USA and elsewhere have lost all faith in statements made by government officials, because of the scientific jargon used to mask the truth.

In the USA, external radiation exposure records (film badge and TLD* readings) are carefully kept for workers, but corresponding health records for workers are not kept and analysed nationally. In other countries, especially those with socialised medicine, excellent health records are kept but accurate radiation exposure records are neglected. Collection and analysis of radiation exposure records together with experience of ill health, including chronic long-term (non-fatal) problems, are required in order accurately to assess radiation-related health problems. Merely recording the first cause of death for workers is not sufficient.

The public is at an even greater disadvantage than the worker. There are no cumulative records of radiation exposures for individual members of the public from nuclear testing, military or commercial nuclear industries anywhere in the world. Because of this record-keeping vacuum, it is difficult, if not impossible, to challenge ICRP predictions.

---

* TLD – thermoluminescent dosimeter, used to measure individual radiation doses of workers. It contains radiosensitive chips and must be carefully screened for the kind of radiation it is meant to detect. In a pilot study done in the US some processors of TLDs discovered that some of their chips were completely insensitive to the type of radiation for which they were purchased. See P. Plato and G. Hudson, 'Performance Testing of Personnel Dosimetry Services: Alternatives and Recommendations for a Personnel Dosimetry Testing Program', US Nuclear Regulatory Commission (NUREG/CR–1593), 1980, p. 9.

Inadequate collection of information on public health by governments makes it difficult for scientists concerned about rising radiation exposure levels to document changes in public health. The problem is not that they are poor scientists, but that they do not have access to detailed information, since governments have failed to collect it. The health changes which can be detected, in spite of poor records, represent only a minute proportion of the undocumented whole.

One key to understanding what priority a country places on the health consequences of national defence and energy choices is the precision of its measurements of resultant health effects. Measurements of health effects can be made through controlled animal experiments or observation of the effects of unplanned human exposures. These measurements serve as an audit of human health effects or as an after-the-fact check on the accuracy of predictions. This technique of controlled observation is normally applied when a new drug or new medical procedure is introduced into general use. A prediction must prove its worth in real life.

As one would expect, predictive dose/response estimates for radiation exposure and specifically chosen severe health effects have been prolific in the USA. Not only has the USA maintained a tight control over and interest in research on the Japanese survivors of radiation exposure from the nuclear bombing of Hiroshima and Nagasaki, it also has a system of government-sponsored research laboratories controlled successively by the Atomic Energy Commission, the Energy Research and Development Administration and the Department of Energy. These bodies have been the source of almost all the original research papers published between 1945 and 1977 on the health effects of ionising radiation. Because radiation-related health effects are the result of the production, testing and use of atomic weapons, military goals and military secrecy have influenced both the selection of research questions and release of findings in the USA. The nuclear age is predicated on public acceptance of its consequences, hence 'proving' that public acceptance is 'rational' has a very high priority for government and industry-employed scientists. They have a vested interest in verifying the status quo.

Prior to the above-ground nuclear weapon test ban in 1963, the USA set off at least 183 atmospheric nuclear tests, more than all the other nations of the world combined. About half these tests were set off near the Pacific Trust Territory of Micronesia, given into US protection by the United Nations after the Second World War, and the other half were set off on the 1,350 square miles at the Nevada Test Site north of Las Vegas. By 1978 the USA had set off an additional 400 nuclear bombs below ground in Nevada, some of which were officially admitted to have 'leaked' large amounts of radioactive chemicals. Some of the tests were of UK weapons since it also uses the Nevada test site. Underground tests are still taking place in the USA,[36] the USSR and French Polynesia. In the Northern Hemisphere, above-ground tests have also been detonated by the USSR, China and India and in the Southern Hemisphere by France and South Africa.

The Nevada nuclear tests have spread radiation poisons throughout central and eastern United States and Canada, and produced in the stratosphere a layer of radioactive material which encircles the globe. They also cause nitric oxides to form in the atmosphere which then descend on earth as acid rain. Radioactive chemicals can now be found in the organs, tissues and bones of every individual in the Northern Hemisphere, and the contamination from past nuclear explosions will continue to cause environmental and health problems for hundreds of thousands of years, even if all nuclear activities are stopped today. Siberian tests affect the north polar region.

Pollution of the Southern Hemisphere, though less than in the North, is progressing along the same path. Although the United States and Great Britain have ceased nuclear tests in the Pacific Ocean, France has not ceased them, and it appears that South Africa has begun to test. Brazil, Argentina and other nations are thought to be developing a nuclear weapon capability.

A 1977 report of the United Nations Scientific Committee on the Effects of Atomic Radiation stated that twenty atmospheric nuclear tests – six in the Northern Hemisphere and fourteen in the Southern Hemisphere – plus unnumbered underground tests, took place between 1972 and 1977. As a result of this nuclear testing radiation doses to the population increased by about 2 percent in the Northern Hemisphere, and 6 percent in the Southern Hemisphere, over the dose estimated in 1970. The nuclear weapon testing carried out between 1972 and 1977 was insignificant when compared to that between 1945 and 1963.

The total global dose commitment for each individual from all nuclear explosions carried out before 1976 ranges from about 100 mrad (in the gonads) to about 200 mrad (in the bone-lining cells). In the northern temperate zone the values are about 50 percent higher, and in the southern temperate zone about 50 percent lower than these estimates.[37]

This estimate does not include the dose from radioactive carbon (carbon 14) which, because it takes time to distribute in the oceans of the world and return to humans via the biological food chain, has not yet taken its human toll. For comparison purposes, 100 mrad is about equal to the amount of radiation a person receives from naturally occurring radiation in one year of chronological ageing. The dose commitment from nuclear weapon testing is spread over a fifty-year period, with most of the dose being delivered in the first year.

There has been no lack of victims of radiation pollution in the West to study both for refinement of predictions of biological harm and checks of adequacy of predictions relative to the real-life situation. Checking *adequacy* of predictions means including all hidden costs which must eventually be paid, including damage to agriculture and the biosphere. Government oversight should also include full disclosure of findings to the public as a test of the acceptability of such costs and as an evaluation of the judgments made for society by the nuclear experts.

Can Health be Measured?

The obvious answer is that we can, of course, find a way to measure gains and losses in health; only the will to do so is lacking. In order to measure subtle changes in health a good reporting and recording system is needed, together with protection of privacy for the individual and ongoing biostatistical analysis of the accumulated data. Whole bodies of statistical theory, such as sequential analysis, used for product quality control, and system analysis, used to predict the outcome of a complicated interaction between inter-dependent variables, need to be used in the public health sector. This could provide a public health technology capable of managing military and industrial technology, able to act as a reality check on predictions and to give an early warning of dangers arising from within the bio-system and threatening survival of the nation or, indeed, the human race. Biostatistical detection of problems needs to be followed by pathological, cytological and other confirmatory studies. No such serious systematic commitment to public health is evident relative to this nuclear issue anywhere in the world.

Governments seem unaware that economic and military policies can be destructive of human health within the nation.

The radiation issue is further confused by statisticians and public health specialists who claim that there are some inherent and insurmountable problems which make it impossible to monitor the public health effects of pollution.[38] These professionals seem to limit themselves, consciously or unconsciously, to current inadequate data collection systems and mathematical tools. This is like deciding that it is impossible to travel to the moon on the basis that the only transportation possible is a commercial airliner. It will very probably require grass-roots scientific initiatives to cause governments to begin to act as strongly in protection of public health as they act to promote their own economic and military strategies.

Many people have become aware that national security strategies, especially nuclear weapon stockpiling, are increasing individual insecurity. Capital-intensive national economic strategies, designed to balance import/export dollar flow, can cause havoc with the individual citizen who is having to cope with the side effects of inflation and unemployment. Government neglect of health monitoring relative to economic and military strategies is, however, not yet perceived by the public as a serious problem.

It should be obvious that pollution of the environment with fission products will cause a wide variety of physiological changes in people exposed to them. There is little disagreement among scientists with regard to this conclusion.

There is also little controversy about the tragedy caused by uncontrolled fission – whether deliberately or accidentally unleashed, whether from a nuclear reactor accident or an exploding warhead.

The question which causes controversy is: which health effects should be recognised as important for fiscal planning? 'Important' may relate to public acceptance of the problem, or to the money which must be paid out for damage compensation, or the productive years lost through premature disability or death of workers. Once the significant health effects are identified, then quantification of these effects becomes the primary societal goal. This gives rise to scientific controversies. Present scientific controversy on low level radiation has to do with estimating the number of radiation-induced 'excess cancer deaths' that are related to a given dose of ionising radiation. Fiscal concern has centered on radiation-induced excess cancers, and scientific concern on predicting this outcome.

These excess cancer numbers are important to planners who wish to show that their development schemes are less harmful than an alternative scheme. They are important to government officials who have to decide whether or not to assume the financial burden of ordering evacuation of a danger zone in a reactor accident like that at Three Mile Island. They are important to insurance companies, since they allow calculation of theoretical liability due to an accident. They are important to legislators who need to balance risks (deaths) against some military or economic benefit. They are important to strategic planners who calculate 'collateral damage', i.e. the number of human deaths, after an atomic attack.

These numbers of specifically selected health effects, 'radiation-induced excess cancer fatalities', predicted on the basis of the 'average man's' reaction to a given average dose of ionising radiation, are of little meaningful use to individuals. Firstly, no one is really an 'average man'. Also, populations may vary in the proportion of people with above-average susceptibility to radiation damage. Secondly, a 'radiation-induced excess cancer fatality' is one of the least likely of the health problems to occur with exposure to low level radiation. More likely scenarios are radiation accelera-tion of a cancer caused by some other factor, such as cigarette smoking,* an earlier clinical expression of cancer, benign tumours, or related non-malignant health problems. Thirdly, even if the individual has a cancer it is almost impossible to present evidence to prove that his or her cancer is the *excess* one which would not have occurred without the radiation exposure. Therefore compensation for damage is almost impossible to obtain. Only one veteran from the USA exposed to radiation in its nuclear bomb programme has ever received compensation: Orvile Kelly. About six months before he died the Veterans Administration admitted that his illness could be attributed to radiation exposure. About 1,000 veteran claims have been refused.[39]

The usual 'rational' approach to risk versus benefit planning by governments is irrational from the point of view of the individual. It undermines the individual's ability to control and understand his or her environment and to hold government accountable to its electorate.

The human body is delicately fashioned and the unique gifts of each person are meant to enrich the human family. Crude quantification of random damage to people which is used to justify

* Many researchers believe that the primary carcinogen in cigarettes is polonium 210, a radioactive daughter product of radon gas which is released from uranium mining and mill tailings.

political or military gains of the nation may be labelled sophisticated barbarianism. It is the decadent thinking of those who have accepted the rule of force and who envision a future earth ruled by a powerful country (the USA or the USSR) with a monopoly of weapons of mass destruction, able to terrorise all other nations into co-operating with some form of global economy and resource-sharing of their choosing.

The Health Physicist

A word needs to be said about health physics, a relatively new academic specialty which has emerged since the dropping of the atomic bomb. Systematic study of radiation health questions began at the University of Chicago when the first nuclear reactor began operating on 2 December 1942. Primarily under the leadership of physicists E. O. Wollan, H. M. Parker, C. C. Gamertsfelder, K. Z. Morgan, J. C. Hart, R. R. Coveyou, O. G. Landsverk and L. A. Pardue, it grew to become a recognised graduate-level discipline.[40]

While this was a much-needed specialty, its bias toward the so-called 'hard sciences' – physics, chemistry and engineering – and neglect of the 'soft sciences' – biology, physiology and psychology – has tended to create radiation safety officers rather than health professionals.

In a message from the President of the Health Physics Society published in the July 1971 issue of the *Health Physics Journal*,[41] Dade W. Moeller stated:

> I think it is interesting to note the results of a tabulation of the records of the 2,862 health physicists who joined the Society from 1960 through 1969. The data showed that although *half* of the new members *with college degrees* had attended graduate school for a year or more, 21.6 per cent of the new members *did not have a college degree*. [Emphasis added.]

Membership of the Health Physics Society is broader than, but includes, licensed health physicists who have passed qualifying examinations. These latter are generally required to have a college degree with a major in physics, chemistry or engineering, and one year of graduate training in radiation measurement and safety practices.

Dade W. Moeller goes on to describe the members who had a college degree:

> by far the greatest percentage (24.0 per cent) received their bachelor's degrees in physics and/or mathematics. Next was chemistry (15.8 per cent) and then engineering (13.6 per cent).

Even members of the Health Physics Society have complained about the pro-nuclear bias of its publication[42] but seldom has this been expressed as clearly as in this address by Dade Moeller. After reporting a need for 2,000 to 3,000 more health physicists by the year 2000 just to support the operation of nuclear power stations, he urged members to be active: 'To paraphrase an old adage, "let's all put our mouth where our money is".'

Unfortunately, the Health Physics Society probably will not be in the vanguard speaking on behalf of workers and members of the public whose health is at risk from nuclear industries. The obvious and outstanding exception to this statement is Dr Karl Z. Morgan who has remained an open, honest and independent student of life. Dr Morgan has spoken out courageously on behalf of lowering worker and public exposures to radiation and avoiding all unnecessary exposures. In so doing he has alienated many of his peers and jeopardised his own research and teaching position. Karl Morgan was a friend of Hermann Müller and he remembers the geneticist's warning about undermining the health of a nation and its children.[43]

The United States, a leading nuclear nation, has failed to provide any reliable human health study either to confirm or to deny its prediction of the human health effects of exposure to chronic low level radiation, or even to provide a systematic health follow-up of the significant groups exposed to radiation so that there will in time be such a reliable study. The predictions of health effects are based primarily on the effects reported at Hiroshima and Nagasaki and the applicability of these estimates to chronic low dose exposure of a normal population has always been doubtful.[21]

The US government has also failed to supply the worker or the public with trained health professionals whose jobs are independent of the nuclear industry and whose training and background would enable them to alert people to a slowly deteriorating health situation. Adequate record-keeping and reporting would force public awareness of the problems, and probably the facing of ultimate questions such as: for what perceived benefit can society sacrifice the health of future generations?

The health physicist, while serving a necessary safety function within nuclear installations, does not fulfil the role of a health advocate in this situation. His or her job is to enforce regulations, not to question them and to support the nuclear plant management even if it is clear that the management is wrong.[44] This is not so much the result of malice as a normal outcome of believing 'permissible' is the same as 'safe', and trust that present regulations are 'very safe'. It thus becomes acceptable to handle radioactive material and to cheat a little on over-exposures.

The first key to understanding governments' commitment to ensuring the survival of individual citizens is its adoption of a verification process for testing its prediction of severe health effects resulting from its economic and military strategies. In the United States, this leads to a preliminary judgment that individuals have been considered expendable. Health damage from radiation associated with military or economic ventures has not been easily traceable to the cause or immediately apparent to the public. No efforts deliberately to trace and make public all the health effects have been made. In fact, when any research has begun to show such effects, the researcher has been 'discredited' and his or her funding discontinued.

On the basis of the US government's neglect of follow-up and record keeping on radiation-exposed people, and its lack of concern for mild genetic effects, the unrest of the US public with respect to further development of nuclear technology is highly rational. Continuance of present government neglect and unconcern is at best irrational and at worst genocidal. We may observe the same syndrome of irrational behaviour in other nuclear nations which are experiencing public unrest.

Although the problems inherent in the production of nuclear weapons and nuclear power reach a climax of scale in the United States, they are experienced in all countries with nuclear technology. Where one country may keep excellent public health records, it has poor records of individual radiation exposures. Where another keeps detailed radiation exposure histories, it has no detailed medical history. As long as part of the information is missing, the worker and general public are forced to rely on predictions made by 'recognised experts' which are not verified by factual studies. This is really a forecast with no audit allowed. The promotion of nuclear technology in developing nations as the industry loses support in the developed world is even more disturbing.

Before moving on, some of the concepts of radiation protection important for nuclear workers, the general public and medical personnel need to be emphasised. First, an assurance of 'no immediate danger' with respect to exposure to ionising radiation is empty when it masks long-term effects resulting from incorporation of radiochemicals in sensitive tissues and/or the results of biological magnification of cell damage or radiation-induced genetic mistakes. Secondly, independent testing of urine, faeces, exhalation, tissues removed in surgery, baby teeth and hair for radioactivity, must become routine laboratory tests for medical diagnostic purposes as we try to cope with the fission product pollution already in the biosphere. Thirdly, when assessing the impact of any leak, abnormal release, normal effluence or waste which is radioactive, it is essential to know the radiochemicals involved: their physical and biological properties, the potential pathways to human beings and the length of time they remain toxic. Fourthly, the health effects of radiation differ with the age of the person exposed, his or her physical status and prior experience.

The second key to governmental priorities in decision-making is found in the historical context of the nuclear development. This is examined later. First we must try to understand the practices of nuclear technology in the military and civil sectors.

PART TWO

The Practices:
Military and Civil Nuclear Technology

Parental Exposure: Ted Lombard's Story

To telescope an individual's whole personal and family medical history into one fact, the first cause of death recorded on the death certificate, as is done in most health research, is a very crude way to measure the quality of individual or collective health. It is like trying to do bookkeeping for a business when you know only the closing balance each month. As an illustration, the story of Ted Lombard, from the USA, is included here. Ted has not yet died of a radiation-induced fatal cancer, nor do his children have officially recognised genetic diseases.

In 1944, during the Second World War, Ted left his young wife and small son to 'make the world safe for democracy'. He was stationed at Los Alamos, New Mexico, and given the responsibility of transporting uranium and plutonium. Because of secrecy and the few records remaining about this part of the Manhattan Project, Ted's level of exposure and the description of his duties are labelled by Veteran Administration (VA) officials as 'unverifiable'. Ted regularly transported enriched uranium and plutonium, extracted from the weapon reactors at Hanford, Washington and Oak Ridge, Tennessee, from a pick-up area, Fort Douglas, outside Salt Lake City, to Los Alamos, New Mexico. He also worked at times near the processing building in Los Alamos and handled the radioactive material freely without gloves or protective clothing and without knowing its danger.

Ted was at Los Alamos when Harry Dahlian, a civilian from Purdue University, died after a criticality accident exposed him to radioactive fission products. Harry lived for three and a half weeks after the accident and one of Ted's friends was sent to photograph him every day in hospital as he slowly died a horrible death. Though

65

The First Nuclear Accident*

Harry Dahlian, an employee at the Los Alamos (New Mexico) laboratory, was working alone during an August night in 1945. A guard was seated on a stool about 12 feet away. Dahlian was in the process of making critical mass studies and measurements, gradually adding tamper material to an assembly of fissionable material and watching an instrument which indicated when concentration would cause fission multiplication, i.e. a chain reaction, to be produced. The last piece of tamper material was quite heavy. Dahlian noticed that the instrument was indicating an impending critical configuration as he brought the piece closer to the already assembled pile of fissionable material. He tried to move the block away, but as he did so he dropped it on top of the assembly.

A blue-glow was observed and the employee proceeded to disassemble the critical material and its tamper. In doing so, he added heavily to the radiation damage to his hands and arms.

Dahlian died 24 days later. The guard was exposed 'beyond the established daily limit'. He was reported as having no immediate injury 'observable', and was never followed up for the purpose of learning the long-term effects of his exposure.

* Information on this accident can be found in 'Radiation Accidents: Dosimetric Aspects of Neutron and Gamma-Ray Exposures', Oak Ridge National Laboratory, ORNL–2748, Part A, p. 3. It is also found in 'A Summary of Accidents and Incidents Involving Radiation in Atomic Energy Activities, June 1945 through December 1955', TID–5360, p. 2.

the men were all forbidden to talk about the incident, Ted never forgot it.

While still at Los Alamos, Ted experienced stomach problems, diagnosed as ulcers, and infections in his feet, diagnosed as fungus, by military physicians. By the time he was discharged from the army, his eyes were changing noticeably, varying at each medical examination. At time of discharge, he was told he might be sterile, and also that if he had any future medical problems the fact that he had worked on the atomic bomb project would guarantee medical compensation. He had no written proof of his participation, however.

Ted eventually became an assistant vice-president at a Boston bank. He and his wife, Ruth, had four children in the twelve years after his discharge from the service. He was not, obviously, sterile.

Barbara, the first child born after the war, appeared normal but was unusually susceptible to measles, chicken pox, and other viruses. It was finally discovered that her body lacked certain antibodies and enzymes, a condition never previously identified in the family history. Barbara later developed neuro-muscular problems, eventually becoming severely disabled.

The second daughter showed the same inability to produce antibodies and enzymes as did Barbara, but managed to live a more normal life without developing severe neuro-muscular disease. She now has two young children of her own, who are beginning to experience physical symptoms which may indicate the same genetic damage she and her sister and brothers experienced.

The third child, a boy, has severe migraine headaches, seizures and dyslexia. He has a chronic susceptibility to respiratory disease and other medical problems.

The youngest son is epileptic, deaf and mute. He has had to be institutionalised as mentally retarded and aphasic (i.e. speechless).

The child born to Ted and Ruth prior to his enlistment has none of these medical problems.

Periodically, when his children began experiencing problems, and when his own ulcer and skin condition grew progressively worse, Ted asked VA officials for his Los Alamos records. They always said the records were not available. Finally in 1960 he filed a claim with the VA. They eventually ruled that there were no medical records which indicated that he had been sick while in the service, and therefore he had no claim. He accepted the ruling without complaint, thinking he was probably wrong and the VA was being just.

As his own and his children's health continued to deteriorate, Ted began to confront VA officials in further attempts to obtain his army medical records. The VA never released his records to him and never honoured his claim.

Finally, in 1976, after appealing to a local congressman and to the Energy Research Development Administration, Ted's efforts began to produce some of the missing records. Ted's claim was turned down again in 1979 but he was given partial assistance to help with his medical bills. Ted now still has the undiagnosed dermatitis and the stomach ulcer as well as severe pulmonary fibrosis and other medical problems. He is totally disabled at the age of 57 years.

The VA board stated:[1]

We understand the anguish which the veteran has experienced as a result of his children's disabilities . . . we note that there is no provision in laws and regulations for compensation based upon birth defects in children claimed to be caused by the genetic injury to the parent-veteran in service.

The VA Board recently granted partial compensation for Ted's ulcer and skin condition, but not as radiation-related. They declined compensation for his lung fibrosis since Ted smoked. The VA argued that they could not be absolutely sure that it was the inhalation of radioactive plutonium and uranium which had caused the lung damage. It is usual for moderate smokers to experience severe lung fibrosis at fifty-five years of age. It seems probable that the VA is fearful of setting a precedent for radiation exposure compensation to veterans because of the 250,000 to 1 million exposed men who would be potential claimants.

In 1980 Ted had a blood chromosome test done. It showed that he had severe bone marrow damage causing a high number of abnormal blood cells. In reporting the test results, Dr Avery Sandberg, who knew nothing about Ted's personal or family history, made the comment that this level of damage to blood chromosomes usually indicated that there was also genetic damage to offspring.[1] Blood chromosome tests on the children have not revealed any gross abnormalities.

In 1983 a physician at the Radiation Oncology Centre, Palms of Pasadena Hospital, examined Ted's hands and arms, which still showed the rash, scaling and fissures he had had since his Alamos days. The physician diagnosed it as 'chronic eczematory dermatitis, consistent with but not pathognomic [exclusively characteristic] of radiation dermatitis'.[1]

In a letter dated 4 September 1981, on official letterhead stationery of the General Counsel of the US Department of State, William H. Taft, IV, tried to persuade US Representative to Congress, G. V. Montgomery, Chairman of the Committee on Veteran Affairs, to delete assistance to veterans exposed to nuclear weapon testing from pending legislation. Taft argued:

We are more concerned, however, with the perceptions among both veterans and the general public which would result from enactment of HR 3499 as passed by the Senate. Section 3 of the Senate passed bill creates the unmistakable impression that exposure to low level ionising radiation is a significant health hazard when available scientific and medical evidence simply does not support that contention. This mistaken impression has the potential to be seriously damaging to every aspect of the Department of Defense's nuclear weapons and nuclear propulsion programs. The legislation could adversely affect our relations with our European allies, impact upon the civilian nuclear power industry, and raise questions regarding the use of radioactive substances in medical diagnosis and treatment.

This carefully worded lawyer's message clearly states the vested interest in low level radiation. Any evidence of its harm would obviously be seen as subversive. Acknowledging that there is a problem would identify the victims and provide evidence – an unwanted outcome.

The Legislative Act before the US Congress in January 1983 was modified to read: 'persons injured by toxic chemicals and other hazardous substances'.

Because no systematic records of the health of men exposed to the military nuclear operations have been kept, each one must now individually try to prove a causal connection between his exposure and subsequent health problems. Records documenting exposure are either non-existent or difficult to obtain. VA procedures are antiquated and unable to deal justly with delayed health effects or the ill health of offspring, which are service connected. Ted Lombard's story is only the tip of an iceberg of stories of records not kept and tales not told.[2]

Besides the direct exposure of military personnel, thousands of unidentified individuals have been exposed at various radioactive dumpsites throughout the USA. For example, in the Four Corners area of New Mexico where intensive uranium mining took place in the 1950s, Navajo Indian children under the age of fifteen were found to have 17 times the expected rate of reproductive organ cancers, 5 times the national rate of bone cancer, 1.7 times the national lymphoma rate and 1.6 times the national rate of cancer of the brain and central nervous system.[3] Many military accidents have gone unreported, shrouded in secrecy for 'national security' reasons. For example, the plutonium fires and spills at Rocky Flats, Colorado, the plant which produces the triggers for all the nuclear bombs in the USA, were not reported to the people of Denver 16 miles downwind of the plant. It was an independent researcher, Dr Edward Martell, who went out and tested the soil, discovered the plutonium pollution and reported the accidents.[4] Dr Carl Johnson,

Jefferson County Health Commissioner, reported increased cancer rates in the vicinity of the Rocky Flats plant.[5] In these and other cases of radiation exposure, neither the federal government agency knowledgeable about the problem nor the industry responsible for the contamination told the public, or tried to determine the actual damage to health.

The Marshallese

The Marshall Islands, a part of Micronesia, have borne the brunt – 66 out of over 200 – of US nuclear bomb blasts in the Pacific Ocean. Usually the radiation doses were small, received via the food chain: fish, water, coconuts and other foods. The small damages to health accumulated slowly, almost imperceptibly, until they gradually began to vie with economic and social matters in the minds of the islanders. One incident, however, was immediately traumatic. A US nuclear test in 1954, at Bikini, caused acute radiation sickness in the people of Rongelap: the symptoms were vomiting, diarrhoea, severe skin burns, irritation of the eyes and throat, bone marrow damage and blood changes. About three weeks after exposure most of the indigenous population and the black US servicemen developed a transient bluish-brown colour on their fingernails and toenails (never satisfactorily explained). During the first two months after exposure more than half of the exposed population suffered in an epidemic of upper respiratory diseases. More than 90 percent of the Rongelap children had epilation (loss of hair on the head) and scalp lesions. This occurred to a lesser degree (about 28 percent) in the adults.[6]

The Pacific Ocean nuclear test blast on 1 March 1954, resulting in the extraordinary fallout in the Marshall Islands which caused acute radiation sickness, was a 17 megaton blast, *about 1,000 times the force of the bombs used at Hiroshima and Nagasaki.* It was the largest detonation of a hydrogen bomb. The navy reported that a shift in the wind caused the radioactive cloud to drift over the inhabited islands but this shift has been denied by several of the US weathermen who were on Rongerik and who reported every four hours to the military command post on Enewetak.

The people of Rongelap, about 100 miles from Bikini, received the fallout first, about four to six hours after the blast. It took another hour to reach the twenty-eight American weathermen on Rongerik and twenty-two hours to reach Utirik (about 280 miles from Bikini). The blast itself was 100 miles off the coast of Rongelap, so no one on Rongelap suffered blast and/or fire effects of the bomb as had been the case in Hiroshima and Nagasaki. Their injuries and diseases were the direct result of radioactive fallout.[7]

There was a cloak of military secrecy over the incident, so none of the acute immediate health effects are attributable to 'scare tactics' by the media, or to public panic. The people had no information about the testing. No precise measuring of the individual exposures was undertaken but it was assumed to be quite high, ranging from a total gamma dose of 14 rem* in Utirik to 175 rem in Rongelap. This could be compared with the 5 to 12 rem permissible exposure for nuclear workers over the course of a year, or the 0.5 rem per year allowed for the general public.

The children of Rongelap had, of course, never seen snow, but after the hydrogen bomb explosion of 1 March 1954 on Bikini, small, white, powdery material began to fall out of the sky. It was one and a half inches thick on the ground in places. No one knew what it was and the children were allowed to play in it. Forty-eight to seventy-two hours later, US military personnel arrived and informed the people that the white fallout was from a thermonuclear device exploded on the Bikini atoll. The inhabitants of Rongelap and Uterik were to be evacuated immediately; no more food or water from the island should be consumed. It was difficult for the people to comprehend that this white rain was poison; that for two days they had been consuming the contaminated water and food and that the children had played in it. The polluted air had already been breathed and now their homes and possessions and land must suddenly be abandoned. Thus was the Bikini hydrogen bomb exploded, escalating the nuclear arms race from the kiloton to the megaton range. With some demented sense of humour, the bikini bathing suit was named for the atoll whose middle was blown away in the hydrogen bomb blast.

During the first five years after the radiation exposure of the Marshallese there was a significant increase in miscarriages and stillbirths among exposed Rongelap women.[6] When the number of offspring deaths increases we can assume that the proportion of damaged but viable children also increases. But these slightly damaged survivors were never identified by the researchers, so there is no count. It took about nine years for researchers to notice

---

* For external radiation, rem and rad measurements are generally interchangeable. A mrad or mrem is one-thousandth of a rad or rem. 1 rem = 1,000 mrem and 1 rad = 1,000 mrad.

the high rate of growth retardation and thyroid abnormalities in the Rongelap children. In 1972, Lekoj Anjain, from Rongelap, died of leukaemia at the age of 19. He had been exposed to the gamma radiation cloud on Rongelap at the age of one year and had played in the white fallout. Lekoj had suffered acute radiation sickness, beta skin burns and epilation (loss of hair) at the time of the nuclear explosion but had seemed to recover. When 13 years old, he had surgery for thyroid nodules. His mother, father and two brothers also had thyroid surgery. Lekoj died in the USA at the National Cancer Institute Hospital in Bethesda six weeks after his admission. He had been the youngest child on Rongelap at the time of the bombing.

Medical examinations carried out on adults in Rongelap between 1970 and 1974 compared exposed and unexposed islanders. The results indicated that there was a higher than average incidence among the exposed population of acne, anaemia, arteriosclerosis, bradycardia (unusually slow heart pulse), cervical erosion, cystourethro-rectocele (urinary tract or intestinal) hernia, leprosy, migraine, prostatic hypertrophy, rheumatic heart disease, benign and malignant tumours. While the incidence of some of these conditions might have been higher by chance, the overall level of ill health could not be explained away. The most apparent and widespread health effect noted among the Marshallese was thyroid disease. This was very probably associated with the inhalation and ingestion of the radioactive iodine in the fallout. Diabetes incidence in the islands was found to be the highest in the world, with the possible exception of certain American Indian groups. It did not seem to be associated with radiation dose, occurring in about 28 percent of all the islanders irrespective of the level of radiation to which they had been exposed. Studies of diabetes among Marshall islanders are still in progress and the relation of this condition to the pollution is not clear.

In 1972 the Japanese Atomic Bomb Hospital in Hiroshima opened its doors to Marshallese radiation victims. There are few other medical centres in the world able to recognise or treat this type of 'sickness'.

With respect to genetic damage, a report issued by Dr Conard states: 'on the basis of incidence of *gross anomalies*, no evidence of inherited radiation-induced mutations' was found.[6] Just as in the case of Ted Lombard, gross changes do not tell the whole human story or encompass the whole reproductive experience. The number of children studied was small and there had been an unusually high number of miscarriages and stillbirths in the first five years after exposure. None of the severe genetic diseases in live-born offspring, which are officially used in the developed world to measure the 'acceptability' of radiation exposure, were reported as having been observed in the Marshallese population by Dr Conard. This may well indicate that the list of criteria is irrelevant. The Marshallese people report gross changes in their offspring. Women in the Marshalls have described their babies as 'a bunch of grapes . . . or a jelly fish baby . . . the heart beats for 2 to 12 hours and then they die.' In the case of a blighted embryo, the uterus can fill with cystic grape-like structures varying in size from microscopic to 3 centimetres in diameter. Sometimes one large hydatidiform mole is formed from a fertilised ovum which has lost its nucleus. These women experience shame and humiliation at not being able to bear normal children, not wanting even their husbands to see the offspring.

Dr Conard has established a body of data on the Marshallese with respect to their genetic characteristics after the nuclear experiment which he hopes 'may be useful as a base line, should genetic changes appear in later generations'. It is obvious that such a base line comes too late to be useful. No other population in the world has a scientific base line of genetic information against which to measure changes in successive generations.

The Bikinians had been evacuated from their island homes before the nuclear tests. The American military governor of the Marshall Islands, Commodore Ben Wyatt, went to the island before the Pacific nuclear tests began and spoke to 167 islanders at the closing of their Sunday religious service on 10 February 1946. According to official navy records, Wyatt 'compared the Bikinians to the Children of Israel whom the Lord had saved from the enemy and led into the Promised Land'. He asked them if they would be willing to sacrifice their island for 'the good of all men'. Chief Juda later reported to the Commodore the decision reached by the Bikini islanders: 'If the US Government and the scientists of the world want to use our island and atoll for further development, which with God's blessing will result in kindness and benefit to all mankind, my people will go elsewhere.' The Bikinians had been awed by the US defeat of Japan, and Wyatt's description of the atomic bomb led them to believe that they were unable to resist any wish of the US military.

Navy records show that Wyatt was asked what commitment concerning reparations he could make to the people. 'He was told that he could promise them no more than the opportunity to submit claims for damage.'

The Bikinians were first moved to Rongerik Atoll, which proved unable to support them. Some of the fish were poisonous and the vegetation was much more sparse than on Bikini. In March 1948, after they began to show visible signs of malnutrition, the islanders were moved to Kwajalein, an atoll being used by the USA as a support base for the Bikini and Enewetak tests. That summer they were moved again, to the island of Kili, about one-sixth the size of the Bikini atoll. The once self-sufficient population became dependent on US food supplies and never really adjusted to their supposedly temporary exile.

Although navy records showed that a 1946 nuclear blast had left 500,000 tons of radioactive debris in the Bikini lagoon, navy press releases were always optimistic and reassuring: 'Scientists now engaged in an intensive six-week survey of the Bikini Atoll can find few visible effects of [the atomic tests]' (1947).

The US Marshall Island testing ended in 1958, but it was not until 1967 that a group of 'prestigious scientists' appointed by the US Atomic Energy Commission declared the atoll 'once again safe for human habitation'. Four to five million dollars were spent cleaning up the atoll; 50,000 new trees were planted and 40 homes built. About 150 jubilant islanders moved back to Bikini in 1971.

The Bikinians old enough to remember the atoll before the bomb tests could hardly believe that some islands had evaporated in the blasts, leaving only blue water and sand bars. One of the leaders wept openly. The people said their islands 'had lost their bones'.

Gradually, warnings were sounded by the 'experts': initially islanders were told not to eat or drink the juice of the coconut; then they were forbidden to eat the coconut crabs; still later they were advised not to eat any but imported foods. Since outside food supplies did not always arrive on time, people were forced to eat some local foodstuffs.

It soon became obvious to the US civilian officials charged with resettling Bikini that the US radiological survey done by the military, declaring the island 'safe', had been very superficial. The US Department of the Interior began requesting a new survey as early as 1972, but the survey was not actually conducted until late 1978, after the Bikinians filed a lawsuit and after their second evacuation. In early 1978, US scientists confirmed that the Bikinians had high levels of internal contamination (with radioactive particulates) due to the consumption of food grown on the

Bikini soil. The islanders had experienced a 75 percent increase in cesium 137 contamination of their bodies in only one year, reaching probably the highest levels of cesium of any population in the world. Cesium does not occur in our natural environment, but is produced in fission explosions.

Alarmed, the US government declared Bikini uninhabitable for another fifty years, and in September 1978 the islanders were moved back to Kili or other islands in the Marshalls. No one has been allowed to live on Bikini since, and no one knows what the future will bring for the Bikini islanders and their children. The people of Bikini are still asking for restoration of their islands, and striving to disengage themselves from US political control.

On 19 September 1979 officials of the US Department of Energy (which replaced the Atomic Energy Commission in the United States) advised the people of Enewetak in the Marshall Islands that what was left of their northern islands would be uninhabitable for at least another thirty years.[8] About forty-three nuclear detonations took place off the shore of Enewetak. The land is more severely contaminated with fission products than is Bikini.[9] Fortunately, the islanders of Enewetak were removed before the blasts, not after. Yet who could escape the distant fallout and the contaminated food?

In 1982 the Enewetak islanders went back to repossess their island home. Five entire islands in the Enewetak atoll, which once blossomed with tropical vegetation, had been 'vaporised', to use the US military terminology. As part of clean-up operations, over 110,000 cubic yards of radioactive sand and debris were mixed with concrete and dumped in a crater which had been blasted in Runit Island by a nuclear test called 'Cactus'. A 370-foot-wide, 25-foot-high and 18-inch-thick concrete dome covers the dump. Conservative estimates are that Runit Island is 'off-limits' for human life for the next 25,000 years. The concrete dome has a 'life expectancy' of 300 to 1,000 years, and reportedly is already cracked.

Micronesia, which includes the Marshall Islands, was given to the USA as a Trust Territory by the United Nations after the Second World War.[10] As trustee, the USA was pledged to protect the people of Micronesia and their resources, and to assist them in achieving economic independence. Since May 1979 the territory has begun to function with its own government. A radiological survey of the islands completed in January 1979 under US Department of Energy and Defence Nuclear Agency auspices has not yet been released to the Micronesian government. Health problems on islands not officially designated 'contaminated' have been ignored by US radiation experts, and all islanders are being assured that low level radiation will pose no 'noticeable' health problems in the future. These serious problems will be turned over to the newly formed Marshall Island nation and disputes over just compensation for injury will be between Marshallese citizens and their Marshallese government. The once self-sufficient island paradise is now dependent on imported food and drink, depleted in health, damaged in genetic integrity and politically divided on how to relate to the US military and civilian bureaucracy.[11]

Learning about the long-term effects of radiation exposure on individuals and on their children and grandchildren is a continued research goal of the US Brookhaven team. A report of June 1958 states:

> Even though . . . the radioactive contamination of Rongelap Island is considered perfectly safe for human habitation, the levels of activity are higher than those found in other inhabited locations in the world.[12]

The level of human exposure on Rongelap is not only greater than elsewhere but the type of chemical/biological interaction is different. Fission products which contaminate Rongelap and other Marshall Islands did not occur at all in the human habitat prior to 1945. Radiological similarity between fission products and naturally occurring radionuclides is no basis for confidence that their migratory behaviour in the environment and biochemical behaviour in a living human being will be the same.

The Japanese Fishermen

A Japanese fishing boat, *The Fifth Lucky Dragon*, was in the path of the fallout from the 1954 weapon test which sent radioactive fallout over the Marshall Islands. The ship and crew took 14 days to return to Japan, arriving on 14 March 1954. All were hospitalised immediately on arrival and were not released until May 1954, with the exception of one man who died on 23 September 1954 from liver and blood damage due to the radiation exposure. Twenty-three men, ranging in age from 18 to 39, were on the ship. Their total external radiation dose was estimated to be 170 to 600 rem. None have yet died of radiation-induced cancer and little is known of their subsequent medical history.

All the men suffered the symptoms of radiation sickness. A special study of their sperm was conducted. The number of sperm decreased about two months after exposure, and those sperm present had reduced mobility and visible morphological abnormalities. It took two years for the sperm count to return within the normal range. 'Most of the patients got healthy children' is the only official comment.[13] Even twenty years after their exposure the men exhibited chromosomal abnormalities in their blood cells. The level of abnormality was found to correspond to the severity of their injuries.

The hydrogen bomb explosion on 1 March 1954 escalated the nuclear weapon race and marked a definitive commitment in the Western world to 'peace' through military strength, regardless of the cost in personal suffering and destruction of the life-support system. It was the Western response to the Soviet detonation of its first nuclear bomb in Siberia in 1949 and was followed, predictably, by a Soviet explosion of a hydrogen bomb in 1955.[14] Hydrogen bombs were later exploded in Australia, on Christmas Island and on the island of Novaya Zemlya in the Arctic Ocean,[15] but we do not know who was exposed to the fallout or what health effects they experienced.

A hydrogen bomb consists of three separately detonated parts: an atomic fission bomb, which creates the conditions necessary for the fusion which follows, and many 'smaller' fission bombs detonated by the neutrons hurled off in the fusion. There is no upper limit to the blast power of a hydrogen bomb except the ability of the planet earth to absorb the blow. It essentially eliminates the concept of war and replaces it with annihilation.

The Military Experience

The Soviet Union has also had accidents and difficulties with nuclear development. A major nuclear accident occurred during the winter of 1957 in the Ural Mountains. The first public mention of the accident in the West came in 1976 when a former Soviet scientist, Dr Zhores Medvedev, referred to it in *New Scientist*.[16] It was later described more fully in a book.[17]

A US study of the accident, carried out at Oak Ridge National Laboratory, estimates that most of the industrial city of Kasli was contaminated and fourteen lakes and about 625 square miles of land were poisoned. About 60,000 people had to be permanently relocated and thirty towns eliminated. It is believed by some that the accident occurred when ammonium nitrate (a by-product of nuclear waste reprocessing) exploded, spewing 10 million curies of strontium 90 and other highly toxic radionuclides on the surrounding countryside. A deliberate weapon detonation has been ruled out because of the high population density in the Chelyabinsk area (approximately 3,400 persons per 62 square miles) and because of the valuable nuclear plant which had been built there. It is known that all other Soviet nuclear tests have been conducted in remote unpopulated areas of Siberia. The most likely scenario for this accident is a low level nuclear waste dump explosion, but this will be discussed later (see pp. 175–184).

American, British and Canadian military personnel and scientists were exposed to ionising radiation during the development and testing of the first atomic bombs, and in the massive programme in the 1950s to develop a pentomic army.* More than 250,000 troops and an unknown number of civilians took part in military manoeuvres to prepare for combat in a nuclear war.

Many of these men reported the same acute radiation effects as were experienced by the Marshallese, but no government study was done until the 1980s to document those immediate effects or the health effects which occurred ten to thirty years after exposure. In fact, it is not uncommon when the men exposed request their military service records, for them to be told that their records have been lost or destroyed by fire in government offices.

After discharge from the service the men returned to civilian life. They never spoke about their military experiences since they had been told that these had the highest classification status for secrecy on behalf of US national security. Their subsequent illnesses and even some deaths went unnoticed for years because no records were kept and the men lost contact with one another. Each assumed his problems were unique.

One veteran dying of myelogenous leukaemia took the whole story to the press in the late 1970s. He attracted the attention of the US Congress, which ordered a study of this veteran and all the participants in the Smoky blast which had been detonated on 31

---

* The US military term for men prepared to fight in a nuclear war was the pentomic army. The derivation of the term is unknown, but it appears to be either a combination of penultimate, 'the last member but one of a series', and atomic, pertaining to the atomic bomb, or a short form of 'Pentagon-atomic programme'.

August 1957 at the Nevada test site. This study related, of course, to only one of the 600 or more US nuclear blasts and only one health effect, myelogenous leukaemia, but it marked a beginning of official US government concern for the fate of the men who participated in the blasts. It signalled a beginning of congressional concern for the activities of the US military.

In August 1980, the Centre for Disease Control in Atlanta, Georgia, published its preliminary findings on the incidence of leukaemia among the 3,224 men exposed to that one nuclear blast.[18] In the thirty years following this radiation exposure, nine men were known to have had leukaemia. This was 3.5 times as many cases as would have been expected based on the experience of other US males their age. The incidence rate for other blood diseases, other cancers, reproductive problems or the ill health of offspring, was not studied. An adjustment for the fact that servicemen have atypical good health at the time of enlistment was also not made. These healthy males could be expected to have less leukaemia, not more, than their peers.[19]

Dr Glyn G. Caldwell published a second follow-up study of about 3,000 participants in the 1957 Smoky nuclear test in 1983. The men showed an excess, not statistically significant, in five cancer types: skin melanoma, genital system, eye and orbit of the eye, brain and nervous system, as well as the previously noted significant increase in leukaemia. Only cancer deaths were counted and the follow-up time of these men was twenty years. Since the average age at exposure was 24.4 years of age in 1957, and they were chosen for military service on the basis of their good health, it is not unusual that as yet these 46-year-old men did not show a statistically significant increase in radiation-related cancer deaths when compared with other men their age who had not been fit for military service. Their experience over the next fifteen years would be more relevant, but there do not appear to be any plans to continue follow-up. In conclusion Caldwell comments relative to his own research:

Furthermore, this conclusion cannot generalise to include participants at other nuclear tests or resolve the low-dose controversy.[20]

Caldwell, in his 1983 study, withdrew support for his own 1980 study, although there were no new findings relevant to the latter.

The military personnel exposed to radiation in the Smoky test, who later were known to have leukaemia, all had less than 3 rem exposure, except for one man whose exposure was unknown. This dose is less than the permissible dose per year for workers in nuclear industries.

No government inquiry into reproductive experience, general health problems, cancerous or non-cancerous blood problems, benign or malignant turmours has been conducted on the US marines who cleaned up in Hiroshima or Nagasaki in 1945, the navy nuclear engineer corps, or any non-military participants or down-wind recipients of fallout from the US nuclear testing, with the exception of the Marshallese. A study of nuclear shipyard workers in the USA has been embroiled in scientific controversy.

The civilian population downwind of the Nevada test site was never warned of the health effects which might result from their exposure. They were never even given the option of moving away from the fallout area. They have reported startling increases in miscarriages, stillbirths, thyroid disorders and cancers, but a serious government-sponsored study of these claims has yet to be undertaken.[21]

Dr Joseph Lyons, examining vital statistics for the State of Utah, found four times as much cancer in children downwind of the site as among other children in the state.[22] Utah had always had a very low cancer rate before testing began in Nevada. The St George area of Utah received the heaviest fallout, but the people were not warned and have not been helped to deal with this gross poisoning of their living space and the serious deterioration of health which they now perceive to have occurred.[23] A court trial brought by the Governor of Utah against the Federal Government for the wrongful death of the people of Utah from nuclear testing revealed that testing of the environment for radiation was deliberately limited to avoid alarm-ing the public. Health studies were suppressed.

The Workers

Many US and Canadian workers were involved in the Manhattan Project (which produced the bombs dropped on Hiroshima and Nagasaki), and the aggressive nuclear weapon programmes of the 1940s and 1950s in the USA. Their exposures to radiation seldom produced the dramatic acute effects noted in the Marshall Islands or downwind of the Nevada test site, but the frequent small exposures nevertheless can be expected to cause cumulative damage which eventually takes a toll on health.[24] The decision or lack of decision by government officials and the armed services involved in the weapon testing resulting in not setting up an adequate medical surveillance of the soldiers and civilian workers at the bases and the exposed downwind population, has masked the real health effects and placed an extremely unfair burden on those affected.

Producing the bombs required uranium mining and milling, enrichment, fabrication, operating nuclear weapon reactors (these nuclear reactors were originally operated to produce spent fuel rods, a source of plutonium for bombs, but their energy was not used to provide electricity), transportation, decommissioning and managing the radioactive waste from each step of the chain. The number of exposed personnel in all phases of the nuclear weapon industry increased rapidly as the world moved into the nuclear age.

The danger of uranium mining had been known for a hundred years before 1945, yet it took the USA until 1967, with many unnecessary miner deaths from lung cancer, to legislate regulations for mine ventilation and subsequent reduction of airborne radioactive particles and gas in the mines. Even this drastic reduction of radiation levels is now reported to be inadequate for preventing the doubling of lung cancer rates.[25]

It is estimated that about 1,100 excess lung cancer[26] deaths will have occurred among early US uranium miners by 1985. The full number of deaths in the USA including non-malignant respiratory diseases and other fatal and non-fatal illnesses induced in miners is unknown. No adequate study of the reproductive experience of uranium miners had been undertaken until it was initiated by the Navajo Indian Community in 1981. This was not a government or industry study. It was funded partially by the March of Dimes Foundation and partially by donated services. The analysis is expected to be complete in 1985. The general public exposed to the mine debris, unsealed abandoned mine and mill tailings have never been studied. There are nearly 100 million tons of such radioactive tailings in mid-western USA, and about 6 billion tons globally. The radioactive debris has been used as land fill, mixed in cement and macadam road surfaces, and also for construction of homes, schools and churches. At no time has a factual audit of the public health impact of these dubious practices of carelessly contaminating the living space been undertaken by governments and hence the vital feedback on health consequences has never been documented, quantified and reported to the public. Some even use this lack of research to back their claim that no harm has ever been proved to result from radioactive pollution, thereby continuing the same cynical attitude towards disposal and 'justifying' it. Other nations use US permissiveness to justify comparable carelessness within their own borders.

After increased public inquiry, US government agencies undertook a study of two populations of nuclear workers. The US National Cancer Institute and then the National Institute of Occupational Safety and Health (NIOSH) funded studies of uranium miners, while the Atomic Energy Commission (AEC) which subsequently became the Energy Research Development Administration (ERDA) funded the study of workers at the Hanford Nuclear Facility at Richland, Washington, and other government installations.

It has been known since 1920 that uranium miners in Eastern Europe have died of lung cancer caused by exposure to radon gas. This experience was repeated in the USA, and the grim story illustrates the tragic consequences of waiting until there is irrefutable evidence that the same death story is being again repeated before action is taken to protect human health.

Waiting for official death certificates constitutes a delay and is an unacceptable means of detecting a cancer epidemic. The time between the initiation of a cancer and the point at which it is diagnosed can be fifteen to fifty years. After death, another one to two years are required to process, prepare, and distribute official federal statistical reports. Besides the problems caused by such reporting delays, there are also problems of underestimation resulting from cancer patients dying of heart attacks, pneumonia or other illnesses, with the actual cancer never being reported. All of these record-keeping problems combine to make it possible for a cancer epidemic to go undetected for ten to fifty years. This type of delayed health evaluation proved tragic for more than a thousand US uranium miners.

Commercial uranium mining in the USA began in 1949. The probability of lung cancer deaths was well known, but an aggressive military programme to 'stay superior' in nuclear weapons seems to have overridden caution. It was not until 1957 that the US Public Health Service made a prediction, based on the early respiratory disease rate in uranium miners, that a significant number of deaths from lung cancer would occur in reality as a result of the US programme. This warning seemed to fall on deaf ears. Dr Victor Archer and Dr Joseph Wagoner were among the pioneer researchers documenting the lung cancer effects among the miners.[27]

The AEC claimed a legal loophole relating to responsibility for the health of uranium miners during the 1950s. When Congress created the AEC in 1946, the Commission's authority included the licensing of 'source materials' containing uranium. In the enabling act, it was stated that the AEC's power was limited to 'source material after removal from its place of deposit in nature'. The AEC relinquished authority over mining regulations by its strict interpretation of the enabling act, 'after removal' being equated with 'after leaving the mine site', and claimed ignorance about the health problems of uranium miners. This left the mines essentially unregulated. The AEC was the sole purchaser of uranium from 1946 until the late 1960s.

In 1965 the US Public Health Service reported that the evidence of a significant increase in deaths from lung cancer was considered to be 'conclusive'. Neither the uranium mining companies nor the AEC initiated remedial action. It became obvious that scientific evidence of 'health effects' was not sufficient to move the military decision-makers and industrial bureaucrats to introduce even minimally adequate ventilation in the mines.[26]

The US Secretary of Labour, William Wirtz, called hearings on the question. At these hearings, in spite of the 450 miners already dead, Dr Robley Evans, expert witness for the AEC, testified that radon gas levels in the uranium mines were below a 'threshold level' for human health damage. It was only at the insistence of Secretary Wirtz, over the protests of the AEC and the mining companies, that safety standards and improved ventilation of the mines were finally initiated in 1967 and further improved in 1972. The 1972 improved uranium mining standards were challenged again in 1980 as not yet sufficiently protective of the miners' health.[25]

It is now estimated that there will be 1,000 to 1,100 excess lung cancer deaths among US uranium miners by 1985 caused by this 'error' in judging radiation effects. One might speculate that military planning justified this cost for national security.

Indigenous People

The exploitation of indigenous populations in the USA and Canada, Aborigines in Australia[28] and Blacks in the Belgian Congo and Namibia[29] comprises an especially disgraceful aspect of the history of uranium mining. The rich uranium veins on indigenous land in north-western New Mexico, supplying one-fourth of the commercial and military uranium needs of the USA, were leased to fourteen major energy companies by the Bureau of Indian Affairs. Exxon alone has 400,000 acres near Red Rock, New Mexico. Other companies with holdings on Indian land in the USA are: Continental Oil, Anaconda, Grace, Gulf Minerals, Homestake, Humble Oil, Hydro Nuclear, Kerr McGee, Mobil Oil, Pioneer Nuclear, Western Nuclear, Phillips Petroleum and Marathon Oil.[30] New Mexico was also the site of the first nuclear bomb explosion, that which confirmed A-bomb 'success' before its use in Japan in 1945. Now, in addition to the bomb fallout, the state is dotted with uranium mines, more than a 100 million tons of uranium mill tailings and an unknown amount of mine tailings. Both mine and mill tailings, unless stabilised, will continue to pollute air and water, emitting highly toxic radioactive gas for at least 250,000 years. LaVerne Husen, director of the Public Health Service of Shiprock, New Mexico, reports increases in both lung cancer and pulmonary fibrosis. Lung cancer used to be rare among the Navajos. Dr Gerald Buker reported in a monograph called 'Uranium Mining and Lung Cancer Among Navajo Indians' that the risk of lung cancer had increased by a factor of at least 85 percent among Navajo uranium miners.[31] Husen reported that the levels of radon gas in the mines are now one-hundredth of what they were in the 1950s. Even allowing for some possible margin of error in these estimates, the incredible lack of responsible government action to protect the public health is shocking. The radon gas released from the mine and mine tailings is heavier than air, and can spread 1,000 miles from the source in winds of 10 to 15 mph, before half is disintegrated. The tailings piles have been left close to schools, homes and churches.

In Shiprock, New Mexico, residents have been unsuccessful in attempts to obtain compensation for the tragic situation created by the uranium mining. In addition to the deaths and debilitating illnesses of the miners themselves, there is the continued hazard from mill tailings which not only lie unattended at abandoned mine sites, but have also been used in New Mexico and elsewhere as building materials for thousands of homes, schools, roads, playgrounds, driveways and other public constructions.[32] These structures, even if identified and condemned, cannot be replaced without federal financial assistance because of the magnitude of the task. Although Pete Domenici, a US Senator from New Mexico, introduced a 'Uranium Miners' Compensation Act' bill during the 1978 Congress, he had little hope of obtaining adequate funding either to compensate the miners or to clean up the environment.

When Congress later authorised a funded study on the effects of radioactive uranium debris on Navajo living near tailings sites. Domenici did not apply for funding because of a slump in the uranium market.[33]

Money spent to clean up some of the radioactive debris contributes to the hidden expenses of nuclear power and nuclear weapons which are never calculated in the cost to the consumer. It is paid in taxes rather than electricity bills; the environmental super-fund rather than the Department of Defence budget.

The ambivalence of government officials is an obvious part of the nuclear problem. For example, Senator Domenici, who did try to obtain assistance for uranium miners, also has reportedly interceded with the US Nuclear Regulatory Commission on behalf of the Bokum Resources Company of New Mexico. This is one of the mining companies responsible for the uranium tailing pollution. The Nuclear Regulatory Commission had refused to license a dam which Bokum proposed to build out of mine tailings. This proposed dam would need to be maintained intact for as long as it contained radioactivity, i.e. for at least 250,000 years – a virtual impossibility – or be removed and stored as radioactive waste. With the help of top-level political figures, a 'conditional' licence to build was granted – the first such licence granted in New Mexico.

Funding for adequate health care for the Navajo uranium victims has been cut at the same time as realisation of the severity and extent of the problem escalates. Dr Leon S. Gottlieb, MD, an internationally known expert in lung cancer and other respiratory disease, had all his funding cut off in the summer of 1980. He was forced to move from Shiprock, New Mexico, to California in order to earn a living. In a jointly written paper, Drs Gottlieb and Husen reported that sixteen of the seventeen people with diagnosed lung cancers at the Shiprock Hospital in the summer of 1980 were uranium miners. Of the sixteen cases, only two had ever smoked cigarettes. The number of other respiratory diseases and tuberculosis cases is also elevated at Shiprock. About 95 million tons of radioactive waste from the mines and mills are piled above ground there.

In 1972 the US Congress authorised $5 million for a mill tailings removal programme in Grand Junction, Colorado. The increased number of children born with cleft palates had drawn attention to the radiation pollution problem there. As of July 1978, $5.4 million had been spent on the clean-up, and Senator Floyd Haskell, from Colorado, had proposed authorisation of another $3 million.

The US Department of Energy has identified an additional twenty-two sites contaminated with radiation which require clean-up, and has estimated that $85 million to $135 million will be needed. Meanwhile, people become ill and die and do not even realise the connection between their illness and their exposure to radiation.[34]

The US Nuclear Regulatory Commission has now established safety regulations to lessen the danger from uranium ore tailings being generated by mining conducted at the present time and in the future. However, the regulations do not apply to the more than 100 million tons of uranium waste already polluting the western USA. The US Senate Energy Subcommittee on Production and Supply held hearings on this problem in July 1978. In a press release from the office of Senator Floyd Haskell, chairperson, it was stated that 'over several million tons of tailings' are not covered by the new NRC regulations. Moreover, Senator Haskell said: 'I am alarmed that there are large piles of tailings throughout the West that won't be covered by the bill [for appropriation of funds for clean-up] before the Senate. More and more medical evidence indicates that exposure to these waste materials causes severe health hazards.' The situation must be equally monumental at the older mining sites in Canada,[35] Australia[36] and Namibia.[29] Without drastic changes in attitude, this rape of the land and people will spread to Saskatchewan, the Black Hills of South Dakota, new areas of Australia and elsewhere where uranium mining is proposed or has just begun. Canadian production of uranium mine tailings according to the Atomic Energy Control Board in Canada, 1984, is about 10 million tons per year.

The economic and human health cost of the so-called 'front end' of the nuclear weapon and fuel cycle has never been counted. If Ian Watson, Member of the Canadian Parliament, is correct, the 120 million tons of accumulated uranium tailings in Canada represent only 2 percent of the global total. This makes the global total about 6 billion tons (US).[37]

In the wake of awareness of the increasing radon gas exposure of the public from carelessly handled uranium mine and mill tailings, it became apparent to some people that the magnitude of the problem of radionuclide contamination of the environment had been masked by the deceptive use of language. Thorium, uranium, radium, radon gas, and its radioactive daughter products are officially called 'natural background radiation' in nuclear jargon,[38] even though they have been removed from their relatively harmless natural state deep in the earth. The term 'natural background radiation' has been used for all radioactive particles not man-made (human-made) in the fission reaction. Blasted out of the ground with dynamite or leached with acids, then pulverised into very small particles, these natural radioactive chemicals have been enabled more easily to enter into the human body with air, food and water, damaging human cells and tissues. Although in a sense they are 'natural', they are not in their natural state.

Another confusing term is 'background radiation', which is a broader category including both natural and 'man-made' radioactivity. Fission products may be called 'background radiation' when they do not emanate from the installation under consideration or when they have been in the environment for a year or more. Thus in the US two nuclear power plants on the same tract of land can be licensed separately, the pollution from one being considered 'background radiation' whilst possible contamination from the other is considered. Similarly, last year's pollution from a reactor becomes 'background' if it persists in the environment longer than a year (as much does). An individual's yearly radiation exposure estimate attributable to nuclear activities is an assessment of a fresh fission dose from a particular radiation source – not a realistic measure of the total dose from all sources, whether external and left over from an earlier year's pollution, or already incorporated into body tissue or bone from previously ingested or inhaled radionuclides.[39] Thus strontium 90 can be ingested and then incorporated into bone where it will continue to give a small dose of radiation to the individual every year for life.

It is also misleading to report pollution in terms of a percentage increase in background radiation levels. Little or nothing is said about the steady increase in background radiation – natural or otherwise – due to human activities. Hence a percentage of background radiation added may stay constant, masking the total accumulation.

A first attempt to deal with this confusing use of language is found in a US Environmental Protection Agency (EPA) report, 'Radiological Quality of the US Environment, 1977'.[34] The EPA introduced a new phrase, 'technologically enhanced natural radiation', to designate the human-made hazards such as uranium tailings, which, although natural, have been rendered more dangerous to humans by being removed from their natural state.

This 'technologically enhanced natural background radiation' (TENR) now ranks first as a source of individual internal radiation exposures in the USA, ahead of medical and radio-pharmaceutical exposures. It ranks a close third, after medical and radio-pharmaceutical exposures, as a source of general population dose.[34] This TENR ranking is due primarily to the lung dose from inhalation of radon dissolved in natural gas.

Hanford, Washington

During the Second World War a semi-desert area in south central Washington, near the Columbia River, was chosen as the site for one of three atomic cities to be built by the US army to support the Manhattan A-bomb project. This military installation, called the Hanford reservation, comprises about 5,700 square miles of land and currently employs about 7,000 workers. It contains conventional nuclear reactors, breeder reactors, reprocessing plants, huge nuclear waste disposal trenches and more than 150 large tanks for high-level liquid nuclear waste. The latter section of the property is sometimes called the 'tank farm'. Hanford's nearest neighbour is the town of Richland, Washington, with a population of about 26,000.

This installation was studied by Thomas Mancuso, Alice Stewart and George Kneale to evaluate the hazards of radiation in the workplace.[40] Unlike most death certificate studies, the Hanford Worker Analysis contained yearly reported readings on radiation monitoring badges for each worker. Hence it was possible to correlate death-cause with radiation exposure level. The findings of the Hanford study provoked a veritable storm of defensive behaviour on the part of both the US government agencies and the nuclear industry.

The Hanford installation is something of a showplace for the nuclear industry. It is well managed and average radiation exposures of workers are about the lowest in the US nuclear industry. If cancer levels at Hanford proved to be significantly higher than was predicted by the radiation research scientists, radiation exposure would be expected to be a still greater problem at other nuclear installations in the USA and around the world.

The cancers did occur more frequently than expected, and the US government agencies moved quickly to counteract the expected impact of this finding on their nuclear weapon programme. Dr Thomas Mancuso, Research Professor at the University of Pittsburgh, Graduate School of Public Health, was both astonished and dismayed when he learned that government officials were terminating the funding for his follow-up study of the nuclear workers at the Hanford Atomic Energy installation in the State of Washington and at other installations, because of his 'imminent retirement'. Dr Mancuso was sixty-two at the time. University retirement policy allowed professors to continue to the age of seventy, but the government officials had not bothered to telephone to inquire. This termination of funding for such a questionable reason raised suspicions that other reasons were involved.

Dr Mancuso wanted to continue the nuclear worker research, realising its international importance, and had so informed the government agency. Mancuso had originally been requested to undertake the research because of his extensive investigative experience in occupational cancer. He is credited with having developed the basic epidemiological methods of studying biological effects of workplace hazards over a span of decades. His techniques have been generally adopted and used by other researchers in this field. The Hanford installation, however, proved to be the most politically difficult worker health situation of any he had tackled.

Funding for the Hanford study initially came from the US Atomic Energy Commission (AEC). Then followed the successor organisation, the AEC promotional arm, called the Energy Research and Development Administration (ERDA). This successor agency was later raised, in the bureaucratic hierarchy of the US government, to the Department of Energy (DOE) by President Carter and its head was given a place in the President's Cabinet.

The real reason for Mancuso's funding cut seemed to stem from more than scientific disagreement. The trouble began with his refusal to publish the Hanford study findings prematurely, at ERDA's request, and his refusal to refute the allegations of Dr Samuel Milham, Epidemiology Director for the Washington State Health Department. Dr Milham reported higher cancer death rates, as designated on the death certificates, among Hanford atomic energy workers than among other industrial workers in the state. Milham's two-volume study contains information on causes of death among males employed in and dying in Washington State between 1951 and 1971. He found excessive rates of pancreas cancer and multiple myeloma, a rare type of bone marrow cancer, among former Hanford employees.

Fearful of the public release of the Milham findings, the ERDA officials tried to get Dr Mancuso to agree to an immediate press release that would refute Milham's findings. Mancuso refused, since he realised that releasing a worker health analysis before the number of follow-up years required for developing the occupational cancer had passed, would lead to false positive findings. Such a press release would have been misleading and probably misused. Mancuso's refusal led to a chain of events ending in a Congressional hearing in which the Congressman, Chairperson of the investigating committee, publicly expressed dismay at the 'attempted government cover-up'.[41]

Milham had direct access to records of recent deaths in the State of Washington which provided information from the 1970s. Mancuso, who had to trace deaths through the US Social Security system on a national scale, was experiencing delay. It was probably evident to ERDA officials that as soon as Mancuso received the 1970 death certificates he would discover the same increased cancer death rate as was noted by Milham. Mancuso's findings would be even more devastating for the nuclear industry than Milham's, since Mancuso was tracing Hanford workers to any place of death in the US, while Milham had records only of those who died in the State of Washington, and Mancuso had detailed information on each worker's occupational exposure to radiation. Mancuso, able to correlate exposure with cause of death and using a larger data base, could produce a much stronger study.

After this request for a press release to refute Milham's findings, and as the Hanford data accumulated to the 1970s and reached the critical stage for analysis, Dr Mancuso called on Dr Alice Stewart to provide an independent assessment of the death certificates and radiation data for the Hanford nuclear workers. Alice Stewart, a medical doctor in Birmingham, England, who had been the first to document the health effects of low level radiation on the human foetus,[42] was on Dr Mancuso's Advisory Committee for the Hanford study. Dr Alice Stewart and her assistant, Dr George Kneale, a mathematician, agreed to come to the USA. They conducted a series of statistical analyses of the Hanford data, and identified an increased risk of cancers among Hanford radiation workers relative to the rate of those Hanford workers who were unexposed.

The government officials were informed of these developments and they were briefed with a preview of findings before they were first presented at the meeting of the Health Physics Society in September 1976. At the briefing ERDA officials suggested that perhaps more research should be done before making the information public. The irony was that during the years in which the data was being developed and the findings were negative, the government officials had urged Mancuso to publish the data. Now, when the findings were positive, the same officials were 'suggesting' that any release of the findings should wait for further years of research. Dr Mancuso was not swayed and the positive cancer findings of the Hanford Study were presented at the Health Physics meeting and made public.[43]

At the Health Physics Society Meeting, by a remarkable and probably planned coincidence, immediately after Dr Alice Stewart had made her presentation of the Hanford data, Dr Sidney Marks of ERDA and Dr Ethel Gilbert of Battelle Northwest, proceeded to present a very different analysis of the health risks of the nuclear workers at Hanford. Both oral presentations were necessarily brief, limited to about fifteen minutes.

Drs Stewart and Kneale prepared their paper for publication, analysing the data in a number of ways, with and without assuming a normal distribution of radiation exposures and controlling for internal radiation. The analyses showed that the main findings were statistically significant. The paper was submitted to scientific peer review and was accepted for publication in *Health Physics* of November 1977.[40]

Dr Thomas Mancuso was a mild man, but having been involved in occupational health problems for many years he was prepared for an angry response when the first written report was released. After all, both government (especially the military) and industry had made a tremendous financial commitment to nuclear weapons and power under the assumption that people could safely handle the radioactive materials.

Dr Alice Stewart, who had waited out the storm which followed publication of her findings of radiation-related cancers in children exposed *in utero* to medical X-ray, was also expecting a battle. They were not wrong.

First criticisms of the study revolved around the 'short time' (about six months) that Stewart and Kneale had spent on the project, ignoring Stewart's long-time role as adviser to the study.

However, neither Mancuso nor Stewart expected the ERDA representatives' attempts to confiscate the Hanford data. An ERDA officer contacted the Computer Centre at Oak Ridge and without getting the university's authorisation and over the objections of Dr Mancuso, got most of the Hanford data. Fortunately, Mancuso had his complete file in Pittsburgh. Dr Mancuso and others found this an unwarranted interference by a government agency in scientific research.

A further large-scale effort at confiscation was made by government officials of all the other research data that Dr Mancuso had developed as Principal Investigator of the radiation study of atomic energy installations. This was strongly and successfully resisted by Dr Mancuso and the University of Pittsburgh officials.

The final insult which completely disillusioned Dr Mancuso in his relationship with federal agencies occurred when he discovered that ERDA had circulated, in the USA and Europe, major critiques by its consultants without making these critiques available to the Principal Investigator for response. In fact, Mancuso was only able to obtain the critiques under the Freedom of Information Act. It was clear that the 'fight' would not be fair and dialogue on a scientific level was not even attempted by the government officials. In contrast, after finally obtaining copies, Stewart, Kneale and Mancuso prepared a scientific response to these critiques and sent the document to each of the government officials and their consultants for their consideration before the congressional hearings on the dispute began.[44]

When the excuse of 'imminent retirement' tendered by ERDA for termination and transfer of the Mancuso, Stewart and Kneale Hanford worker research project floundered, ERDA advanced another reason – 'negative peer reviews' during the congressional hearings. US Representative Tim Lee Carter asked whether the two negative reviews produced by ERDA were the only reviews received by ERDA prior to its decision to remove the funding from Mancuso for the Hanford and related studies. After an affirmative answer from Dr Liverman, of ERDA, Congressman Carter produced four additional reports, all complimentary towards Dr Mancuso, which Representative Carter's aides had obtained. They had been received by the ERDA peer review panel at the same time as the two negative reviews. The four positive critiques received by ERDA had been ignored, and their existence denied under oath at a congressional hearing.[41]

Although the number of death certificates for former Hanford workers has significantly increased since the first published Mancuso, Stewart and Kneale report, and the increases in radiation-related cancers are now more strongly confirming the researchers' original conclusions,[45,46] the verbal battle is still raging. Numerous critiques of the methodology of the first 1977 paper have been written and published. These are being widely quoted without noting the authors' response, and without reference to the more recent confirming additional data.

As noted before, the Hanford data is superior to ordinary vital statistics information in that each worker's death certificate could be matched with complete radiation exposure badge readings since date of employment. This meant that more research questions could be asked and answered than would have been possible with only a death certificate. For example, one could ask if individuals dying of cancer had, on average, more radiation exposure than those dying of non-malignant diseases. The answer was yes. In response, critics pointed to the fact that workers now alive averaged more exposure than those who were dead. The researchers' answer to this question was little publicised. The Hanford workers now living are younger, and, while young, tend now to get jobs with higher exposure. Their life story is not yet over. They have not reached the critical age for cancers, nor has their total work average yet been determined.

In another test within the process of continuing analysis of the Hanford data undertaken by Mancuso, Stewart and Kneale, average radiation exposures for workers with cancer types designated as 'probably radiation-related' by the International Commission on Radiological Protection (ICRP) were compared with average radiation exposures of those workers with cancer types never shown to be radiation-related. The ICRP classification was derived from observation of cancer types of higher than expected frequency among A-bomb survivors and radiotherapy patients.

A Hanford data re-analysis by Mancuso, Stewart and Kneale showed that average radiation exposures were higher among workers dying of known radiation-related cancers, and lower among workers with other cancers. In other words, the radiation-related tumour types occurred among workers with higher exposure levels.

Another test of the validity of the relationship between worker exposures to radiation and excess cancer deaths had to do with the timing of events. If, for example, the excess radiation exposures received by those with radiation-related cancers had occurred at a time not consistent with the induction time needed for the tumour to develop, the findings would have been suspect. The Mancuso, Stewart and Kneale findings passed this test also.

It is possible to use statistical techniques which reduce the possibility of finding health effects. Some of these include: first, putting all cancers together for analysis so as to 'dilute' the clear evidence of increase in the sub-group of strongly radiation-related cancers; second, extreme fractionisation of the data, considering each cancer type individually, so that the number of cases became too small to give statistical significance; or third, eliminating portions of the data rather than controlling for possible confounding variables.[47] A final tactic is the use of a general US population as a control group for the medically screened healthier population at the Hanford installation. Mancuso, Stewart and Kneale avoided this last pitfall and used (appropriately) internal controls, i.e. workers with the usual Hanford medical screening and health care, who had lower radiation exposure badge readings. They also avoided the other three pitfalls in technique.

The large number of criticisms of the Hanford worker analysis seems to be related more to its perceived political importance to the foundation of the whole nuclear industry than to its scientific merit. Criticisms are easily generated from the large establishment of national nuclear laboratories in the USA. Some questions raised have required more extensive analysis, and Mancuso, Stewart and Kneale have tried to give forthright answers to these scientific questions, but the answers have received much less notice and distribution than the questions. Other problems raised with respect to the Hanford study seem to be merely a smoke-screen to continue the impression that the scientists disagree. Some heated remarks have come from physicists or engineers unable to understand either the biomedical reality or the biostatistical analysis. The most quoted criticism, emanating from Sidney Marks and Ethel Gilbert, deserves special mention.[48]

The Battelle re-analysis of the Hanford data was done by Dr Sidney Marks with the assistance of Dr Ethel Gilbert, whose only other radiation research seemed to be a re-analysis of the work of Dr Milham who had first pointed out the problem of increased cancer among Hanford employees.

The re-analysis of Mancuso, Stewart and Kneale's research data by Marks and Gilbert used all the techniques for masking effects mentioned above. Their lowest radiation exposure category, i.e. workers with less than 2 rad dose, included 70 percent of the workers. This category was assumed to be unexposed for the purposes of examining trend. The remaining 30 percent were divided among three higher exposure levels, the highest being 15 rad or more. Then they looked at thirteen different cancer types separately. This made 3 times 13, or 39 different categories in which to place a small number of exposed workers. In contrast, the Mancuso, Stewart and Kneale analysis used seven radiation-dose categories based on log scale divisions, with the lowest cut-off at

0.08 rad (about natural background level) and the highest at 5.11 rad, providing categories of exposure better suited to the workers' experience.

If radiation exposure is associated with increased cancer rate, then a trend ought to be present, i.e. the cancer rate should progressively increase as the radiation dose increases. The Mancuso, Stewart and Kneale analysis which began with natural background exposure levels and identified seven levels of increase over background was more likely to elucidate such a trend if one were present, than was the Marks–Gilbert analysis.

Marks and Gilbert also eliminated all Hanford workers with less than two years' employment and all those whose employment did not extend beyond 1 January 1960. Hanford went into full production in 1944. The Marks–Gilbert elimination of workers from the study reduced the sample size to 820 certified death certificates, of which 171 (20.9 percent) were cancers. In contrast, Mancuso's first analysis included 3,520 male certified death certificates, 670 (19.0 percent) of which were cancers, and 412 female death certificates, of which 126 (30.6 percent) were cancers. Instead of eliminating whole groups of workers, Mancuso, Stewart and Kneale used the usual method of statistical control for these possible confounding variables. The first Mancuso, Stewart and Kneale analysis controlled for calendar year of exposure, since badge types and skill at measuring worker radiation dose changed over the years; it also controlled for employment year of exposure, for pre-death year of exposure, for age at the end of each radiation badge year, and for age at death. The later Mancuso, Stewart and Kneale analysis, which included death certificates received between 1972 and 1977, was based on 4,033 deaths, 832 (20.6 percent) of which were due to cancer. This analysis was done using standard Mantel–Haenszel techniques and controlled for five possible sources of bias.[44]

In contrast, the smaller sample used by Marks and Gilbert, with only 171 cancer deaths, plus their odd choice of exposure categories, with 112 cancer victims assigned to the lowest radiation exposure category, left only 59 cancer victims to be divided among 13 cancer types and three higher levels of exposure. Marks and Gilbert, of course, then complained that no firm conclusions could be based on such small numbers of 'high exposures', claiming that it would require another twenty years to reach a definitive decision on whether or not radiation was a workplace hazard. In spite of their poor statistical analysis, Marks and Gilbert found two cancer types significantly increased among Hanford workers. They then suggested this might be due to the exposure of workers to some toxic chemicals other than radiation in the workplace. They have made no effort to resolve this doubt; hence, one can say the re-analysis served only to muddy the water and paralyse constructive action towards protection of radiation workers.

Dr Sidney Marks was chosen by ERDA to represent the USA at the International Atomic Energy Agency symposium on the late effects of ionising radiation exposure, in March 1978. Attendance at IAEA meetings is open only to government designated scientists from member states. It is not a scientific conference in the usual professional sense of contributed scientific papers and free scientific participation. The US participant was designated by ERDA, an agency which was not likely to certify attendance at the IAEA symposium for Dr Thomas Mancuso or other scientists rejecting the politically expedient ERDA position on radiation health effects. However, to its credit, the British government sponsored the attendance of Dr Alice Stewart at the meeting.

The same Marks–Gilbert IAEA paper was later presented by Ethel Gilbert at numerous nuclear-industry-sponsored scientific meetings. In 1980 she received an award from the American Statistical Society for her Hanford worker analysis. One cannot but wonder at the award process.

It becomes very difficult for physicists and engineers to sort out scientific fact and illusion in such a climate of scientific corruption. The general impression of scientific disagreement encourages reliance on the status quo decisions made on behalf of other scientists and the public by 'prestigious' experts from the military-industrial complex.

Since the publicity about the Hanford study became widespread in Britain, the UK Atomic Energy Authority has distributed a popular-level pamphlet called 'The Effects and Control of Radiation', by P. G. H. Saunders. It mentions the Hanford study with the remark that 'the data used have been recently reanalysed by independent expert groups using correct statistical techniques and no significant radiation effect was found.' The names of the disputing researchers were not given, no fair rebuttal of charges and countercharges was made, and no way of following up this significant research was provided. The very clear message was 'trust the experts'.

US military control over radiation-related health research, seen by it as vital for the continuance of US military nuclear strategy, was severely threatened by the Mancuso Hanford worker analysis. Projections of health effects for workers exposed to small fractionated doses of radiation could no longer reasonably be based on A-bomb survivor data when direct information on healthy workers exposed at this low level had become available. The house of cards was in danger of falling since human willingness to handle radioactive material is basic to all military nuclear planning.

Direct human experience gained by 35 years of nuclear technology should logically replace the optimistic predictions and comforting myths which have supported nuclear growth. An audit always replaces a forecast. However, resistance to this logical change is strong.

In order to give an estimate of the magnitude of the change in thinking about the cancer risk associated with radiation exposure demanded by the Hanford worker findings, we can look at the estimate of radiation dose assumed to double cancer rates in a population. The generally accepted nuclear industry estimate of a cancer-doubling dose from radiation is 500 rad exposure. The Mancuso, Stewart and Kneale estimate is 33.7 rad. An independent analysis done by Dr John Gofman, MD, PhD, former director of health research at the Lawrence Livermore Laboratory, USA, on only those Hanford workers who survived for more than fifteen years after their radiation exposure (i.e. those not dying before the cancer developed) put the estimate at 43.5 rad.[49] Since nuclear workers are permitted to receive 5 to 12 rad doses of radiation per year, they could double their general cancer risk in three to nine years. Because most workers do not receive the maximum permissible dose each year, it has been argued in the past (by nuclear proponents, relying on the 500 rad doubling-dose estimate), that it is unlikely that workers would accumulate enough radiation exposure to double their cancer risk in a lifetime of work in the industry. However, the lower doubling estimate significantly increases the probability of doubling cancer risk even for workers with exposures well below standards, making this employment more hazardous than the government and nuclear industry has stated.

Put another way, in work generally recognised as hazardous, such as in quarries, mines, railroads or construction, fatal accidents occur on average to between 1 and 3 per 10,000 workers per year. Prior to the Hanford worker study it had been argued that nuclear workers averaging 20 percent of the permissible radiation exposure level per year (1 rem or 10 mSv) would experience no more than one or two fatal cancers per 10,000 workers per year.[49,50] The implications of the Hanford worker study are that the actual nuclear worker cancer fatalities will be 10 to 16 per 10,000 workers per year at 20 percent of the permissible dose level. This is clearly unacceptable employment.

Since not all human tissue appears to be equally sensitive to radiation-induced cancer, cancers in radiosensitive tissues such as breast and bone marrow would be expected to have a lower doubling dose than that for all cancers. The average cancer-doubling-dose estimate also fails to take into consideration those more sensitive to radiation – young children, women, the elderly, or those with certain diseases – for whom the doubling dose would be lower.

The uranium miner studies resulted in a dramatic lowering of permissible exposure levels for workers in the late 1960s, but only after strong opposition from the mining industry and the US Atomic Energy Commission. The results of the study of first cause of death for Hanford workers have yet to make an impact on the industrial or military nuclear world.

The rather mild, well-documented statement that Hanford worker cancers were a little more than 10 times higher than predicted, based on Hiroshima and Nagasaki studies, caused a furor of defensive actions on the part of the military and industrial nuclear establishment, including the termination of Mancuso's funding.[41] After considerable difficulty Dr Mancuso was able to continue his research for two more years with the aid of the Environmental Policy Centre, a public interest organisation in Washington, DC, and NIOSH. He obtained still more death-certificate evidence to substantiate his original charge.[51] However, publication of the updated later reports became difficult because of the continued severe attempts to 'discredit' the original report.

Meanwhile the international nuclear establishment resumed (or really continued) relying on its predictions of the number of fatal cancers expected to be induced in workers due to exposures within permissible levels, ignoring the Hanford death audit as 'not yet absolutely certain proof'.[52]

A project report from Oak Ridge Associated Universities and the University of North Carolina was published in May 1984, containing summaries of the US Department of Energy's (DOE) in-house studies of its nuclear-related activities. One study[53] reported excess mortality due to leukaemia, lung cancer, brain cancer, digestive tract cancer, prostate cancer, Hodgkin's disease and non-malignant respiratory disease among workers exposed to uranium dust and/or radiation from other internal and external sources over a forty-year period. In a study of leukaemia deaths of employees of Oak Ridge National Laboratory a trend towards increased leukaemia with increase in radiation dose was noted for those with long-term employment in maintenance and engineering jobs.[54] An overall significant risk of cancer was found among workers with maintenance, construction, janitorial and labourer job titles.[55]

Not only has an inadequate breadth of health effects research been considered, but even selected *severe* effects like death due to radiation-induced cancer seem to have been underestimated for nuclear workers. Although in financial affairs an audit would replace a prediction, the opposite occurs in health. The burden of proof that predictions are wrong is placed on the victim and such proof must be researched by unfinanced independent scientists.

Other Nuclear Weapon Testing

The suppression of health information consequent to nuclear testing is not peculiar to the USA. It appears rather to be related to military preoccupation with short-term goals and power diplomacy. Prior to beginning nuclear testing in Polynesia in 1966,[56] the French discontinued publishing vital statistics there.

John Teariki, the French Polynesian Deputy, received an early warning of the nuclear dangers from Dr Albert Schweitzer: 'long before I received your letter I was worried about the fate of the Polynesian people . . . those who claim that these tests are harmless are liars.' (Personal letter from Schweitzer to Teariki, April 1964.)

In May 1966 the Centre d'Expérimentation de Pacifique (CEP) announced the first nuclear test and indicated the 'danger zone'. When it was pointed out that this danger zone included seven inhabited atolls, the CEP acknowledged that it had made a mistake. It drew a new 'danger zone' of smaller radius, with one inhabited atoll, Tureia, still included.

After June 1966 public health statistics were not published on a regular basis in French Polynesia, and persons requesting such statistics were reported to the secret police. Prior to that, public health records were available monthly in the *Journal Officiel*. The problems of information-gathering on health by the Polynesian people themselves were compounded by the fact that only army doctors were allowed to practise in the Papeete civilian hospital.

Four days after one nuclear test blast in September 1966, the New Zealand National Radiation Laboratory measured 135,000 picocuries per litre radioactivity in rainwater in Apia, Western Samoa, 2,000 nautical miles downwind of the test site. No measurement of local fallout was reported to the French Polynesians.

In August 1968 scientists at the University of Baja, California, reported increasingly high levels of radioactivity in fish caught along the Baja Peninsula in Mexico. They blamed the French tests.

According to the French army colonel in charge of the French Polynesian Department of Health, however, the nuclear tests were harmless. He had serving under him fifty military doctors 'who had never reported to him any detrimental effects on the health of the population in the territory that might be ascribed to the nuclear tests'.[57]

France has declined to co-operate with a World Health Organisation study of cancers among Pacific Islanders initiated in 1982.[58] According to the French Secretary of State about fifty Polynesians were sent to France for treatment for brain tumours in 1976, seventy in 1980 and seventy-two in 1982. There is no information on other years. These patients were under thirty-five years of age according to a physician at Val-de-Grâce Hospital near Paris where they were treated. Other Polynesian cancer patients are also being treated in New Zealand at a rate which appears to observers to be abnormal. However, since physicians are not required to report cancers when diagnosed, and since islanders on the smaller islands frequently die without medical care or of a pre-cancer infectious disease, it is difficult to document the French Polynesian experience.[58]

Louis Gonzales Mata, a journalist who undertook an investigative tour of Polynesia, reported abnormally high miscarriage rates, especially on the island of Mangareva. There are also young children, under the age of six, needing open heart surgery because of untreated rheumatic fever. Until record-keeping is fully reinstated and the reporting of diseases and death causes becomes compulsory, it will be difficult to estimate the full human toll due to the French nuclear weapon programme.

The secrecy and the sparse revelations of trouble are consistent with the known exploitation of human health which has taken place in connection with US weapon testing in the Pacific. These Polynesian people are among the early victims of the Third World War.[59]

As early as 1971, Peru, a long-standing friend of France,

threatened to break off diplomatic relations because of the tests in the Pacific. The ocean current sweeps through French Polynesia and then makes its way north along the South American Pacific shore, affecting Peru and Peruvian fisheries.

In June 1972 the UN Conference on the Human Environment in Stockholm formally condemned the French tests and criticised the inadequacy of the French reports to the UN. In the words of the New Zealand representative to the UN, Mr Temple, the French reports gave 'a partial picture of incomplete data'. Peru had every right to be uneasy.

Finally, in July 1973, Peru broke off relations with France, judging that the atmospheric tests were endangering her marine life and her citizens. These fears were confirmed in August 1973 with the actual measurement of nuclear fallout in Peru by its Atomic Energy Control Commission.

On 23 June 1973 the World Court urged France to avoid nuclear tests causing radioactive fallout on Australia, New Zealand and Fiji. France officially informed the court that she did not recognise its jurisdiction. After another series of seven atmospheric tests, France finally announced in September 1974 that she had finished atmospheric testing and would now test underground.

'Underground' meant in basalt rock beneath the coral island or the lagoon. On 5 June 1975 the Australian government reported that an underground explosion had occurred in French Polynesia. The French government admitted that it had exploded an 8 kiloton bomb, 623 metres below the Fangataufa atoll. A concrete slab placed over the test pit had cracked. Underground tests at Moruroa began in April 1976. France admitted that some radioactive gases might leak out but claimed they were 'harmless'.

Two indirect effects of this blasting beneath the water should be noted: one is climatic change and the other is ciguatera fish poisoning. Prior to the French nuclear testing there had been a long period in the South Pacific with no strong cyclones – at least sixty years. The dangerous cyclones usually occurred in the Fiji Islands and had never occurred in Tahiti. The first cyclone, on 28 November 1980, hit Moruroa, site of the nuclear tests. During the night of 11 to 12 March 1981, a second strong cyclone hit Moruroa. In early 1983, five cyclones passed over Moruroa causing 10-metre-high waves and considerable damage throughout French Polynesia. Even French military scientists are reporting a 'heating of the water', with ocean temperatures between Tahiti and the equator reaching 31° to 32°C even in the cold season. The threshold for cyclones occurs at about 28°C.[60] This warmer ocean current, eventually reaching the Peruvian coast, is called 'El Niño'. It is blamed for the 1983 Peruvian rains and mud slides, for reversing the equatorial trade winds and for causing drought in Australia and Fiji. The 'El Niño' effect is not new, since it was reported in Peruvian history a hundred years ago. However, there has been no underwater volcanic activity to explain the heating of the current in 1983.

In the 1983 Peruvian mud slides, people died and precious seed varieties were completely wiped out.

Although France denies any connection between the warming of the ocean and its nuclear tests, one cannot help but wonder about the injection of hot gases into the water and the laying bare of bedrock in the lagoon after a blast. Changes in these ocean currents and trade winds can alter weather in all Pacific rim countries.

Fish poisoning has always been common in the Pacific, and was reported by Captain Cook even in 1774. However, after the nuclear testing began, the indigenous population reported a startling increase in the phenomenon. The rate per 100,000 people in French Polynesia ranged between 366 and 633 cases per year between 1977 and 1980. (The actual number of cases reported each year was between 502 and 937.) By contrast, the Solomon Islands reported two or three cases per 100,000 and the Cook Islands reported only one case in the entire four-year period.[61]

There is also a high rate of ciguatera fish poisoning in the Micronesian Trust Territory, which includes the US military testing ground in the Marshall Islands.[61] The missiles are released from the Vandenberg Air Force Base in California and come crashing into the lagoon at the Kwajalein atoll at some 8,000 miles per hour.

Victims of ciguatera poisoning start with vomiting, headache, fever, pain in all joints, and chills, and eventually the victim can be paralysed or die. After this poisoning, a recovered person can never again eat fish without a recurrence of the trauma. Fish is, of course, the main protein source for the island nations.

As carefully documented by Dr Takeshi Yasumoto, Professor in the Faculty of Agriculture, Tohoku University in Japan, a fish becomes poisonous after it ingests a micro-organism which grows only on broken or damaged coral.[62] The main source of these micro-organisms in French Polynesia is in the Gambier Islands. (It is forbidden to eat fish caught in the Morora lagoon.) These islands were used as a safe harbour for military vessels employed in observing the nuclear tests. The lagoon was dredged for sand and in its harbour radioactive fallout was washed off the observation ships after the tests. Other harbour construction and alterations took place to facilitate military uses. Again, the French deny a connection between military alterations of the natural environment and the dramatic increase in ciguatera fish poisoning, but to the observer, the evidence appears to be overwhelmingly against their claim.

As already mentioned in connection with ciguatera fish poisoning, the Kwajalein atoll in the Marshall Islands of Micronesia is the testing ground for US intercontinental missile systems. Beginning in January 1983, MX missiles have been fired from the Vandenberg Air Force Base in California to this atoll, 4,200 miles to the west, apparently to test the accuracy of the military delivery system.

About 8,000 to 10,000 people from Kwajalein, or related to a worker at the US base, are now living on the 66-acre island called Ebeye. Almost all the vegetation on that island has been destroyed and the once self-sufficient islanders are forced to import 95 percent of their food. About 6,000 of these Micronesians are living in four-room cinder-block flats, with 30 to 40 people per flat, sleeping in shifts, using one kitchen and bathroom. Another 2,000 Micronesians are homeless, living on the beach in shacks, without kitchen or sanitary facilities. The lagoon water near Ebeye has a bacteria count about 15,000 times above the World Health Organisation's emergency level. There are serious epidemics of TB, malaria, dysentery, leprosy and other infectious diseases. The health and life of Kwajalein islanders is clearly being sacrificed for US military aims.[63]

These are some of the non-radiation-related tragedies making up the pre-Third World War climate. Many other victims who die of starvation, surrogate war-fighting and unemployment might be named. But I want to focus on the radiation victims since the Third World War, if allowed to occur, will be a nuclear war. They often fail to recognise their own victimisation, or feel the meagre comfort afforded by solidarity in suffering. For them, the Third World War has begun.

When the Soviet Union experienced a major nuclear accident in 1957–58 at its nuclear weapon production installation near Chelyabinsk,[17] thousands of people were seriously injured and many died. The victims have not been allowed to contact other radiation victims so that they could have at least this small shred of human comfort in their suffering. Soviet military policy keeps outside concerned persons from assisting these people, sharing with them the experience of other radiation victims and learning from their tragic experience.

But accidents and ignorance are not the only source of victims. More than 1,200 nuclear bombs have already been detonated, and the number of victims globally must already be in the range of 13 million. Great Britain has set off bombs in the Christmas Islands and Maralinga, South Australia; China and India have exposed their own citizens to radioactive fallout; and the United States has polluted and endangered the whole Northern Hemisphere with more than 600 atmospheric nuclear blasts. Soviet tests at Novaya Zemlya have endangered the people and fragile life-support system of the arctic region. The Lapps, for example, are estimated to have more nuclear material in their bodies than any other people of the world, with the possible exception of the Pacific populations of Bikini and Enewetak. Canada sends scientists into its northern territories every summer to measure levels of radioactivity in animals and plants.

The following is an attempt to estimate the numbers of 'early' victims of the Third World War, beginning with Hiroshima and Nagasaki in 1945:

155,521	immediate civilian fatalities;
2,140	pregnant women with their children killed;
400	aborted embryos and foetuses;
147,033	civilians who died between September 1945 and January 1950 from bomb injuries;
1,523	children born with severe congenital malformations;
200	microcephalic and severely mentally retarded children;
1,384	children with milder congenital malformations;
1,350–4,090	cancer victims among survivors;
1,000–21,600	genetically damaged offspring each generation (after equilibrium) until death of the family line.

The Hiroshima and Nagasaki victims number about 322,000. Genetically damaged children will continue to be born and in their turn produce damaged offspring for generations to come.

The production and testing of nuclear weapons since 1945 has resulted globally in even more deaths and casualties:

68,000–95,000	embryonic, foetal and infant deaths;
2,252,000–6,620,000	cancer victims (some of whom die in pre-cancer states);
18,000–22,000	children with severe congenital malformations;
7,000,000	children with milder congenital malformations;
420,000–8,900,000	genetically damaged children each generation (after equilibrium) until death of the family line.
10,000,000–22,000,000	(calculated to the year 2000)

The global victims of the radiation pollution related to nuclear weapon production, testing, use and waste conservatively number 13 million. The current rate of weapon production globally (1985) generates between 7,000 and 15,000 victims yearly (between 20 and 40 a day) even without further nuclear weapon testing.[64]

These estimates include the miners and nuclear workers whose radiation-related illnesses have never been acknowledged by either government or industry. Most victims are unaware of the toxic substances added to their air, water or food. Even if aware of exposure, most victims are unable to prove that their sickness is related to their exposure.

The Nuclear Power Industry

The idea behind the nuclear generation of electricity is quite simple. A chain reaction is initiated when a neutron strikes an atom of U_{235}, which then splits, releasing more neutrons to strike other nearby U_{235} atoms. In the process a great deal of heat energy is released. This heat can be used to boil water, creating steam, which in turn can be used to turn a turbine, producing electricity in the conventional steam generator. To slow down or stop the fissioning, one can use water or insert control rods which interfere with the chain reaction by absorbing the neutrons and preventing them from striking other U_{235} atoms.

The heat generated by fissioning is very intense. About one-third is converted into electricity, and about two-thirds is wasted as thermal pollution. The excess heat cannot be used directly (cogeneration) in manufacturing processes requiring steam or in space heating because it is polluted with fission and activation products which are radioactive. The excess heat generated by fossil fuel electrical generation does not have these limitations, and can be captured for useful purposes. It is speculated that the waste heat from several nuclear generators located on one geographical site can generate a localised storm or small tornado. The waste heat can also be life-threatening for certain plants and animals.

American light water reactors are of two types: boiling water (BWR) and pressurised water (PWR). The boiling water reactor has only one loop, i.e. the same water which reaches boiling point as it circulates through the reactor core is used as steam to turn the turbine. It is then condensed and pumped back to be reheated in the core. Small particulates and some of the radioactive gases escape from the fuel rods into the air or water within the reactor. This water, as either steam or liquid effluence, needs to be periodically released into the environment. These releases are normally held back until the short-lived radio-isotopes have decayed. It is not possible to operate a nuclear plant without any releases of fission fragments and activation products (also called radionuclides or radioactive chemicals). Typical release figures for one operating nuclear generator would be between 10,000 and 100,000 Ci of radioactive gases, 1 to 10 Ci of halogens and particulates, and 100 to 10,000 Ci of tritium per year, assuming no abnormal problem.[34]

The fission process is violent, causing normal molecules in the air, water and solid structures in near proximity to become radioactive. Eventually the pipes, machinery and even the containment building at a nuclear reactor become radioactive. This effect is euphemistically called 'turbine shine'.

Pressurised water reactors (PWR) have a two-loop system, so that the water in the first loop, under higher pressure, boils at a higher temperature and then can be used to boil the water in the second loop. The water in the primary loop does not mix with the water used in the secondary system, provided there are no leaks. Actually, all such generator systems leak at times. PWRs are also subject to dramatic loss of water, which is under high pressure, if there is a break in the pipe or a faulty valve. The Three Mile Island accident involved such a sudden loss of water.[65]

The gas-cooled reactor is used extensively in Great Britain and France. It is considered by many to be an engineering failure. Great Britain subsequently began planning (1981) to build or buy light water reactors in the future, perhaps to help consolidate design in a failing commercial nuclear industry. On 14 October 1979 the *Observer* announced Prime Minister Margaret Thatcher's plans for the purchase of twenty PWRs from Babcoc and Wilcox in the USA (the company which built the Three Mile Island reator). Recently there has been promotion of gas-cooled reactors in the USA, 'so that a Three Mile Island type accident cannot occur' again. The public is beginning to understand that this reassuring statement merely means that the type of accident will change.

Canada pioneered and built the successful CANDU reactor, a pressurised heavy water moderated and cooled reactor. By using an isotope of water (deuterium) which is heavier than the more abundant form of natural water, the CANDU reactor achieves the two-loop advantages of the American PWR, without necessitating as high a pressure. The CANDU reactor, while it has many safety advantages, is also one of the reactors most vulnerable to theft of fissionable materials. Spent fuel rods from all types of nuclear reactors contain plutonium which can be used for nuclear bombs. A CANDU reactor, however, can be refuelled while operating, making theft of plutonium without detection possible. Other reactors must be shut down for refuelling. It was spent fuel rods from a CANDU reactor which provided the plutonium for India's nuclear bomb exploded in 1974.

The Canadian reactors, Pickering 1 and 2, began generating electrical power in 1971 and 1972 respectively. These two reactors, plus five other Ontario reactors of the CANDU type, were rated among the ten most efficient nuclear reactors in the world, i.e. until 1 August 1983. On that date a pressure pipe in the 12-year-old reactor burst, pouring 240 gallons of heavy water into Lake Ontario. The second Pickering reactor was shut down shortly after that, also for pressure tube leaks. A third CANDU reactor at Douglas Point, Ontario, leaked 594 gallons of heavy water into Lake Huron on September 3rd and 4th, 1983. The Douglas Point reactor is being permanently closed down after only 17 years of service, and the Pickering reactors will be outfitted with hundreds of new pressure tubes. The intense heat and radiation from the nuclear fuel rods made the pressure tubes brittle, leading to a major (6-foot) and some minor cracks. The cost of retubing a reactor is about $250 million (Canadian) and a year's loss of operational time. Canadian taxpayers bear the costs of repair and dismantling of these reactors in addition to paying their electrical bills and a more than $500 million (Canadian) federal subsidy of the nuclear industry each year. They also, of course, have borne the research, development and construction costs.

The commercial nuclear industry began after recommendations

for radiation exposure standards were made by the International Commission on Radiological Protection (ICRP) in 1959. Most nations accepted these recommendations, and incorporated them into national standards and regulatory guides. There was never any pretence that nuclear power plants would operate without releasing radiation,[66] although the public generally assumed this to be the case. Rather, releases were planned to be well within the 'permissible' standards set on the basis of military experience and the desirability of the technology. Most physicists and engineers took for granted that the standards were adequate and that the scientific challenge was to design a nuclear plant to generate electricity, to be reasonably safe and efficient, and to be cost competitive with coal. 'Safe' meant having a low probability of accident scenarios and keeping pollution within 'permissible' standards. 'Safe' is therefore a value judgment, not an objective evaluation.

A new scientific discipline was developed to provide personnel trained to measure levels of radioactive pollutants in the workplace, or in the gaseous and liquid effluence of the nuclear reactor. These pollution measurements were converted into probable radiation doses to workers and members of the public. Elaborate computer models for predicting dispersion of radioactive gases and airborne particulates, including the effects of wind currents and washout with rainfall, were developed. Estimates of pollution of groundwater, leafy vegetables, grasses and secondary foods such as milk from cows eating contaminated grass were generated by means of mathematical models. The probability of living plants concentrating radioactive chemicals in their cells at a higher level than the concentration in the air or soil, and the further concentration of radionuclides in animals eating the plants had to be estimated. This concentration of pollutants is a well-known property of the food chain. Humans, of course, are at the top of the food chain, and they receive the most concentrated doses of all environmental pollutants – pesticides, herbicides, toxic chemicals and radioactive chemicals. Radiation dose estimates to the 'Standard Man' had to be made on the basis of some specific pathways to humans by which radioactive chemicals might travel, and an individual's probable 'average' diet of these foodstuffs.

The 'Standard Man' was expected to react physiologically to the radiation dose in a statistically predictable way, regardless of race, sex, geographical and environmental differences.[67] The Standard Man is a white temperate-zone male, in his twenties and in perfect health. There is no modification of the Standard Man's reactions to allow for synergistic effects between radiation exposure and other environmental or occupational hazards, individual medical history and medications, prior radiation exposures, or even changes in future generations relative to susceptibility to radiation damage. There is no adjustment for the smaller size or differing food habits of the Asian nuclear worker. Although a systematic audit of environmental pollution in the USA was initiated to test predicted radionuclide levels around light water reactors, no audit of the predictions of human health effects of radiation exposure for the general public, or the validity of using a 'Standard Man' approach has ever been undertaken. Statistical adjustments to take care of the differences in the foetus and growing child when exposed to radiation have been introduced, but this is the only recognition of human variability.[68] The needs and weaknesses of the elderly or of disabled adults are not recognised. Similarly ignored are variations within the same person at different times in life. For example, an individual with a broken bone will take up more radioactive strontium than a normal person. Strontium acts chemically like calcium. It is a radioactive chemical always emitted in a nuclear weapon test. It or its radioactive precursor is also sometimes emitted from nuclear power plants – namely when the system of containment does not operate as perfectly as planned.

During the Second World War, long before nuclear reactors were used to generate electricity for the civilian economy, nine General Electric nuclear reactors were built along the Columbia River on the Hanford reservation to produce plutonium for the weapon programme. Eight of these reactors are now 'mothballed', i.e. retired from service because of plant deterioration and the build-up of radioactivity in pipes, walls and stationary equipment. In the USA, a nuclear plant is considered functional for between fifteen and thirty years. This large range of estimates reflects a dispute over how much building contamination with radioactive chemicals and structural deterioration makes the cost (radiation exposure to workers and the public, danger of a major accident, and so on) outweigh the benefit (production of plutonium for bombs or generation of electricity). When cost outweighs benefit in the minds of planners, the installation is shut down. Cost and benefit are usually calculated with respect to present, not future, generations, and with respect to 'worst' outcomes such as radiation-induced excess cancer deaths. Cost often refers to money paid in worker compensation or legal claims.

Retirement of a nuclear generator from active service is called decommissioning, and the decommissioning process has economic as well as radiation exposure costs. According to nuclear industry estimates in 1976 US dollars for three decommissioning options of a 1,100 MWe pressurised water reactor, the costs would be as follows:[69]

> *Entombment* (filling the containment with concrete and leaving it in place): $7.39 million plus $58,000 annually for security inspection and maintenance in perpetuity. This method would involve approximately 130 man rem (the bone marrow dose equivalent to 26,000 chest X-rays* distributed over an unknown number of employees).
> *Mothballing* (guarding the structure in perpetuity): $2.31 million with annual costs of $167,000 in perpetuity. This method would involve an estimated radiation exposure of 150 man rem (the bone marrow equivalent of 30,000 chest X-rays).*
> *Removal*: $26,860 million with no annual cost in dollars at the generator site. The cost of perpetual storage of the decommissioned generator in a guarded repository elsewhere is not estimated. This process would involve an estimated 630 man rem (the bone marrow dose equivalent of 126,000 chest X-rays).

The decommissioning dollar cost for a reactor, assuming 'mothballing' for 100 to 180 years in order to allow radioactive decay of some of the activation products, followed by removal of the structure rubble to a permanent repository, may eventually become equal to its construction cost, with the added human radiation exposure which was not involved in building the plant. The cost of decommissioning a nuclear installation and constructing a permanent repository for the radioactive rubble is not included in rate costs for electricity or in weapon budgets, and will probably become a permanent national expense for nuclear nations in the years to come. Like fuel enrichment and nuclear insurance, the cost of nuclear waste disposal is charged to taxes rather than electricity rates or military defence.

---

* Only the external radiation dose is counted in both man-rem estimates and X-ray dose. In the decommissioning but not the X-ray, internal doses might also occur. Man-rems is the product of the number of people exposed times the average exposure of each person, for example, 100 man rem could mean 400 people each receiving an average dose of one-fourth of a rem, or two people each receiving 50 rem. I assumed 0.005 rem bone marrow dose per chest X-ray.

The long-range plans for removing the mothballed US military reactors at Hanford have not yet been made public.

Nuclear Technology's Other Waste

The Hanford nuclear reactors were operated solely to obtain the spent fuel rods, i.e. the fuel rods after the nuclear fissioning has taken place in the reactor.

Fuel rods are composed of uranium pellets about 0.4 of an inch in diameter and 1 inch long. The pellets are placed inside a 12-foot-long cylindrical rod of zirconium. A fuel assembly consists of between 50 and 200 fuel rods, and can weigh 0.2 to 0.5 metric tons* (200 to 500 kilograms).

A 1,000 MWe (megawatt electric) US light water generator, of the pressurised water type, requires about 180 fuel assemblies, or about 90 metric tons of enriched fuel, to operate. 'Enriched' means that the fissionable fraction of the uranium fuel, U_{235}, has been concentrated from 0.7 percent to 3.0 percent by weight. About one-third of the enriched fuel, 30 metric tons, is removed and replaced in an American power reactor each year. Replacement of fuel would be more frequent in a reactor used only for producing plutonium for weapons because this maximises the proportion of fissionable plutonium 239 relative to the non-fissionable plutonium 240 and 238 in the rod. When plutonium 239 incorporates a neutron into its nucleus instead of fissioning, plutonium 240 is produced.

The Canadian 542 MWe heavy water reactors at Pickering each use about 80 metric tons of uranium fuel per year. CANDU technology does not require enriched fuel.[70] Assuming that a 1,000 MWe CANDU reactor would use twice this amount of fuel yearly, the fuel requirement estimate would be about 160 metric tons of uranium concentrate. The 160 metric tons of uranium concentrate would require mining of 145,264 tons of ore.[71] In the table on p. 112, estimates for the CANDU reactor are given in parentheses under the corresponding figures for the light water reactor.

Only approximate estimates of direct waste are given in the table. There are also large quantities of radiation-contaminated tools, equipment and clothing; radioactive water pipes, walls, buildings, barrels, trucks; condensates and all manner of objects which have

* A metric ton is 1,000 kilograms or 2,240.6 pounds. Sometimes ore is measured in long tons, 2,240 pounds avoirdupois, as in the UK, or in short tons, 2,000 pounds avoirdupois, as in the USA or Canada.

Nuclear Fuel Cycle Products and Waste Associated with a One-year
Operation of One 1,000 MWe Generator

| Product | Amount in metric tons | Waste | |
		Type of waste	Metric tons of waste
Uranium ore	180,000 (145,000)*	Uranium mill tailings	179,728 (144,790)
Uranium refining	272 (210)	Uranium refinery waste	242 (50)
Reactor fuel	30 (160)	High level waste (spent fuel rods)	29 (159)
Mixed plutonium isotopes[46]	1	Plutonium waste	0.2
Weapons grade plutonium	0.8	50 to 90 Nagasaki-type bombs	–

* CANDU reactor plutonium yield would be about the same as for the light water
reactor. The British gas-cooled Magnox reactors use natural uranium fuel rods
similar to those used in the CANDU reactor. About 1 percent of the fuel is recovered
as plutonium.

become radioactive by proximity to the fissioning or which are
contaminated by direct contact with the fuel or fuel products.

Only solid fuel rods and the high level liquid waste resulting from
the chemical dissolving of fuel rods are stored. All radioactive gases
and contaminated liquids (other than dissolved fuel rods) are
eventually released to the environment. The only purpose of
dissolving the fuel rods, producing the high level liquid radioactive
waste – the type which needs to be re-solidified into a glass or salt –
is plutonium extraction. The plutonium is needed not only for
weapons, but also for fuelling the breeder reactor, which 'breeds'
more plutonium. Plutonium extraction from commercial fuel rods
has become the rule rather than the exception in most nuclear
nations, as, for example, France, Great Britain, West Germany and
the United States prior to 1972. In 1982, President Reagan
announced that spent fuel from US commercial reactors would

again be needed to fulfil the proposed weapon build-up of the 1980s. Other nations try to keep the 'peaceful atom' programme separate from the military connection.

Gaseous wastes are filtered and after a short hold-up time to allow decay of radioactive gases with short half-lives, released directly to the air. Under normal conditions some radioactive particulates such as plutonium and cobalt 60 are emitted with gases, carbon 14 and tritium. They are routinely released to the environment; the solids in small quantities and the gases in huge quantities. Some of the radioactive gases decay into radioactive solid particles after their release into the atmosphere.

Filters are used in nuclear installations to retain some of the airborne radioactive particulates, such as plutonium and other heavy metals. However, there is a routine amount of such particulates which escapes the filters and enters the atmosphere, and the contaminated filters themselves become radioactive waste. In spite of being contaminated with plutonium the filters are classified as 'low or medium level waste'. Only spent fuel rods and reprocessing waste are called 'high level'. Waste is classified by the concentration of radioactivity, not by its potential hazard to humans or to the biosphere. Plutonium particles whether from low level waste or high level waste can induce cancer.[5]

The continuance of commercial nuclear technology in the USA and elsewhere depends on the creation of state, regional or national nuclear waste repositories to prevent the solid fission waste, especially strontium 90, cesium 137, cobalt 60, carbon 14 and the trans-uranic elements like plutonium, from contaminating the environment. The repository must be stable for millions of years. A state, regional or national high level waste repository has yet to be constructed and tested anywhere in the world with the possible exception of the USSR. There is no known way to prevent the escape of radioactive gases, tritium, or the uranium by-products.

Plutonium Breeders

The uranium fuel used in a nuclear generator is a mixture of uranium 238 and uranium 235. The fissionable fraction, uranium 235, is about 0.7 percent in the CANDU and Magnox reactors, about 3 percent in US light water reactors, and as high as 23 percent in French reactors. Weapons grade uranium is about 95 percent uranium 235.

When the uranium 235 fissions, i.e. breaks down into two lighter chemical elements, it releases neutrons which if they strike other uranium 235 atoms cause them to fission also. This is called a chain reaction.

Sometimes, however, a neutron strikes a uranium 238 atom at a rapid speed and the uranium captures it, incorporating it into its nucleus. This new uranium isotope, called uranium 239, quickly transmutes into plutonium 239 which is fissionable.

Occasionally a plutonium 239 atom, instead of fissioning, will capture a neutron, becoming the non-fissionable plutonium 240. It is the fissioning which produces the energy release used either as an explosive or as a heat source to boil water and steam-generate electricity.

The breeder reactor capitalises on these reactions by using as fuel a core of plutonium surrounded by a 'blanket' of uranium 238. The fissioning of a plutonium atom releases two or three neutrons – one to strike another plutonium atom and continue the fissioning, and one or two to 'breed' more plutonium by striking the uranium 238.

By using plutonium 239 and uranium 238 to produce more plutonium 239, the amount of usable fuel obtained from the original uranium ore is increased. Whether or not the fuel advantage warrants the increased health and safety risk of breeder technology, or the biohazards and weapon proliferation risks caused by deliberate production of plutonium, is being debated internationally. Countries which are actively developing breeder technology include the United States, the Soviet Union, Great Britain, France, West Germany, India and Japan, although it should be mentioned that in 1983 the United States halted its principal breeder reactor programme (called the Liquid Metal Fast Breeder Reactor) at Oak Ridge, Tennessee.

'Breeding ratios' in the USA are about 1.24, that is for every ton of plutonium burned 0.24 tons of plutonium are produced by the surrounding 'blanket' of 'fertile' uranium 238 (in industry jargon). The water coolant used in non-breeding nuclear generators (light water reactors and the CANDU reactor) slows down neutrons not favouring the uranium 238 capture and transmutation into plutonium 239, although some plutonium is always produced in these reactors.

Water slows down the neutrons to about the speed of a rifle bullet, while neutron capture, which is maximised in the breeder, requires about 100,000 times that speed. The breeder reactor uses liquid sodium instead of water because this liquid does not slow down the neutrons as much as water does. Since sodium is a metal, this type of breeder is often called a Liquid Metal Fast Breeder Reactor (LMFBR) in industry jargon.

Being highly inflammable, the liquid sodium is subject to spontaneous fires if exposed to air and to violent explosions if exposed to water, problems not experienced when water is used as a coolant.

Because of the difference in coolant, i.e. sodium instead of water, breeder reactors behave differently during a loss-of-coolant accident. The uranium 235 fissions best in the water moderated slow neutron environment. If the water is lost, the neutrons speed up, the fissioning rate slows down somewhat and the reactor does not explode. A disastrous core melt-down, steam explosion or other catastrophic event can, of course, occur, but the reactor cannot explode like an atomic bomb.

When a breeder reactor suffers a loss-of-coolant accident, the fissioning increases and an explosion can occur. It is not the same as an atomic bomb, since the reactor core is surrounded by the whole containment structure which can absorb and contain some blast and fission fragments. Even more important, the fuel blows apart before it has time to fission. However, the blast can distribute large quantities of plutonium aerosols a considerable distance downwind. Nuclear advocates avoid the term 'explosion' using instead phrases like 'rapid disassembly' or 'a deposit of considerable energy in a short period of time'. The reality is not mitigated by such language manipulation.

A breeder reactor has a characteristic breeding time. This is the time it takes to build up a quantity of new plutonium sufficient for extraction and use to fuel a reactor. The uranium 238 blanket design proposed for a small 375 MWe breeder reactor at Clinch River, Tennessee, USA had a breeding time of thirty years. Proponents hope to reduce the time to eighteen years with a new heterogeneous design which inserts the blanket of uranium 238 directy into the plutonium fuel core.

The International Atomic Energy Agency estimates that electricity from breeders will cost twice as much as from conventional reactors, even with the customary nuclear subsidies. Sir Francis Thomas, Chairperson of the British Electricity Council, has called for a shelving of England's four reactor, $10 billion breeder programme. The French and Japanese have virtually admitted that their aggressive programmes to develop breeders are an 'insurance' against manipulation of the uranium supply, most of which must be imported into these countries. The breeder reactor is viewed by nuclear technologists, however, as a necessary technology for extending uranium supplies, regardless of financial costs. It seems to be a combination of international resource insecurity and technological myopic vision which keeps breeder technology among global energy options. The general public hardly enters into the debate, either being manipulated by public relations information, as in the USA, or having protests suppressed through government action, as in France.

In the USA, the Clinch River reactor, originally estimated to cost $669 million and to be operational by December 1979, was, in 1983, scheduled to be operational by February 1990, with an estimated cost of $3.2 billion (US). Both the economic and political costs of this technology are heavy and appear to be increasing with time.

Although President Reagan's Budget Chief, David A. Stockman, deleted the commercial breeder reactor from the 1982 Reagan budget, the White House restored funding at a level of $254 million for Clinch River. There has, however, been little or no national debate on any question other than the economics of the US commercial breeder. The military breeder reactors operating at Hanford and Idaho Falls were constructed without public hearings or, in many cases, public awareness of their existence. Even the economic problems fail to be discussed when political leaders claim that 'national security' requires plutonium production for weapons.

On 16–18 September 1981, a debate on breeder reactor policy in France took place at the French National Academy of Science in Paris. The debate was held at the request of the Groupe de Bellerive, Swiss scientists from Geneva.[72] The French breeder reactors, Phoenix and Super Phoenix, are dangerously close to the cities of Lyons and Geneva. Hearings were closed to the public.

Following a resolution of the Governing Board of the World Council of Churches (WCC), adopted in August 1980, the French debate was designed to initiate scientific–public dialogue on the implications of a plutonium breeder economy. The WCC resolution reads as follows:

a. The Central Committee endorses the call to the Heads of Governments by the WCC Conference on Faith, Science and the Future for an immediate five year moratorium on the construction of new nuclear power plants to enable the overall risks, costs and benefits of this energy option to be properly evaluated in public debate; and asks the member churches of the WCC to study all the recommendations on energy adopted by the Conference;

b. the Central Committee urges Churches to encourage a debate in all countries and to discover for themselves the most effective ways to implement these recommendations in their own activities;

c. the Central Committee takes note of the series of consultations being planned in Third World countries on the theme: 'Just Energy Policies for Sustainable Societies';

d. in the light of these discussions the Central Committee requests Church and Society to present an assessment of the energy debate to the Central Committee in 1981.[73]

The WCC resolution was built upon two resolutions of the US National Council of Churches, one opposing a plutonium economy in 1975 and another calling for the phase-out of nuclear technology in 1978. The National Council of Churches' (NCC) plutonium statement developed along the following logical path:

The principle of ecological justice provides ethical guidelines for human choices and is a basis for decisions about energy policies.

The statement articulates values that must be the criteria for the development of energy technologies:

sustainability: requiring that 'biological and social systems be neither depleted nor poisoned';

equitable distribution: requiring that 'the earth's resources be fairly distributed among the present generation on the basis of need, and economically used in deference to the needs of future generations'; and

participation: requiring that 'decisions be made by all members of the community'.

Because of these guiding principles, the Council urged

the United States government to call a moratorium on the commercial processing and use of plutonium as an energy source, and on the building of a demonstration plutonium breeder reactor pending further study.[74]

The NCC plutonium statement was developed by a study

committee co-chaired by Dr Margaret Mead and Dr René Dubois.

There were strong objections to the NCC statement opposing a plutonium economy, especially since the statement was drafted without input from the nuclear technology community. It also became obvious that nuclear power needed to be considered as a part of the whole energy strategy of the USA.

The NCC responded by forming a consultative committee of 100 people – nuclear physicists, oil company executives, trade unionists, homemakers, economists, biologists and environmentalists – together with twenty-eight ethicists and theologians, requesting that they draft an energy policy statement. I was an ethicist on this committee. As a result of the work of this Energy Task Force, headed by Katherine Seelman and Chris Cowap, the following policy was adopted by the NCC governing board in May 1979:

> We support a national energy policy which will not need to utilise nuclear fission. Secure handling of nuclear wastes over thousands of generations and safe operation of nuclear plants require that humans and their machines operate without endangering human beings or the environment. Human beings are not infallible; they will make mistakes, and machines will fail. The result may be irreversible damage to the environment and to the human genetic pool.
>
> We support a continued ban on the commercial processing and use of plutonium as a fuel in the United States, and stringent efforts to reach worldwide agreement banning such use of plutonium. Commercial use of plutonium can result in proliferation of nuclear weapons. The potential misuse could result in pressure to curtail civil liberties in order to prevent such a threat.
>
> We support the rapid development of enforceable regulations to require a social and environmental impact statement of a technology before it is widely used and the monitoring and control of its use to prevent social and environmental damage.
>
> We support US policy which seeks to share technologies internationally without imposing capital-intensive energy technologies on other countries.
>
> We support full US co-operation in international efforts to ensure equitable distribution of necessary energy supplies and the rapid development and deployment of appropriate technologies based on renewable energy resources such as solar energy, including wind and water.[75]

The commercial nuclear industry, not realising the shaky ethical, biological and social foundation of the whole military nuclear dream, has inherited the nuclear waste dilemma, the citizen protest movements and the unrealistic projections of the health consequences of radiation exposure. The very attempt of this industry to separate itself conceptually from the nuclear military seems to have blinded it to the problems. Although it can be argued that nuclear power does not need the nuclear weapon industry, it cannot be argued that the nuclear weapon industry does not need nuclear power. The convenient 'peaceful atom' façade, created in 1954, was promoted by the major nation-states of the world in a manner that no other nascent commercial venture has ever been. In spite of this support no country has been able to develop a commercially viable nuclear industry. The industry is kept alive by the will of governments, through taxpayer subsidies. Even Canada, the latest nation on the scene with nuclear power development, is wavering. A Gallop Poll in early 1983 showed that more than half the population favoured *not* developing nuclear power.

Reprocessing Plants

'Weapon' reactors do not differ from the nuclear reactors which generate electricity except that the steam they produce is not used to turn steam turbines. They are fired simply for the sake of obtaining the 'spent' or waste fuel rods.

After firing in a reactor, the rods are removed and the zirconium outer sheath is dissolved in nitric acid, releasing the uranium fission products. This process routinely releases nitrous oxides to the air causing acid rain. Radioactive iodine and other fission products are also released in the plant's gaseous effluence. A chemical process is used to separate out plutonium and, sometimes, strontium 90.

Those fission products which are airborne at the temperature of the 'hot' fuel, such as krypton, tritium, iodine and carbon 14, are routinely released into the atmosphere. The radioactive solution left after separation of the plutonium and release of the radioactive gases and volatile products is still highly radioactive, corrosive (acidic) and thermally hot. It contains a mixture of fission products. As the fission products with shorter half-lives decay, the predominant waste products are strontium 90, cesium 137 and actinides. The actinides, elements with an atomic weight greater than that of uranium, remain radioactive and toxic to humans for millions of

A Dangerous Choice

In 1985, India will become the only third world country to commission a fast breeder reactor – a 40 MWe test reactor on the shores of Kalpakkam in Tamil Nadu . . . Nuclear planners are already designing a 500 MW prototype commercial reactor.

Given our [India's] scant resources and our vast ambitions, nuclear designers are economising both on safety and on research. Thus, while in other countries nuclear designers are compromising on containment [the concrete and steel building containing the reactor], our nuclear planners have, for the 500 MW prototype they have designed, done away with containment altogether in order to cut costs.*

The Indian breeder reactor is a replica of the 40 MW French test reactor Rapsodie, which had to be shut down in 1983 after fifteen years' operation, because of a sodium leak. The French fast breeder operated with a fuel of plutonium and uranium carbide, however, India will have to use plutonium and uranium-oxide, an as yet untried fuel. The state-of-the-art for fast breeders leaves technical gaps in several major areas including lack of equipment to detect local blockages in sections of the reactor, spread of fuel meltdown if control rods fail, sodium fires or vapour explosions and emergency core cooling systems.

* *The Illustrated Weekly of India,* 16 October 1983

years. Some plutonium with a half-life of 240,000 years is emitted with the waste gases.

In the early days of the Hanford and Chelyabinsk plutonium factories, the liquid acid waste was neutralised and then stored in a liquid form. Depending on the amount of dilution, the volume of this liquid radioactive waste varies from 1,000 to 100,000 gallons per ton of fuel reprocessed. A one-year supply of fuel for one reactor is 20 to 30 tons, producing 20,000 to 3 million gallons of high level, liquid waste, if reprocessed.

By 1975 the Hanford Reservation contained about 70 million gallons of high level radioactive liquid reactor waste. The US Atomic Energy Commission had been slowly trying to evaporate the waste down into solid cakes of salt to be stored in steel tanks. This task now falls to the Department of Energy. There are more than 150 high level radioactive liquid waste containers at Hanford.

Nuclear industry estimates of the dimensions of the nuclear waste 'cubes' which will be required to provide each member of the public with electricity needs during a lifetime are misleading with respect to the volume of space and centuries of care which the human community will have to commit to storing this nuclear waste. If actually packed closely, the fuel would form a critical mass. The industry also implies that only reactor fuel rod waste should be of concern to the public. It was low level waste, not reactor fuel, which apparently exploded in Chelyabinsk and almost exploded in Hanford, Washington. It is uranium mine tailings which have bathed the North American continent in radon gas. It is tritium, carbon 14, and stratospheric pollution from weapon testing which is already in the biosphere and which will slowly continue to pollute the food chain, undermining the life-nurturing power of the planet earth. Even if nuclear pollution were to stop today and all the controllable waste jettisoned into space,* the uncontrollable waste already released into the biosphere would cause devastating damage over time.

The official ICRP recommended limits of uptake by living plants, animals and humans of radioactive material released into air and water have also been challenged as scientifically unreliable.

In a paper presented at the Annual Meeting of the American Public Health Association in Washington, in January 1982, Bernd Franke, a scientist from Heidelberg, West Germany, pointed out that the search for the lowest possible estimate of human radiation dose from milk contaminated by radioactive airborne particles had led to the use of estimates for American cows eating English plants grown on Russian soil. He also pointed out an error which resulted in an estimate of human uptake of cobalt 60 from milk which was more than 2,000 times lower than it should have been. This and similar errors can result when chemicals present in the workplace as inorganic molecules are converted to organic molecules in the environment, having much greater bioavailability in the latter form.[76]

The water effluence from the nuclear plant or waste facility was treated in the same theoretical way. Models estimated the rate of emission of radioactive particles, the settling rate on the bottom of the nearby river or lake, and the incorporation into flora and so-called 'bottom-eating' fish. These fish are then eaten by other

---

* The dream of sending nuclear waste into space is utopian because of its extremely heavy weight, and the volume required to prevent formation of a critical mass.

fish, and eventually the polluted food makes its way to the human dinner table. All these steps must be quantified and averaged.

Other sources of radiation dose to humans are through swimming, bathing, or other recreational water activities where the skin is exposed to radioactive particles, or the lungs exposed to radioactive vapours rising from polluted water.

Besides these pathways involving air, land, water and food pollution, humans can be exposed to radioactive materials transported by truck or rail into and out of the nuclear power plant. The spent (or used) fuel rods are especially hazardous as they give off penetrating gamma radiation which passes through most shielding material, at least to some extent. Obviously, an accident situation involving spent fuel rods would cause the most serious exposures. All other parts of the nuclear fuel cycle, from uranium mining and milling to waste disposal, involve transport of radioactive materials by truck, train, ship or plane, through and over areas of high population density. For an extreme example, between July 1979 and 30 May 1980, the San Onofre nuclear power plant in southern California applied to the Nuclear Regulatory Commission for route approval of 52 truckloads of spent fuel rods each to be transported 2,200 miles to Morris, Illinois.[77] This is only one small power plant.

The fourth source of radiation exposure for the general public from the ordinary operation of a nuclear power plant is direct gamma radiation from the reactor building, euphemistically called 'turbine shine'. This problem is especially serious with BWRs.

The Averaging Syndrome

By the time we calculate the total loss of radionuclides expected in the average 30 to 40-year lifetime of a nuclear plant, average this loss over time to determine approximate yearly effluence in curies, and then determine average air pollution and desposition on living plants and groundwater, average uptake of radioactive chemicals by plant and animal, average diet and recreational habits of the average person living in the vicinity of the plant, we have a (fictional) estimate of human radiation exposure over the normal plant lifetime. Sometimes the dose is averaged over the whole population of the country rather than the population downwind or downstream from the plant, making the individual average doses seem smaller. Of course, some people receive much higher than the average dose.

In the 1972 edition of *Biological Effects of Ionising Radiation* (BEIR I) there is an estimate of the average whole-body gamma radiation dose to the US population from commercial nuclear reactors.[78] The estimate, 0.002 mrem* per person, is often used by nuclear industry spokespeople to show the relative harmlessness of the industry. This very small number is compared to 100 mrem per year natural background radiation and 70 mrem per year medical X-ray exposure estimated for each person. The argument is very convincing until one realises that: (1) only the external gamma dose (i.e. where the source of radiation is external to the body) is included in the estimate for the reactor, thus omitting alpha or beta external doses and all internal doses; (2) only the generator phase for the nuclear fuel cycle is included, omitting radiation exposures related to the uranium mining, milling, enrichment, fabrication of fuel, transportation and waste management; (3) the estimate was made as if each person in the USA shared the radiation dose, while in reality the greater burden is on persons living downwind or downstream from the installation; and (4) at the time of the estimate (1970) there were only 11 commercial nuclear generators operating. There were 74 operating in 1979, with 120 more nuclear generators planned for the 1980s. Also, this BEIR estimate of dose is limited to the US population, while it assumes the global distribution of some of the radioactive pollution.

When presented to the public, a fictional and outdated average radiation dose is often compared with natural fluctuations in radioactivity due to varied rock content of soil, height above sea level or rainfall (laced with fission products from weapon tests). If the average dose is within that expected range of natural variability, the dose is considered 'acceptable'. What this sophistry actually implies is that with present record-keeping on the health of the community at risk from radiation, natural or man-made, the changes in health due to the changes in radiation exposure will not be 'noticed' or traceable to the cause. People will blame their cancer or asthmatic child on 'hard luck' or 'fate', and not be able to connect it causally with their exposure. At least in theory, the radiation dose is shared equally by everyone so that the harm is shared equitably.[79] Moreover, constant exposure above the average level shifts the average, hence changes the environment.

Of course, the radioactive effluence from a nuclear plant is not

---

* 'Mrem' is an abbreviation for millirem, one-thousandth of a rem (0.002 mrem = 0.000002 rem). 1 mrem = 0.01 mSv in the new terminology.

released uniformly as is assumed in the computer model but in discrete batches; it does not immediately spread homogeneously in food but tends to clump in some foods more than others; people do not all eat average diets but have varied tastes; and there is no such thing as an average response to radiation. Some individuals are much more susceptible to damage than others.[80] The elaborately fabricated assurances are a sophisticated self-deception and rationalisation.

Much serious research has been done to test and document radiation pathways to humans through the food chain; however the standard-setting bodies systematically choose to accept the research which minimises official dose estimates for the public, regardless of the quality or appropriateness of the research.[81] Little research has been done on differential uptake in humans with metabolic diseases, broken bones or other disorders. There has also been little research on identifying and protecting humans who are especially susceptible to damage from radiation, such as those with asthma, heart disease, diabetes or other chronic diseases.[82]

Perhaps because the early nuclear technology attracted people trained in physics, chemistry and engineering, there was from the beginning an emphasis on the measurement of radiation in the environment rather than the measurement of changes in public health. Physicists tend to 'spread' radiation stress in a population in the same way they would spread stress in a steel beam when building a bridge. This is an unacceptable methodology in biology. Perhaps also the military cover-up of the nuclear bomb testing effects made it necessary to continue the pretence that 'low level' radiation is no harm to human health, it presented 'no immediate danger'.

Radiation exposure is one more cause tipping the delicate balance between illness and health for the already vulnerable individual exposed to many environmental carcinogens. Because no one can be isolated from all pollutants except radiation, people are led to believe no cancers can be 'proved' to be attributable to the radiation. Nourishing this myth is political rather than scientific. It is like asking that the human body be completely free of all bacteria and viruses before it can be accepted that a particular bacteria causes a particular disease. Just as one can sort out which virus or bacteria is responsible for a disease, so one can learn to discern the effects of different environmental hazards.

Radiation, because it increases mutations, is both an initiator and a promoter of cancer. The cancer initiator is any physical or chemical agent which disables the cell's growth-control mechanism. Uncontrolled growth of white blood cells produes leukaemia. Uncontrolled growth of red blood cells produces polycythemia vera. Uncontrolled growth of tissue cells produces a solid tumour. These erratically growing cells are 'damaged' and 'different' from normal cells so the body's immune system, designed to attack 'foreign' cells, tries to destroy them. Everyone at some time in his or her life has such tumour processes within the body destroyed by the body's own protective mechanism.[83]

In a radiation environment, the evolving malignant cells can mutate more frequently, increasing the number of slightly different malignant cell strains present in the body. The greater the variety of cell strains, the greater the probability that one strain will be 'successful' and avoid destruction, going on to form a clinically detectable malignant disease.

Once this is realised, the presence of other causes capable of initiating cancer is recognised as enhancing the ability of radiation to produce cancers, not fully explaining away the problem. This synergistic tumour production was important in the occurrence of lung cancer among the uranium miners. Those miners who smoked cigarettes were the first to experience the lethal effect of the radioactive gas, radon, which polluted the air in the mines.[84]

The policy of 'ruling out' all other environmental causes of increases in leukaemia and cancer before radiation can be 'blamed' totally ignores the synergistic effects of combined hazards, the discerning ability of the human mind and the predictability of the living systems exposed.

Record-keeping

The well-publicised nuclear industry public relations statement that 'there is no record of a member of the general public having died from a commercial nuclear power plant operation' has given many people the illusion that this industry is very safe. There is little public understanding of the pro-nuclear philosophy of associating cause and effect and the grossly inadequate health record-keeping done in public health sectors. The latter is, of course, limited by government funding policy. There is little public awareness, also, of the limiting factors used in the carefully worded statement. No *record* of a death does not prove that a death has not occurred.

The gathering of public health records began in an era of

Family Survey Proposal

Introduction: Since it has been my experience that environmental health monitoring proposals are limited in scope, i.e. addressing the problems of measuring toxic releases into air or water, assessment of biological uptake and dose to the 'average' members of the public, together with predicted risk of cancer, or other genetic effects, I would like to submit an auxiliary proposal to complete this information. This proposal is not a substitute for the usual predictive approach, which is a 'best guess' or forecast, but rather serves as an audit of that guesstimate. Due to the excessive use of averaging, and the inability of the predictive methodology to identify the victims of pollution, it can never meet the expectations of those at risk from an industrial accident. The proposed methodology is designed to identify major types of health effects resulting from any hazardous exposure, to enable determination of the proportion of each health effect attributable to the exposure and to determine an acceptable compensation for the victims and their families. It is important that such problems be anticipated, and not merely left to chance solution after an accident.

Phase I: Prior to the operation of any hazardous facility a three-year baseline data Family Survey should be developed for the population at risk. The Family Survey has three parts:
1. a Family Health Questionnaire;
2. an individual physical examination; and
3. a death certificate study.
These instruments should be appropriate to the hazard.

Phase II: Assume a population of 1 million within a fifty-mile radius of the facility. In a population of 1 million, there would be about 200,000 families of child-bearing age. These families could be categorised as:
1. along the transportation route for hazardous material;
2. within the 90 degree segment of predicted maximum airborne pollution;
3. at risk from downstream pollution;
4. other families.
The names of families in each of the above four categories can probably be obtained from census information, and families where husband and wife are between twenty and forty years of age selected out.

All eligible families along the transportation routes will be included in the survey. This number is subtracted from 20,000 and the remainder divided by three. Assume that there are 2,000 families with parents aged between twenty and forty years along the transportation route. Then 6,000 families randomly chosen from each of the other three risk categories would be included in the survey population. The total survey population would be 20,000 families, roughly 80,000 people. These families would be expected to give birth to 3,000–4,000 children during the survey period. The death certificate study would include age at death and underlying cause of death for grandparents, children or spouses (former) of the surveyed families.

Budget for the three-year baseline study:

£		
	6,000	Determination of study population (six weeks)
	500,000	Administration of and data processing for 20,000 Family Questionnaires
	2,000,000	Physical examinations for 80,000 persons
	6,000	Death certificate analysis
	90,000	Data collection on the 3,000–4,000 babies born to the survey population over the three-year period
	10,000	Professional consultants and travel
	260,000	Overheads and staff support
£2,872,000		Total

This cost, less than 0.1 percent of the cost of constructing a nuclear reactor (US experience), should be added to the stockholder's investment for the facility. It is an assurance of industry concern for public health, and its determination to assume responsibility for and provide legal evidence of any damage to health which might occur. This step would undoubtedly increase public confidence.

Phase III: After the completion of the baseline study, and after five years' operation of the facility, the family surveys would be updated. The estimated cost of updating would be about £500,000 (or £25 per family). This cost, about 50p per person or £2.00 per family, could be charged to the entire 1 million persons in the fifty-mile radius.

The updating would be repeated for each five-year period thereafter.

Phase IV: In case of an accident at the facility the four sub-populations in the survey would be designated 'exposed' or relatively 'un-exposed'. For example, if an accidental liquid spill occurred, the downstream families would be at highest risk. Depending on the nature of the hazardous material, its water solubility, family dependence on gardening or farm animals, time of year, etc., an appropriately designed health study could be undertaken using the baseline families.

If a statistically significant health effect is noted, say, for example, miscarriages, the risk of such an event for the 'exposed' relative to the 'unexposed' could be determined. An attributable proportion* of miscarriages can be obtained using the formula: [a–b]/[1–b], where *a* is the percentage of cases exposed, and *b* is the percentage of controls exposed.

The attributable proportion would serve as the basis for compensation of individuals.

Agreeing on a cash settlement of, say, £1,000 for personal trauma plus [a–b]/[1–b] × 100 percent of medical costs, the company would so compensate every couple in the downstream population (whether surveyed or not) for miscarriages which occurred during the time period of identified risk.

The cost of the financial settlement and of the attributable proportion assessment should be paid by the insurance agency. The assessment would always involve examination of an identified, limited, population of previously known size. The persons holding the survey population data bank could estimate the range of costs suitable for various accident scenarios.

* Attributable proportion measurement was first proposed by Morton L. Levin in an article 'The Occurrence of Lung Cancer in Man', *Acta Unio Internationalis contra Cancrum*, 9:531–41, 1953. I suggested the simplified formula and Dr Levin and I sent a letter jointly to the *American Journal of Epidemiology* setting out the methodology (14 March 1978). The attributable proportion is a number between 0 and 1.

infectious disease epidemics and they are designed to handle this problem best. If a person dies of the plague, it is important to know where he or she died and who else might have been exposed to the disease so that quarantine or other health measures can be enforced. With respect to cancer, knowing where people died can be irrelevant and useless information. Their exposure fifteen to forty years before death will be more pertinent. Many servicemen exposed to nuclear bomb tests returned to their hometowns to live and die. In no way can the records of their cancers now be distinguished from those of their neighbours. Their death records contain no indication of their experiences. Because of inadequate military record-keeping policy, the US government seems unable to provide even the names or service experience of military men who participated in nuclear tests. Such record-keeping would allow comparison between their health today and that of other men of the same age. Similarly, workers in nuclear installations, people along transportation routes for radioactive material, or those whose main food supply comes from areas around nuclear power plants, cannot now be identified and their experiences compared with those of persons not so exposed to radiation. Non-record-keeping is hardly a commendable public health policy or the way to 'prove' the safety of an industry.

The subtle limiting factors in the public relations statement, 'there is no record of a member of the general public having died from a commercial nuclear power plant operation', so frequently quoted, include:

'member of the general public' – which excludes all nuclear workers, even part-time employees and those involved in transportation of radioactive materials;

'commercial' – which excludes government-operated or military nuclear plants;

'nuclear generator' – which excludes all other parts of the nuclear fuel cycle such as mining, milling, fabrication, transportation, reprocessing and waste disposal;

'died from' – which excludes non-fatal cancers and other diseases directly caused by the radiation exposure. It also excludes genetic damage such as blindness, deafness or chronic diseases which occur in a child of the victim.

Death Rates in Populations Near Nuclear Power Plants

Martha Drake, wife of Dr Gerald Drake of Petasky, Michigan, USA, whose story will be told later, pp. 206–208, was distressed at the official rejection of her husband's study of health problems in the vicinity of the Big Rock nuclear power plant. She undertook graduate studies at the University of Michigan and her Master's thesis, submitted in summer 1976, was entitled 'An Analysis of Death Rates in Populations Near Boiling Water Nuclear Plants'. She chose three of the earliest US boiling water reactors and the six counties closest to and containing them as her population exposed to the nuclear reactors. There were four different control groups: three randomly chosen samples of eighteen counties each, none of which had nuclear installations, and the USA as a whole.

The eighteen counties exposed to nuclear plants contained nearly

1 million people (959,904). They had essentially the same leukaemia rate as the four control groups prior to start-up of the first reactor. During the second time period, after a lapse of five years, the exposed population averaged nine 'extra' leukaemia cases per year. This number was both a significant increase over the leukaemia rate in these counties containing and surrounding the reactors, before start-up date, and a significantly higher rate than was found in any of the four control groups during the second time period. It was also significantly higher than the US rate as a whole.

Assuming that the average exposure to people in the counties at risk from a nuclear power plant is 0.0001 rem, and multiplying this by 1 million people, this implies that 100 person-rems of exposure has resulted in nine 'extra' leukaemias. The official estimate is 100 'extra' leukaemias for 1 million person-rems, making the number observed about 900 times higher than was expected. That is a rather big error to explain away by assuming changes in socio-economic factors in the area around the plants. One has the option of assuming real nuclear emissions were much higher or that the health effect estimates were too low, or both. Perhaps the 'extra' leukaemias were workers at the nuclear plants, a fact which would not be registered in death certificate summary data. Another explanation would be that an unexpected synergetic effect of radiation exposure and other environmental, age or hereditary factors were causing the increased leukaemia rate. This situation again strongly points to the recklessness of insufficient health monitoring of persons at risk from the whole nuclear undertaking.

In Hiroshima and Nagasaki the increase in the leukaemia rate occurred four to six years after the bombing. The increase in solid tumours occurred after fifteen years, and the ratio between extra solid tumours and extra leukaemias was more than 10 to 1. Should this be repeated around nuclear reactors the cancer situation can be expected to worsen. By the time it is noticed, it will be too late to avert the tragedy.

It is difficult to extrapolate from the Martha Drake study to the broader nuclear power situation. One might argue that improvements in design have reduced radioactive effluence from nuclear generators, and that pressurised reactors are less polluting than boiling water reactors. These factors are offset by the number of power plants being built – more than 100 in the USA alone – and the size of the newer models – 1,100 MWe instead of 75 MWe. There is also an accumulated level of radioactive nuclides in the environment due to thirty-nine years of weapon testing and commercial nuclear plant operations. Nuclear waste is accumulating, there is more transportation of nuclear materials on the highways and airways, and the nuclear industries are hiring more people. The numbers of people being exposed and the levels of exposure are increasing. These factors argue for even greater leukaemogenic and carcinogenic effects.

In spite of many negative health indicators, the official position of the US Department of Energy is that there is not yet 'definitive proof' of harm. A nuclear reactor is presumed innocent until such 'definitive proof' is acceptable to the nuclear community of scientists who make their living from this industry.

Record-keeping for nuclear workers also leaves much to be desired. Although nuclear worker radiation exposure records are kept in the USA, corresponding adequate health records are not. These latter might provide useful information to resolve some of the questions basic to the radiation health effect predictions. Worker health records belong to the employer, sometimes they need to be kept current for only five years and can be destroyed when the worker terminates employment. Solid tumours caused by radiation take fifteen to fifty years to manifest themselves clinically.

Important information on worker reproductive experience and genetic damage to offspring has not even been collected in commercial nuclear installations. From the dearth of information it can be assumed that military 'employers' keep no better health or reproductive experience records than those in the commercial nuclear sector. The alternative assumption would be that such information is classified as secret for national security reasons. An 'in-house' study of the workers' reproductive experience at Oak Ridge National Laboratory, which has been involved in nuclear weapons since the Manhattan Project, has recently been initiated. No details of how this study is being conducted are available to the scientific community. The conflict of interest involved in Oak Ridge studying the genetic effects of radiation is obvious, but the admission of the need for such a study is welcome. The question of why such a study has been so long in coming remains unanswered.

Public concern for health effects from plutonium and other transuranic elements has resulted in the establishment of a Transuranic Registry in the USA for autopsy information on former plutonium workers. Although this is also a welcome step, the project lacks strong scientific credibility from the start because co-operation with the registry is voluntary. Some nuclear installations, such as Keer McGee in Oklahoma and Nuclear Fuel Services in western New York, which have consistently high worker exposures, have refused to co-operate with the registry.[85] There have also been changes over the years in the interpretation of plutonium monitoring data and the quantification of dose to tissue.

The death certificate information on the Hanford nuclear workers, together with their radiation exposure, has provided the first direct information on causes of death among workers exposed to low level radiation.[40] The results of this study, which was first published in 1977, have already been discussed (pp. 88–99). This investigation constituted the first long-term cohort study of all radiation workers, those who were employed and those who left the industry at different points in time. All were traced to determine whether they were alive or dead. If dead, their causes of death were correlated with their radiation exposure. This led to the positive finding that cancer was associated with higher levels of exposure to radiation even within doses routinely permitted to workers.

In contrast, for years government officials had been assuring the public and workers that the health records of atomic workers exposed to radiation did not show a higher rate of cancer or leukaemia than would normally be expected. What the public was not told by the government officials was that the government statements related only to current employees. No study had ever been done prior to the Mancuso study to investigate and evaluate *all* the workers including those who had left the atomic energy plants. Obviously, the workers with cancer are not likely to be at work. The sick and the disabled have left and, most important, the cancers generally occur after a long latent period, years after the workers have left employment. Promoters of the nuclear industry and their related government experts, however, conveniently have not explained this when they have given all these assurances to the public over many years.

Work-related diseases have traditionally served as early warnings to the general population of the relative toxicity of various pollutants. It is always a tragic way to learn. In a nuclear age, this advance warning system is not even operating, perhaps because the nuclear products are too irrationally desired and/or feared.

The present US and Western bloc countries' civil energy strategy, as well as their military strategy, have been heavily nuclear. The USA leads the world in commercial nuclear generators, having in operation or projected 26 percent of the total world number of reactors and 39 percent of the total megawatt electrical capacity according to 1984 estimates.[86]

The expansion of military and commercial nuclear projects in the USA required the development of new university courses and programmes in nuclear physics and engineering. This in turn meant research reactors on campus and more persons exposed to low levels of ionising radiation. Each reactor, whether used commercially or for research, emits radioactive gaseous and liquid effluents during its lifetime.

Nuclear fission by-product utilisation programmes led to the development of radioactive tracers for experiments in biology and nuclear medicine. Commercial uses of fission or uranium by-products include components of smoke detectors, stabilisers in airplanes, components of eye glasses and false teeth, welding materials, and X-ray and gamma-ray sources for measuring stress or imperfections in metals.

Radioactive waste has been dumped in sinks or flushed down toilets, left in dirt trenches or land fills, incorporated into construction materials and trucked through almost every town or hamlet in the nuclearised world.

Many would expect that the USA, noted for its ability to measure physical variables precisely and its ability to carry off successful manned space flights, would also be the nation best able to assess the health effects inseparable from such total commitment to the nuclear age. The USA has financed scientific research institutes well, attracting the best young minds. It enjoys a democratic government with free speech and ample opportunity for professional dialogue and exchange of ideas. It would seem to be the ideal situation for assuring that military and technological planning would not be allowed to jeopardise civilian health or national survival. Indeed many countries, rich and poor alike, are relying on the USA and other major nuclear nations for information on the health impact of the nuclear industry.

Review of professional radiation health literature makes several facts clear. First, numerical projections of health effects have been made primarily for selected causes of death and ill health, namely malignant solid tumours, leukaemia and serious transmittable genetic diseases. Analysis and reporting of more generalised ill health, earlier occurrence of chronic diseases, and most especially the mild mutations in offspring, have been superficial or non-existent. The measurements of fatal radiation-induced cancers and severe congenital malformations or disease syndromes in offspring are highly imprecise and probably underestimate the problems. The prestigious US National Academy of Science

Committee on the Biological Effects of Ionising Radiation, when deadlocked on this issue in 1979, asked Dr Edward Radford and Dr Harold Rossi, the two principal contenders for opposing estimates, to leave the committee. In their absence the committee decided on what the press described as a 'marvellous compromise' estimate of the expected number of excess cancer deaths per rad exposure to ionising radiations.[87] The prediction, the 'marvellous compromise', is used, of course, as a basis for legal liability in case of accidents such as Three Mile Island or for environmental impact statements prior to licensing a new nuclear installation. The prediction is also used as a basis for deciding risks versus benefits, and the level of ill health which is deemed 'acceptable' to the public. It forms the basis for denying veterans' claims and worker compensation cases.

Actual deaths, and radiation-related illnesses other than those officially selected in exposed individuals and their children, still go unmeasured. No major study has been undertaken to resolve the scientific controversy and no public debate has demonstrated the human acceptability of the value judgments made by the experts. A compromise between two estimates of the number of radiation-induced fatal cancer deaths reached by a committee will have little or no effect in the real world of sickness and health. These estimates only affect the legal and political world. It is a bizarre way to solve a problem which has such tragic human consequences.

In the Soviet Union also the global geological process set in motion by human industrial activities is being recognised. In the words of I. V. Patryanov, an academician: 'This process started long ago, and is developing at such a rapid rate that it cannot but evoke anxiety.'[88]

The Cover-ups:
Nuclear Policy-making in
the First Two Decades

Beginning of the Nuclear Age

The War is Over!

The nuclear age began with the dropping of two atomic bombs by the United States on the cities of Hiroshima and Nagasaki in Japan. It is necessary first to focus on these two countries in order to understand how the concept of national security changed with the advent of nuclear war and to grasp the interrelatedness of this concept with the peaceful atom and the human experience with 'radiation sickness'.

In March 1945, prior to Truman's inauguration, an American fire-bomb raid on Tokyo had killed 120,000 civilians. Many other Japanese cities had been bombed that spring and the US ground forces had invaded Okinawa. In April, land fighting claimed 90,000 Japanese soldiers and 100,000 civilians. Japan was close to defeat. Its military leaders were trying to rally enough 'spirit' for a decisive bloody battle on the Japanese mainland. The allied forces had already judged direct and massive killing of civilians as 'acceptable' in wartime.

The decision to drop the atomic bomb came shortly after Harry Truman was inaugurated as President of the United States on 12 April 1945 following the death of Franklin D. Roosevelt. On 4 July 1945, in joint agreement with the British and Canadians, he authorised a test of the bomb in preparation for its use against the Japanese. The Manhattan Project, which developed the bomb, was so secret that even the US Congress did not know about it. It is reported that Truman, who had been Vice-President under Franklin Roosevelt, was told about it only after he was sworn in as President. On 7 July 1945 the Emperor of Japan asked the Soviet government to mediate peace between Japan and the United States. Molotov replied that they would 'study' the question.

At 5.30 a.m. on 16 July 1945, a plutonium bomb was detonated in the US desert near Almagordo, New Mexico. Plans to explode it went ahead in spite of the reservations and warnings of men like General Dwight Eisenhower and the atomic scientists who had designed and built the bomb. Those who witnessed the first atomic explosion explained it in religious terms. A *New York Times* reporter likened it to the second coming of Christ, and a semi-official report went so far as to say: 'Lord, I believe, help Thou my unbelief.' The military operation was called 'Trinity'.

Truman received a full report of the detonation on 21 July 1945 while at the Potsdam Conference, and sent a written message to Winston Churchill, Prime Minister of Great Britain, which read: 'Babies satisfactorily born.' On the same day, 21 July, sixty atomic scientists signed a petition saying the bomb should not be used against Japan unless a convincing warning was given and there was an opportunity to surrender. Truman, from Potsdam, uttered his ultimatum to Japan on 26 July: 'Surrender unconditionally or be destroyed.' Pamphlets were dropped all over Japan threatening 'an enormous air bombardment', but the nature of the new bomb was not described. 'We do not intend that the Japanese shall be enslaved as a race or destroyed as a nation', said the Potsdam Declaration,[1] but this message was not conveyed to the Japanese people.

The bomb, referred to as 'Little Boy', was assembled on 1 August. On 2 August the Japanese Foreign Minister was sent to Japanese Ambassador Sato in Moscow with the message: 'It is requested that further efforts be exerted . . . Since the loss of one day may result in a thousand years of regret, it is requested that you immediately have a talk with Molotov.'

On the morning of 6 August 1945, Colonel Paul W. Tibbetts, Jr, of Miami, Florida, flying in the *Enola Gay* (named for his mother in Iowa) gave the order to drop 'Little Boy' on Hiroshima. Major Thomas W. Ferebee of Mockville, North Carolina, released the bomb. It was a sunny day; no fighter planes rose to oppose them. The people had taken shelter, but when only two planes were seen, the all-clear signal sounded. The bomb exploded about 630 yards above the Shima Hospital, near the centre of the city. At the time of the explosion, people had resumed their morning chores or made their way to work.

The fireball was 18,000 feet across, with a temperature of about

100 million degrees Fahrenheit at the centre. People who looked at the fireball were blinded. More than four square miles of this city of 130,000 people were completely destroyed.[2] Houses collapsed from blast pressure or were destroyed by the fire which burned for two days. About 88,255 buildings – stores, churches, hospitals, fire stations, police stations, schools, offices and blocks of flats – were destroyed.[2] Some people near the centre of the explosion literally evaporated; others were turned into charred corpses. Survivors experienced burns which caused their skin to peel off immediately; some had skin hanging from the ends of their fingers. Those who lived to tell the story said that the victims looked like ghosts.

Japanese newspapers did not fully report the bombing or the immediate fatalities of about 70,000 people. One short item read that Hiroshima had been hit with incendiary bombs and 'It seemed that some damage was caused to the city and its vicinity.'

Molotov arrived back in Moscow after the Potsdam Conference and after the bomb was dropped on Hiroshima. On 8 August 1945 he called Japanese Ambassador Sato and announced that Soviet Russia was declaring war on the Japanese, thus fulfilling his promise made at the Yalta Conference and making a Japanese–USA negotiated peace impossible.

On 9 August 1945 a second atomic bomb was dropped on Nagasaki, and on 11 August 1945 the Japanese Emperor accepted the Potsdam peace terms. This second bomb was constructed from plutonium, while the Hiroshima bomb had been constructed from uranium. There was, apparently, a deliberate plan to study the effects of the two different types of bombs. The Nagasaki bomb exploded about 555 yards above the ground, near the University of Nagasaki Medical School.

The surrender was announced to the world on 14 August 1945.[1]

A small book was published in Japan in 1950. In it 164 A-bomb survivors told their stories and made a strong poignant plea for 'No more wars.' The book was suppressed by the occupation forces, and not rediscovered until April 1981.[3] The survivors wrote of the human bodies they had seen, swollen and charred, with skin peeling off and hanging down; the silent, suffering blind people sitting by the river or throwing themselves into the river to relieve the intense pain. They wrote of those who were trapped in collapsed structures, children hurled for hundreds of feet by the fierce winds, family members evaporated or forever scarred and disfigured by keloids. They described their skin as bean curds and their mouths as gelatin. Yet they could bear no grudge against the Americans who had dropped the bomb. Many blamed their own emperor for causing the horror. Most A-bomb victims became radically pacifist, realising very quickly that the presence of nuclear bombs made war – any war – now unthinkable. Unfortunately, the people of the world were denied access to the Japanese experience. No healing could take place because the wounds were denied and hidden.

The US occupation forces used over 85,000 feet of 16 mm film to document the tragedy. This film lay unused in the US archives, not available to the public until 1980. The Japanese YMCA has raised about $375,000 (US) since 1980 to purchase this film for peace documentaries in the hope of preventing any future use of nuclear bombs.

The complete atomic death toll has never been tallied. The 30-year report to the United Nations in 1975 estimated immediate fatalities in Hiroshima and Nagasaki at 240,000. Even as late as 1978 more than 2,000 cancer deaths of survivors during that year were attributed to the bombing in 1945. Civilian casualties in Hiroshima and Nagasaki exceeded the total number of US military killed during the Second World War, in both Europe and the Pacific sector.

President Truman summed up the military philosophy behind the bombing: 'We found the bomb and used it.' He no doubt thought the show of 'power' would end all wars, but in fact it ushered in a period of unprecedented military spending on the arms race and a policy of 'mutually assured destruction' among nations with nuclear capability. A surprise nuclear attack would totally destroy a nation, thus retaliation, if it was to occur at all, must take place in the time interval between enemy launching of weapons and their arrival at the target. This new situation demands constant mobilisation for war, the escalation of weapon design as a 'deterrent', and the disappearance of peacetime economy as experienced prior to 1939. Since 1945, the world has been in hostage to a nuclear war truce.

The people of Hiroshima and Nagasaki were not at first aware that this bomb was 'different'. They were bewildered when people not visibly injured in the blast suddenly died, when their hair fell out by the handful, when black spots dotted their skin as capillaries broke, and when muscles contracted leaving badly deformed limbs and hands. Slowly they began to learn that these were not ordinary bombs. Even those who escaped the blast, fire and heat were not safe from the penetrating radiation which had the power to invisibly destroy without the usual wounds so familiar to war.

It was not until 1970, when some army radio messages were

declassified, that Professor Barton J. Bernstein, a researcher from Stanford University, found a Red Cross report on Americans who had died in the first atomic bomb blast. The report, dated 23 September 1945, stated that Airman Ralph J. Neal of the 866th Bomber Squadron had been 'wounded by the atomic bomb on 6 August' and that he

subsequently died on 9 August. Apparently two or possibly three more airmen of the same crew – the *Lonesome Lady* – who are now missing possibly died as a result of the Hiroshima bombing.

Professor Bernstein's research showed that the US Sixth Army, then at Hiroshima, was ordered to investigate the Red Cross report. They confirmed on 9 October 1945 that Neal, Norman, Roland Brisset and eighteen other Americans were being held as prisoners of war in Hiroshima at the time of the bombing. The eighteen were instantly killed. Neal and Brisset died later, probably of radiation exposure.

This event was officially classified as secret information in the USA, on the grounds of 'national security'. Not even the families were notifed. Cleo Neal, mother of Airman Ralph Neal, was told he died of wounds suffered when his B-24 bomber, the *Lonesome Lady*, was shot down over Honshu Island on 28 July 1945. The US Army went so far as to have a memorial service in St Louis, Missouri, in 1949 for eight men including five crew members of the *Lonesome Lady*, supposedly all killed on 28 July 1945.

This is the first recorded deliberate suppression of information in the United States relating to the 'health effects' of the nuclear bombing. Professor Bernstein speculated that the government was not interested in revealing facts that could embarrass the army and raise moral questions in the USA about the dropping of the atom bomb.

When asked in 1978 about the suppression of Red Cross information in 1945, Pentagon spokespeople responded that a 1973 fire in St Louis destroyed many army personnel records and made it impossible to investigate the matter further.

The Effects of Nuclear Bombs

On 19 September 1945, a little over a month after the dropping of the bombs, an American research team moved into Hiroshima, set

National Security, Energy and Health

14 September 1945

The following message represents the views of the Los Alamos Association of Scientists. It was signed by the majority of the Staff Members in the laboratory.

The message has been transmitted to the Interim Committee of the Secretary of War and it is our hope to release it to the press in the near future. We are distributing it at Los Alamos to acquaint others with our views and to invite comment. Until permission has been received to release the statement to the press the document is *restricted* and should not leave the hill.

W. A. Higinbotham, Chairman
Temporary Executive Committee
Bldg. U – Room 105 – Tech. Area

* * *

In the years of unrelenting wartime emergency, we, a group of scientists, worked on the development of the atomic bomb. We were aware of the fact that our labours were directed to an end which was certain of realisation in the not too distant future. We worked in the fear that our enemies might be first to create the atomic bomb and then would use it to subjugate the world.

From the beginning, we were aware that the scientific and military success of our work would bring both new dangers and new possibilities of human benefit to the world. Until recently, our work was clothed in secrecy. Now that success is achieved and the nature of our work is no longer secret, we believe that we should speak publicly of the profound consequences of this development. If these consequences are to be for the better, if disasters are to be avoided, it is necessary that every citizen come to appreciate the potentialities of our new mastery over nature. Only through such understanding can we hope that our democracy will be able to make wise use of the recent discoveries.

In this effort to make our considered judgement available to the public, we have attempted to raise questions which undoubtedly occur to everyone who gives serious thought to the coming age. Our answers to these questions are of necessity largely predictions concerning the future, rather than proved facts. They are, however, the best answers that scientific experts can give at this time.

The following statements are supported by those who have worked on the construction of the atomic bomb and know its technical details:

1. What would the atomic bomb do in the event of a future war?
The extent of the damage inflicted by atomic bombs was demonstrated in the two drops on Hiroshima and Nagasaki. In the event of future wars, use of such bombs would quickly annihilate the important cities in all countries involved. We must expect, furthermore, that bombs will be developed which will be many times more effective, and which will be available in large numbers. Although all combatants would suffer, these weapons would be most disastrous for countries whose industry is tied to congested centres of population.

2. What defence would be possible against it?
Even with great advances over present radar detecting and directing devices, one hundred percent interception of bombs should be considered impossible within the foreseeable future. A surprise attack in which relatively few bombs could destroy our largest cities, might so cripple our industry that we could not effectively retaliate. Therefore, were there a possibility of attack, we could not afford to gamble on defensive means alone, and would have to make certain drastic changes in our country. These changes would include the abandonment of all large cities, the decentralisation of communications, and the location underground of all important factories.

3. How long would it take for any other country (besides the United States, Great Britain, and Canada) to produce an atomic bomb of its own?
The time between the discovery of the fission of uranium and our use of the bomb was six and one half years. The first three years saw no intensive development and many of the results of this period have been made public. It is true that during the war, we, together with the British Empire, have put an unusually great concentration of scientific talent and industrial capacity on the development, but the knowledge that the bomb is feasible may have a great enough stimulus, in some countries, to more than compensate for this advantage. It must be pointed out that the development of the atomic bomb has involved no new fundamental principles or concepts; it consisted entirely in the application and extension of information which was known throughout the world before intensive work started. Furthermore, deposits of the basic materials for atomic bombs have been found, even before the war, in many parts of the world and new deposits will undoubtedly be discovered. It is, therefore, highly probable that with sufficient effort, other countries, who may, in fact, be well under way at this moment, could develop an atomic bomb within a few years.

4. What would be the effect of an atomic bomb armament race on the development of science and technology?

During the war it has been necessary to turn a major portion of our technological and scientific heritage to the development of weapons and the improvement of defence. Such a policy, with the attendant secrecy, if continued, would seriously interfere with the teaching of new scientists and engineers. Future concentration of effort on military weapons would not only impede the peacetime applications of atomic power and radioactivity, but it would also reduce the progress of fundamental research which has, in the past, led to such vast technological advances.

5. Assuming the international control of the bomb is agreed upon, is such control technically feasible?

From a scientific point of view we assert that international control of the atomic bomb is feasible and that such control need not interfere with free and profitable peacetime research and development. A minimum requirement is free access for the agents of the world authority to all the laboratories, industries, and military installations of the world. We do not believe that any simple or easily hidden production methods are likely since manufacture and research on the atomic bomb requires recognisable instruments and materials and large laboratories. It is, therefore, almost certain that a controlling authority could be organised which would be able to discover and control any secret attempt to produce atomic bombs.

Because the atomic bomb provides a weapon so many times more powerful than any other single military force, we who have developed the bomb have held the hope that proper control of this new weapon will provide the opportunity to prevent all future major wars. We now

up headquarters on the top of a nearby hill and began the systematic estimation of early and delayed 'health effects' caused by exposing a human population to nuclear fission. The Atomic Bomb Casualty Commission (ABCC) thus began its research work which continues even as this book is being written.

On 19 September 1945 the Allied Headquarters issued a press code restricting reference to A-bomb matters in speech, reporting and publication. The Allied Economic and Scientific Bureau later announced that surveys and studies of A-bomb matters by the Japanese would require permission from the General Headquarters and publication of A-bomb data was prohibited. These restrictions remained in effect until the San Francisco Peace Treaty of 8 September 1951 took effect on 28 April 1952.

share this hope, both with our government and with all humanity.

The foregoing summary of arguments has, to the best of our ability, been presented to provide the public with the technical knowledge which must be taken into account in formulating the policy of the United States on the future control of its new weapon.

The information which we have presented, coupled with a desire for peace and for our own safety, leaves us, we are convinced, but one course of action. We must share the benefits of atomic energy with the rest of the world and the use of atomic energy as a weapon must be controlled by an international authority. A continued armament race would be disastrous.

During the course of the next few years it will be necessary for the United States and the Soviet Union, together with Great Britain, France, and China, not merely to co-exist and trade with each other; it will be necessary for them to co-operate on the solution of many political and economic problems which the end of the war has left the world. We have seen the beginnings of co-operation during the war. But already there are signs that it can be replaced by mutual suspicion. However, the terrible nature of future war, in which both sides use atomic bombs, makes the need for international co-operation imperative. A world wide control of these bombs also provides the opportunity for preventing present suspicions from growing into tensions and ultimately into war.

We, therefore, believe that it is essential that America take steps for such international control at once. Today the world is still united in its desire for peace. If we in America fail to recognise the opportunity afforded by the development of the atomic bomb, or if we encourage a world wide armament race by delaying even a few months the use of this opportunity, we will be setting the stage of unprecedented destruction, not only of other countries, but of our own as well.

Dr Fumio Shingetō, himself a Hiroshima A-bomb victim, discovered that hermetically sealed X-ray plates in the basement of his hospital had been exposed to radiation in the bombing. He was among the first to know that Hiroshima had experienced radiation warfare as well as fire-bombing. It was he who violated the research prohibition, painstakingly collecting evidence on the bomb health effects. In 1952 at a meeting of the Japanese Haematological Society he presented his thesis that leukaemia was connected with exposure to the atomic bomb. When news of the physician's concern about leukaemia was reported in Japanese newspapers, he was severely criticised by the Atomic Bomb Casualty Commission. The US Army surgeons had, after all, issued a statement in late 1945 that all people expected to die from the radiation effects of the bomb had already died. No further cases of physiological effects due to residual radiation were acknowledged in the subsequent seven years.

Not to be intimidated by these official pronouncements, Dr Shingetō interested a young medical student, Yamawaki, in the problem. There were no statistics available on leukaemia, so they wrote letters to university hospitals all over Japan to ascertain the general incidence rate of this disease in the population. Yamawaki also examined the medical records of about 30,000 people who died in Hiroshima after the war. For each recorded leukaemia case he visited the physician who had diagnosed it and independently collected specimens and confirmed the diagnosis. As his work progressed, the Atomic Bomb Casualty Commission began to take notice of it and made their data available to him. He completed the work in two years, carefully documenting the statistical connection between the leukaemia and radiation exposure. Dr Yamawaki is now a paediatrician in Hiroshima, but he suffered greatly over the political, journalistic and scientific attacks on his work when it was first publicised. It was called 'merely statistical' and 'medically weak' even by the Japanese Haematological Society. Only later was it accepted and even used by the ABCC as if it were its own accomplishment.[3]

Structurally, the ABCC has recently been reorganised, with the Japanese assuming overall direction of the work under a new name: Radiation Effects Research Foundation (RERF). Health effects research, however, is still under the direction of an American, Dr Edward Radford, Chief of Research. Funding continues to be partially assumed by the US National Academy of Science, which in turn receives its money from the US Department of Energy (present name for the US Atomic Energy Commission). Understanding these US research and financial involvements is helpful for grasping the centrality of health questions in relation to nuclear options. In the USA, the Department of Energy (DOE) is mandated to develop both nuclear weapons and nuclear energy.[4] DOE also controls most funding for basic research on the health effects of both nuclear technologies.

The US National Academy of Science was established by President Abraham Lincoln on 3 March 1863. It is funded by those government agencies which request research, so, for example, it was the US Federal Aviation Agency which funded a NAS study on sonic booms and the US Department of Defence which funded the NAS study on the impact of defoliants in Vietnam. The Academy

served to quiet citizens' protests about sonic booms, scoffed at
Rachel Carson's warning about pesticides in *Silent Spring*,[5] and
supported the US military position on defoliants in Vietnam. The
record is not much better with respect to the hazards of human
exposure to ionising radiation. The Academy endorsed the Atomic
Energy Commission's plan to bury radioactive waste at Lyons,
Kansas, a plan later abandoned because of safety questions raised
by local geologists. It never addressed the problems of nuclear
fallout from above-ground weapon testing. It has never played a
lead role in nuclear reactor safety or occupational exposure
problems in this industry and its support facilities.[6] The US
National Academy of Science convened the BEIR I, II and III
committees to silence dissenters from the prevailing opinion on the
hazard of low level radiation.[7] It has also managed the Hiroshima
and Nagasaki data since 1945.

The nuclear bomb was a quantum leap in destructive weapon
capability, having about 1,000 times the blast force of the
blockbuster which was the major bomb used in the Second World
War. On 27 December 1945, US President Truman and the Prime
Ministers of Great Britain and Canada met with Joseph Stalin, head
of the Soviet Republic, and issued a communiqué popularly known
as the Moscow Declaration. In this they

> agreed to recommend for the consideration of the General
> Assembly of the United Nations, the establishment by the United
> States of a Commission to consider problems arising from the
> discovery of atomic energy and related matters.[8]

In January 1946 the General Assembly of the United Nations
unanimously established the Atomic Energy Commission, which
was to consist of the UN Security Council plus Canada. Canada, not
a member of the council at that time, was included in recognition of
its part in the US and British development of the first nuclear
bombs. This agency was later disbanded, and in 1956 the United
Nations Conference on the Statute of the International Atomic
Energy Agency approved its replacement, the International Atomic
Energy Agency (IAEA). The IAEA statute entered into force on
29 July 1957 with its principle objective 'to accelerate and enlarge
the contribution of atomic energy to peace, health and prosperity
throughout the world'. The IAEA is a specialised agency with
autonomy and fiscal independence within the UN system.[9] It is
directly responsible to the UN General Assembly. All other UN

specialised agencies and autonomous organisations report to the UN Economic and Social Council rather than the General Assembly.

At the first meeting of the AEC on 24 January 1946, US Representative Bernard Baruch introduced the plan which later became known as the Baruch Plan, for control of atomic energy. This plan was further developed and more formally expressed at a meeting on 14 June 1946. It called for international control 'of the entire process of producing atomic weapons: from uranium mining to completed weapon'. In view of the uranium tailings and nuclear waste problems of today, this characterisation of the problems reveals incredible naiveté about the handling of radioactive materials and ignorance of nuclear waste problems. The Baruch Plan also called for accountability for all fissionable material used for 'peaceful and humanitarian ends'.

The Moscow–US accord on nuclear policy was short-lived. On 19 June 1946 André Gromyko presented a Soviet Plan to the AEC calling for immediate prohibition of 'production and employment of weapons based on the use of atomic energy'. The Soviet Plan called for destruction of all nuclear weapons then in existence, but at the same time rejected, as inconsistent with the sovereignty of the nation state, verification and control systems needed to assure the absence of all nuclear weapons. It was well known at the time that Soviet conventional military power exceeded that of the United States. Therefore US possession of nuclear weapons was generally considered to have created parity between the two nations.

Peace Through Law

The International Court of Justice was specifically included in the United Nations Charter, Articles 7 and 92, in the hope of providing a rational alternative to the use of force – and nuclear weapons – for the resolution of international conflict. Article 93 of the UN Charter provides that 'all members of the United Nations are *ipso facto* parties to the International Court of Justice'. The court was intended to be an extension and improvement of the Permanent Court of International Justice, previously established under the League of Nations.

The concept of world peace through world law was to be embodied in the UN judicial organ which could actually create law, based on the precedent of its own decisions, for the entire international community. It was to be limited to legal disputes concerned with matters specified in the UN statute and, if a question arose as to whether or not a case was appropriate for court action and within its jurisdiction, that question was to be determined by the court.[10]

Many UN members accepted the Charter and World Court as it was set up by the founders. The USA, however, which had never been a member of the League of Nations or of the Permanent Court, by Senate vote on 2 August 1946 added a reservation in its ratification statement. In effect, this reservation, known as the Connally Amendment, provided that jurisdiction of the International Court would not extend to

> disputes with regard to matters which are essentially within the domestic jurisdiction of the United States of America as determined by the United States of America.[11]

Just as the Soviet Union wanted to outlaw nuclear weapons but feared loss of sovereignty through allowing armament inspection by an international agency, the USA desired peace through law but feared loss of sovereignty through judicial censure.

Great Britain (1 January 1969) and Canada (7 April 1970) circumvented compulsory jurisdiction of the International Court of Justice by adding the reservation:

> also reserves the right at any time, by means of a notification addressed to the Secretary-General of the United Nations, and with effect as from the moment of such notification, either to add to, amend or withdraw any of the foregoing reservations, or any that may hereafter be added.[12]

By 1972, the latest date for which a summary of 'Declarations recognising as compulsory the jurisdiction of the International Court of Justice'[13] was available, forty-six nations had formally accepted compulsory jurisdiction. Most indicated reservations.

France added to its declaration:

> With the exception of disputes arising out of a war or international hostilities, disputes arising out of a crisis affecting national security or out of any measure or action relating thereto, and disputes concerning activities connected with national defence.[14]

This reservation prevented action by Australia and New Zealand to end French testing of nuclear weapons in Polynesia, which was causing radioactive fallout on their territories and damage to Pacific people's health, environment and food resources.

The Soviet Union has never made a declaration accepting compulsory jurisdiction of the International Court of Justice.

The superpowers, each fearing for loss of national sovereignty, effectively prevented formation of a strong International Atomic Energy Commission and a strong International Court of Justice.

The US reservation had strong opponents within the USA as well as within the international community. The US Senate Committee on Foreign Relations had rejected the Connally reservation as contrary to the nature of law. Public approval of compulsory jurisdiction was reported as strong, with Senator Thomas remarking to President Truman: 'Mr President, rarely in my legislative experience have I encountered such an important measure with the people so unanimously in support of it.'[15] The senate acted against this strong consensus of the people. The American Bar Association has consistently opposed the reservation since 1 November 1946 when its Assembly officially adopted a stand calling for repeal of it. President Truman stated explicitly in February 1946 and March 1948 that he favoured compulsory jurisdiction of the court. Similar statements have been made by all the subsequent presidents.[16, 17]

The history of the International Court has clearly demonstrated the self-defeating nature of a major nation's refusal to back compulsory jurisdiction. While the role of the US reservation on the jurisdiction of the International Court of Justice should not be overstated, it was and is a major impediment to replacing war with conflict-resolution skills and law. It contradicts the fundamental principle underlying the United Nations as expressed in Article 1 of the Charter, which states its purpose to maintain international peace and security through a number of peaceful means and 'in conformity with the principle of justice and international law, adjustment or settlement of international disputes, or situations which might lead to breach of peace'.

Perhaps more importantly, it prevents the International Court from acting on behalf of the public good in times of international crisis, thereby causing people to lose faith in government by justice and law rather than power. In a word, people begin to believe that rational resolution of difficulties on an international scale is impossible to human beings, i.e. there is no alternative to war.[18] Given our human history of developing ever more complex networks of co-operation, this pessimistic conclusion seems unnecessarily defeatist.

Senator Javits, in May 1971, introduced a resolution in the US Senate which offered the USA a graceful and secure way out of the predicament which its 1946 stand had caused. Senator Javits proposed that self-judgment with respect to court jurisdiction in legal disputes involving the USA be ruled out in all questions concerning:
1. the interpretation of a treaty;
2. any question of international law;
3. the existence of any fact which, if established, would constitute a breach of an international obligation;
4. the nature or extent of the reparation to be made for the breach of an international obligation.

This would have allowed for a gradual turning over of responsibility from a national to a world court as international law became more developed. The Senate has yet to act on this resolution. Such action is unlikely because of general public ignorance in the USA of the 1980s of the problem caused by the 1946 decision. It is interesting to note, however, that repeal of the Connally Amendment was part of the Democratic Party Platform in Jimmy Carter's unsuccessful bid for the Presidency in 1980.

In the absence of an effective international control agency for nuclear weapons and fissionable materials, and with an ineffective international judicial system, the USA organised nationally in preparation for the nuclear age.

In his closing address to the nation before stepping down as President of the United States on 17 January 1961, Dwight D. Eisenhower noted:

> Our military organisation today bears little relation to that known by any of my predecessors in peacetime, or indeed by the fighting men of World War II or Korea . . . We have been compelled to create a permanent armaments industry of vast proportions. Added to this, 3.5 million men and women are directly engaged in the defense establishment. We annually spend on military security more than the net income of all United States Corporations.[19]

A US system of national scientific laboratories was formed to support research on, production of and testing capability for nuclear weapons. Hanford, Los Alamos and Oak Ridge were started during the Manhattan Project prior to 1945. They were augmented later by Idaho Falls, Brookhaven, Argonne and Lawrence Livermore. Some of these laboratories have university connections. For example, Los Alamos and Lawrence Livermore have undertaken work for which the University of California was defence contractor. The defence contracts indirectly aided the universities' finances and prestige. The military sector also directly aided universities by providing fellowships, research funding, teacher training programmes, equipment and job opportunities for graduates. In this way the university system was drawn into close co-operation with military defence efforts.

Most of the work at the government laboratories was military in nature, requiring secrecy. This need, together with the demand for scientific excellence, gave rise to a group of elite scientific advisers and decision-makers who would necessarily have to 'manage' nuclear technology and guard its secrets. An independent group of elites, the RAND Corporation, formed the 'think tank' for the operation of the new military industrial complex.

The overall supervision of the growing nuclear military scientific venture was placed with the US Atomic Energy Commission and the Joint Atomic Energy Committee of the US Congress. All nuclear-related matters were channelled to these agencies and their personnel became experts at questions ranging from radiation health effects on humans to the waging of nuclear war.

Although the United States assumed a leadership role in setting international policies, it was not the only nation involved. As previously mentioned, the original Manhattan Project included representatives from Canada and Great Britain. These two countries maintained scientific pace with the United States as both weapon capability and technoligical spin-offs developed.

Centrality and Importance of the Health Question

The US population, although aware that a new and terrible atomic weapon had ended the war, did not receive full details of the human suffering it produced in Hiroshima and Nagasaki. In 1945 Americans were still smarting under the outrage they had experienced when Japan bombed Pearl Harbor, on Sunday, 7 December 1941, in a surprise attack which left six civilians dead at first count. One navy battleship, the *Arizona*, was sunk and four others set on fire.

Final official estimates of Pearl Harbor casualties were: army – 228 killed; marines – 109 killed; navy – 2,004 killed; civilians – 68 killed. Some Americans, their memory probably distorted by shock and by differing emphases in media reporting, equated this military attack on Pearl Harbor with the bombing of civilians in Hiroshima and Nagasaki.

The American psyche, then, justified the total destruction of two Japanese cities, about 200,000 civilians, as 'just retribution' for Pearl Harbor.

Ironically, historian John Costello, using intelligence material in the US National Archives and British Security Co-ordination Documents, has determined that a warning of the Pearl Harbor attack was submitted to both the British and US governments at least two weeks before it took place. A secret Japanese intelligence message from Budapest to Tokyo, with a copy to Nazi Headquarters in Berlin, stated that the British had warned Budapest government officials that war between the United States and Japan would break out on 7 December. Churchill's records for that period are marked: 'Closed for 75 years' in the Public Record Office. It has been speculated that the national leaders deemed the surprise element of the attack 'necessary' to rouse the US public sufficiently to enter the Second World War.[20]

The American people's attitude towards atomic bombing was affected not only by the surprise attack on Pearl Harbor, but also by Japanese suicide squads during the war. Americans did not understand the human vulnerability of the Japanese people and many thought that only the atom bomb could have broken Japanese will and caused them to surrender. They knew nothing about Japanese surrender attempts prior to the atomic bombing. They were led to believe that only a bloody invasion of the Japanese mainland could have ended hostilities.

Emotionally, most Americans saw the bomb only as a means to bring their sons and husbands home from war. They never questioned the illegality of the Manhattan Project, or the lack of congressional approval of the spending on the bomb or the decision to use it. Few knew enough of the facts about this atomic attack on human life to protest about the immorality. Most Americans wanted their loved ones home and the war ended. Public opinion was manipulated by the control of information released by government officials and by the overall elation at 'victory'.

It is probably true to say that just as the atomic bomb was developed, tested and deployed without the knowledge or consent

International Herald Tribune, 11–12 September 1982

An Early Vision

Thirty-seven years ago – on 11 September 1945, soon after the United States had compelled the Japanese to surrender by exploding atomic bombs over the cities of Hiroshima and Nagasaki on 6 and 9 August – Secretary of War Henry L. Stimson wrote privately to President Harry S. Truman suggesting proposals for the awesome weapon's control.

The following letter and excerpted memorandum, quoted by the *Los Angeles Times* with permission of the Yale University Library, were virtually the last official documents Stimson wrote. He retired a few days later after 40 years in government service.

* * *

Dear Mr President:

In handing you today my memorandum about our relations with Russia in respect to the atomic bomb, I am not unmindful of the fact that when in Potsdam, I talked with you about whether we could be safe in sharing the atomic bomb with Russia while she was still a police state and before she put into effect provisions assuring personal rights of liberty to the individual citizen.

I still recognize the difficulty and am still convinced of the ultimate importance of a change in Russian attitude toward individual liberty, but I have come to the conclusion that it would not be possible to use our possession of the atomic bomb as a direct lever to produce the change. I have become convinced that any demand by us for an internal change in Russia as a condition of sharing in the atomic weapon would be so resented that it would make the objective we have in view less probable.

I believe that the change in attitude toward the individual in Russia will come slowly and gradually and I am satisfied that we should not delay our approach to Russia in the matter of the atomic bomb until that process has been completed. My reasons are set forth in the memorandum I am handing you today.

Furthermore, I believe that this long process of change in Russia is more likely to be expedited by the closer relationship in the matter of the atomic bomb which I suggest and the trust and confidence that I believe would be inspired by the method of approach which I have outlined.

Faithfully yours,

Henry L. Stimson
Secretary of War.

Memorandum for the President:

Subject: Proposed Action for Control of Atomic Bombs

The advent of the atomic bomb has stimulated great military and probably even greater political interest throughout the civilised world. The temptation will be strong for the Soviet political and military leaders to acquire this weapon in the shortest possible time.

Unless the Soviets are voluntarily invited into the partnership upon a basis of co-operation and trust, we are going to maintain the Anglo-Saxon bloc over against the Soviet in the possession of this weapon.

Such a condition will almost certainly stimulate feverish activity on the part of the Soviet toward the development of this bomb in what will in effect be a secret armament race of a rather desperate character. There is evidence to indicate that such activity may have already commenced.

If we feel, as I assume we must, that civilisation demands that some day we shall arrive at a satisfactory international arrangement respecting the control of this new force, the question then is how long we can afford to enjoy our momentary superiority in the hope of achieving our immediate peace council objectives.

Whether Russia gets control of the necessary secrets of production in a minimum of say four years or a maximum of 20 years is not nearly as important to the world and civilisation as to make sure that when they do get it they are willing and co-operative partners among the peace-loving nations of the world.

I consider the problem of our satisfactory relations with Russia as not merely connected with but as virtually dominated by the problem of the atomic bomb. Those relations may be perhaps irretrievably embittered by the way in which we approach the solution of the bomb with Russia.

For if we fail to approach them now and merely continue to negotiate with them, having this weapon rather ostentatiously on our hip, their suspicions and their distrust of our purpose and motives will increase. Our objective must be to get the best kind of international bargain we can – one that has some chance of being kept and saving civilisation not for five or for 20 years, but forever.

The chief lesson I have learned in a long life is that the only way you can make a man trustworthy is to trust him; and the surest way to make him untrustworthy is to distrust him and show your distrust.

If the atomic bomb were merely another though more devastating military weapon to be assimilated into our pattern of international relations, it would be one thing. We could then follow the old custom of secrecy and nationalistic military superiority relying on international caution to proscribe the future use of the weapon as we did with gas. But I think the bomb instead constitutes merely a first step in a new control by man over the forces of nature too revolutionary and dangerous to fit into the old concepts. I think it really caps the climax of the race between man's growing technical power for destructiveness and his psychological power of self-control and group control – his moral power. If so, our method of approach to the Russians is a question of the most vital importance in the evolution of human progress.

My idea would be a direct proposal, after discussion with the British, that we would be prepared in effect to enter an arrangement with the Russians, the general purpose of which would be to control and limit the use of the atomic bomb as an instrument of war and so far as possible to direct and encourage the development of atomic power for peaceful and humanitarian purposes. I would make such an approach just as soon as our immediate political considerations make it appropriate.

of the American people or their elected representatives in Congress, so were both the American people and Congress kept generally ill-informed about the effects of atomic radiation on humans and their life-supporting habitat. The same has been true in Canada, Great Britain, France and most probably in the Soviet Union.

This tremendous 'power' was to be managed by those who knew its secrets. Lacking strong international safeguards for weapon control or international conflict resolution, nuclear experts set out to find the limits of toleration for human exposure to radiation as they resigned themselves to life in the nuclear age. Some scientists cognisant of the danger, such as Karl Z. Morgan, travelled around the world trying to convince decision-makers that there could be no more war, but the effort seemed futile.

Both the earth's crust and the sun have long been sources of ionising radiation exposure for humans. The mutation caused by such exposure probably helped along the evolution process by which the complex and delicately balanced human body was formed. As mutation occurred, those most 'fit' survived. Would further exposure to radiation improve or destroy the work which had occurred so slowly over thousands of centuries?

Military leaders – convinced that there was no alternative to preparing for war in the nuclear age – and scientists who envisioned nuclear power as a source of energy for industrial and economic growth were asking essentially the same human health questions. Could armies function in a residual radiation environment if they were protected from the initial blast of a nuclear detonation? Could nations survive a nuclear war? What would be the risks entailed in using nuclear fission as an energy source? How much pollution of air, water and land with nuclear fission products would be considered 'safe' or 'permissible' for the general population? All these questions demanded some prediction of health effects relative to increased use of nuclear fission.

During the late 1940s, most of the qualitative effects of ionising radiation were known but the quantification of these effects was imprecise. It was also difficult to fit the newer complicated model of the post-nuclear bomb world into the traditional historical wisdom and categories of risk versus growth. Industrial growth and fossil fuel generation of electricity had produced many health and environmental problems in the past. The public and government alike asked: was this just another problem to be accepted in the name of progress? Progress had become associated with economic growth, jobs and human well-being and its assurance seemed to require continued access to global resources and markets. This in turn meant to most people the strength to protect oneself against human enemies. This was the historical wisdom. But what about radiochemical destruction of life itself? What if polluted air, food and water became 'enemies' within one's own country? What if the common global enemy was the human choice of electrical power over life?

The two fears – national survival and species survival – influenced complicated international strategies formulated at this time for 'balancing' superpower influence in the United Nations and the world, often without public understanding within the nations involved. A closer look at Soviet and US diplomacy during this time of formation for most UN agencies reveals that both nations experienced fears for survival and ambivalence in dealing with these fears. The many unknowns connected with nuclear military technology must have greatly influenced political actions. Eventually the decision-makers seem to have opted to deal with survival of the nation state as the prime governmental priority, risking random killing of their own citizens as a necessary consequence of this decision.

Power Struggle in the United Nations

By June 1947 the Soviet government had adopted in theory the need for international control of nuclear weapons. It introduced a plan to this effect in the UN, but the plan proved ineffective because of veto power in the Atomic Energy Commission. The international control organ responsible for the regulation of atomic energy was composed of the nations then in the UN Security Council plus Canada, and the permanent members of the Security Council could exercise veto power over any punishment meted out by the Atomic Energy Commission on countries failing to abide by regulations for nuclear materials. Bernard Baruch, the US representative, succeeded in obtaining a renunciation of the veto by the Security Council members on atomic matters, but at the same time he was vague on US plans to eliminate atomic weapons.[21]

The Soviet Union, which was then in the process of developing its 'Iron Curtain' concept, seemed to be afraid that if the USSR once lost national sovereignty by creating effective international control of nuclear materials, it could never re-establish it. It apparently feared that the vagueness of the US plans for eliminating nuclear weapons meant that, after the Soviet Union gave up sovereignty, the US would refuse to give up nuclear weapons. The question of national sovereignty so influenced Soviet foreign policy at this time that the USSR, struggling both to rebuild its cities and to develop its own nuclear capability, refused US Marshall Plan aid because of the national co-operation with the US it required. It is well to remember Truman's public statement after Hiroshima and Nagasaki: 'Now I have a handle on Uncle Joe.' The reference was to Joseph Stalin, and the bombs were apparently meant as much as a threat to the Russians as a punishment to the Japanese.

During the Second World War the Soviet Union lost 20 million people. About 70,000 villages and 1,700 towns were destroyed, leaving 25 million homeless. The devastated area included the site of 60 percent of Soviet industry, with 31,850 industrial plants destroyed. About 40,000 hospitals were wiped out; 84,000 schools and almost 100,000 farms were lost.[22]

In the autumn of 1948, in spite of Soviet dissension, the UN General Assembly accepted the Baruch Plan which then became known as the UN Plan. International tensions mounted in the 1950s with the outbreak of the Korean War. The Soviet Union withdrew from the AEC (supposedly because of its failure to seat a representative from Communist China); China became involved in the Korean War on the side of the North Koreans; and, finally, near the end of the year, the first nuclear bomb tests in Siberia took place. The Soviet Union demonstrated to the global community that it also had the ability to wage nuclear war.

Sensing an approaching crisis between the superpowers (the United States and the Soviet Union), and realising the destructive power of nuclear weapons, millions of people throughout the world signed documents appealing for peace. It was popularly known as the campaign to 'Ban the Bomb'. In a meeting held in Stockholm on 19 March 1950 the Stockholm Appeal was formally delivered to the World Peace Council.[18]

This tension between political manoeuvering to maintain national sovereignty and ordinary people's intuitive sense of the threat to survival, erupting in the form of public demonstrations, has marked the history of nuclear weapons.

Inter-relationships between the Military and the Peaceful Atom

In the normal course of events, the implementation of a commercial nuclear power option in the United States would have proceeded cautiously with some openness and with the usual health and safety watchdogs monitoring each step of the way. However, its military origin, secrecy, and other historical events have significantly altered this pattern of development. Nuclear generators were a necessary part of nuclear weapon production and they had operated in the USA from 1943 onwards at the Hanford Reservation in the State of Washington. Fuel rods were burned and then plutonium for bombs was extracted. The 'peaceful atom' programme meant allowing the reactor steam to be used for the commercial generation of electricity.

After the Second World War, the USA embarked on an intense nuclear weapon development and testing programme in order to 'stay ahead' in military capability. In 1950, during the Korean War, US military leaders feared the loss of bomb-testing sites in the Pacific. By December 1950, this fear prompted a military decision to choose a bomb-testing site within the continental USA.

Although the full impact on health relative to the population at risk from this military venture (whether Pacific islanders or mainlanders) was not at all well known, the military desire to assure national military survival of the USA (especially with the possibility of retaliation for Hiroshima and Nagasaki) seems to have prevailed over all caution. Sensitive documents relating to President Truman's decision have been released recently in connection with a lawsuit brought by the State of Utah against the US government concerning the excessive rate of leukaemia, cancers, infertility and other health problems among the people of Utah who were exposed to fallout from the Nevada bomb testing. Parts of the newly released documents were quoted in an article in the *Deseret News* of 9 December 1978, an excerpt of which reads as follows:

> Under pressure of the Chinese entry into the Korean War, President Harry S. Truman approved use of the Nevada site just 34 days after the National Security Council directed the Atomic Energy Commission to make a selection study and recommendations. Truman signed the order on December 18, 1950, even before a Corps of Engineers' survey of the Nevada site was complete and without specific examination of radiological factors.[23]

Under pressure of a national security emergency, as it was perceived, the decision was made to risk random deaths from cancer and other illnesses among the US and Canadian mid-west populations. These deaths were spread out in time (because of the long-lasting pollution of the land and the long time between individual exposure and development of a tumour or other health effects) and in space (since the radioactive clouds passed over most of northern central and north-eastern USA and extended into Canada). This was considered by the military as 'acceptable' to prevent nuclear attack on a major US city. The US Atomic Energy Commission's voluminous public relations material following the establishment of the Nevada test site contained more wishful thinking than science.[24] Indeed, the health and safety studies were never completed at the Nevada test site since they were judged to be unnecessary once the decision was made and implemented.

Paul Cooper got up on the morning of 31 August 1957, in Yucca Flats, Nevada, USA, feeling a little squeamish about his decision to participate in the voluntary nuclear blast manoeuvres. He reviewed his instructions and hurried out, taking his place with the others with very mixed feelings. It was to be a 44 kiloton blast, about four times as powerful as the Hiroshima bomb.

In a trench, with his back to the bomb and his hands over his eyes, he waited tensely. At the detonation – a flash of light – he was startled to see the bones of his fingers even through closed eyelids. He remembered being told that many of the Hiroshima survivors were blinded by the initial blast, which was ten times brighter than the sun. It seemed as though an eternity had passed as he crouched there waiting, feeling the intense heat and being pounded by the wind and dust storm. It was difficult to breathe and he could taste the dry desert soil.

Suddenly the command came to rush ground zero and begin manoeuvres. He had no more time to think until he lay in his bunk that night. He had survived fear, vomiting and diarrhoea. He could really be 'combat ready' after a nuclear attack! The thought brought a feeling of security and his body relaxed into a much-needed sleep, calmed by a feeling of victory. His commanding officer had assured him that he would have no lasting ill effects.

His other nuclear manoeuvres were relatively uneventful. The men recovered from the initial sickness; if they didn't, they were moved off base. Apparently, not all participants in these experiments had had a choice such as Paul's. Some military personnel remember being asked, others have no recollection of either volunteering or being given an option.

After leaving the service, Paul kept a strict silence about his experience, as he had been instructed. It was considered a top national security secret. He never hinted to his family or physician that his later internal bleeding or the fact that his hair and teeth had fallen out might be related to his nuclear bomb experience. These problems subsided with time and he forgot about them. It was a small price to pay for the safety of his family, and for national security.

Twenty years later, in 1977, Paul Cooper of Emmett, Idaho, decided to speak publicly about the military nuclear tests. He was dying of leukaemia and had come to the conclusion that his illness was related to his radiation exposure. He was concerned about his wife, Nancy. His story was startling and hard for most Americans to believe. Through pressure from the news media and the Disabled American Veterans Association, Paul finally received a 100 percent disability allowance, but the government would not admit that his leukaemia was connected with his participation in the nuclear blasts. Paul died about a week after he testified before a US congressional committee about his experience.

After Paul Cooper publicised his story, many other veterans began reporting similar tragic human experiences. There were 192 announced US above-ground nuclear bomb tests between 1946 and the final 1963 above-ground testing ban. Estimates of the number of US servicemen deliberately exposed to the deadly fallout range from 300,000 to 1 million. About 1,000 Canadian military personnel and an unknown number of Canadian and American civilians were exposed. Today, ex-servicemen who request their radiation and medical records, or verification that they were present at an atomic test, are frequently told by the Veterans Administration or the US Nuclear Defence Agency: 'Your records were destroyed in the St Louis fire' (which occurred at the Military Personnel Records Centre in 1973). Many of these men now have malignant or non-malignant blood disorders and tumours and genetically dam- aged children. Some of the exposed servicemen lost their teeth and suffered other mouth and gum problems, others were blinded or developed cataracts at an early age. None of their immediate or early 'health effects' had been recorded in their military records and no attempts were made to follow up or compensate them for health problems which developed after this experiment 'in the line of duty'. The official reason for not compensating the military victims as they request assistance is lack of 'proof' that the illness developed was caused by the radiation exposure. The 'proof', of course, depends on the non-existent records which the military should have provided.[25]

Stories told by former servicemen who participated in the nuclear blasts are consistent with expected health effects of individual exposures above 30 rads. Many men have described accurately physical reactions not likely to be known by persons not actually experiencing such exposures. On the other hand, government officials have consistently claimed that military exposures were all less than 5 rads although they admit that no estimate was ever made of radionuclides inhaled or ingested during or after the military exercises. Film badges, which men sometimes wore, measured only external gamma radiation. It seems unlikely that military officials were so ignorant that they did not know that inhalation of the desert sand and air and ingestion of the local water, both heavily polluted with the radioactive chemicals released by the nuclear detonations, increased actual radiation levels by at least an order of magnitude (that is, 10 times) above that measured externally by film badges.

The beta dose to skin and lungs could have been 5 to 500 times the dose registered.[26] One can only speculate that the denial of compensation to servicemen is part of a larger denial system by which the military handles the fear of not being 'prepared' for nuclear war. Perhaps also if the true horror and consequences of nuclear war were admitted, governments could no longer dare to declare a war because no one would come to 'fight'. Governments would lose some of their sovereign power. The bullets are now microscopic, the dying can be long and painful, and the wounds are carried by our children and grandchildren. This is the fate of those who survive the blast, heat and intense radiation of the initial explosion.

Of more than 500 claims for radiation damage filed with the US Veterans Administration prior to April 1980, only eight claims were granted and only one of these was admitted to be radiation-related.[27] That one admission of radiation as a possible cause was to Orville Kelly who had taken his story to the public via the press. Orville Kelly died of cancer on 24 June 1980, a few months after he began to receive financial assistance. He was 49 years old. His young son spoke for an end to all wars at the Religious Convocation for Peace at the Second UN Special Session on Disarmament in June 1982. He brought tears to many eyes, even to those of the media cameramen.

After Paul Cooper's death in 1978 a limited federal investigation of the plight of atomic veterans was initiated. A preliminary report on this study was published in October 1980 by researchers at the Centre for Disease Control in Atlanta, USA.[28] There were 3.5 times as many leukaemia cases among the participants of this test than would have been expected based on the experience of males of the same age in the general population. Actually, since the military men were selected for good health, one would expect their leukaemia rate to be lower than the general rate. The reported radiation exposures of these military men were quite low – well below the level permitted yearly to nuclear workers.

Unfortunately, only one type of fatal cancer was studied in the small group of 3,224 survivors, less than one-tenth of the total number involved in the weapon testing.

The average time interval between the men's participation in nuclear tests and their diagnosis of leukaemia was 15.6 years. Diagnosis of multiple myelomas has occurred twenty or more years after exposure. Solid tumours take twenty to fifty years to develop to a clinically detectable size. In spite of these known facts, an official in a local office of the US Veterans Administration told the author that he had instructions to discourage any veteran claims for radiation damage if more than fifteen years had passed since the alleged exposure and no claim could be processed if more than twenty years had passed. There is, moreover, no provision for claims of damage to offspring.

On 16 July 1983, the US National Research Council, a sub-group of the US National Academy of Science, released a study of multiple myeloma reported by US veterans stationed in Hiroshima and Nagasaki for clean-up duty after the Second World War. The press-release headline read: 'No excess cases of multiple myeloma found among Hiroshima/Nagasaki veterans'.[29] The research was funded by the US Defence Nuclear Agency (DNA), the agency responsible for funding veteran claims. At the request of the Honourable Paul Simon, US House of Representatives, the Office of Technology Assessment (OTA) of the US Congress undertook a review of this research. Their findings were released on 17 December 1983, and the summary states:

the OTA finds that the evidence and analysis presented by the [National Research] Council do not permit definite conclusions about the rate of occurrence of multiple myeloma among the veterans, and therefore no meaningful comparison can be made with the expected rates of occurrence.[30]

The study was faulted for not including scientists with relevant scientific expertise, as well as for methodological errors. None of the persons on the National Research Council task force were listed as expert in either epidemiology or statistics in *American Men and Women of Science*. Multiple myeloma has a high fatality rate, approximately 80 percent dying within five years of diagnosis, therefore veteran call-ins on a 'hot line' were not a reliable source of information. The surviving family of the veteran might not know where he had served. Moreover, the National Research Council request for call-in specified neither Hiroshima/Nagasaki veterans nor the occurrence of multiple myeloma. Responses constituted less than 4 percent of the total US occupational force.[30, 31]

Some of the veterans who claimed to have been in either Hiroshima or Nagasaki, and who had multiple myeloma, were dropped from the National Research Council list because their military records either did not show them there or were just ambiguous. Others were eliminated because either they or their physicians failed to reply to the National Research Council's written solicitations. All of these methodological problems tended to bias the study towards underestimation of the problem. In addition to these scientific problems with the National Research Council study, the OTA discovered that the committee had estimated the number of multiple myeloma cases expected among a group of civilians over the 35 years between 1945 and 1980, and compared this expected number with the observed number of veteran cases (remaining in their study after almost half were eliminated) which had occurred between 1976 and 1980, a five-year span.

Investigation of the veterans' experience is obviously only the tip of an iceberg with respect to health damage from nuclear testing. Between 1945 and 1958, global nuclear tests released energy comparable to setting off 30 million *tons* of TNT. Prior to 1959, there were 207 atmospheric test explosions, 131 carried out by the United States, 55 by the Soviet Union, and 21 by Great Britain.[32] The number of atmospheric tests conducted by the United States eventually rose to 192, according to Major Sherker of the US Department of Defence. The official estimate given is 183.[33] After the United States, Great Britain and the Soviet Union ceased above-ground nuclear testing, after a test ban finally implemented in 1963, France, India, China and probably South Africa began testing in the atmosphere. Brazil, Argentina and Pakistan are expected by many to begin weapon testing in the atmosphere later in the 1980s.

The number of underground tests set off by the United States on the 1,350-square-mile Nevada site reached 399 by June 1978. The Reagan Administration has more than doubled the budget for the testing, allocating about $354 million a year. The number of nuclear blasts per year in Nevada increased from about 8 in 1972, to 16 in 1975 and 17 in 1982. The Soviet Union set off 20 in 1978, 15 in 1979, 10 in 1980, 9 in 1981, and 4 in 1982, in an opposite trend.

All underground tests have leaked radionuclides into the air. All have left deadly pollutants in the land.[34] Underground nuclear blasting continues even as I write, despite the warning of an ad hoc committee set up by former Nevada Governor O'Callaghan that 'Nevada is in a region of seismic activity and its citizens face a very real and growing earthquake hazard.'[35] It would be difficult to prove that the underground nuclear tests were the cause of the Guatemalan, Columbian or San Francisco earthquakes – but, equally, it is not unlikely that they contributed to these occurrences because of their proximity to the San Andreas fault, which extends from Alaska to Peru.

Besides the military personnel exposed at nuclear tests, many civilians were also exposed and their land contaminated. The American military were forced to evacuate the people of Bikini and Enewetak,[36] as described earlier, and probably should have also evacuated other Marshall islanders and the people of Utah and surrounding states.[37] Yet evacuation just delays the exposure because the air and water carry pollutants around the globe. In a book called *The Biological Peril to Man of Carbon 14 from Nuclear Weapons*[38] there is an estimate that *atomic testing during the month of September 1958 alone would result in 100,000 major birth defects of a physical or psychological nature; 380,000 stillbirths or infant deaths; and 900,000 foetal deaths in the thousands of years to come because of long-lived radioactive chemicals.* However, most of the victims of the testing will never know the cause of their tragedy. Those children who survive with minimal brain damage or other 'minor' defects will never be acknowledged as part of the national price paid for 'superiority' in the nuclear age. They are victims of what Einstein described as 'the slow torture of radioactive dust and rain'. Although not officially admitted, radioactive fallout must be a major contributor to the decline in Scholastic Aptitude Test (SAT) scores in the United States from 1968 on. Children born in the first nuclear test year, 1951–52, were seventeen to eighteen years old in 1968–69 when they began to take this test.[39]

The sharpest decline recorded in any one year of SAT scores was in 1975, eighteen years after the largest nuclear test series ever conducted at the Nevada test site (1957). Moreover, the decline slowed in 1977–78, eighteen years after the testing moratorium of 1959 to the fall of 1961.

There were regional differences in the amounts by which SAT scores dropped, differences hard to explain with conventional wisdom and without considering the radioactive pollution pattern. For example, the steep drop in scores for 1975–76 was smaller (9 points for verbal ability) in the US mid-west region which includes the industrial centres of Detroit, Chicago and St Louis, and greater (19 points for verbal ability) in the more rural western region which includes Nevada, Utah and Alaska where fresh fallout was heaviest.[40]

This trend towards poorer scholastic achievement in the US generally reversed in 1982–84, eighteen years after the ban on above-ground nuclear testing. However this general trend was not followed in Nevada, Utah, Arizona, California, Idaho, Oklahoma, Washington and Alaska – all states in the west.[41] The US government admits to venting of radioactive gases and particulates for some underground tests, and these states would also have received fallout from the Chinese nuclear testing.

This same poisoning will, of course, also extend to plants and animals – the foundation of biological life on earth as we know it and of the human food chain. In one 1953 bomb test in Nevada, 4,300 sheep were killed.[42] Documents originally marked 'Restricted, For Official Use Only' were declassified in 1979 and released to the radiation committee appointed by Governor Scott Matheson of Utah. Michael Zimmerman, a member of the committee, examined more than 400 documents showing that the sheep received up to 1,000 times the maximum allowable dose of radioactive iodine which would have been permitted for humans. Zimmerman charged the Atomic Energy Commission with 'wilful refusal' to investigate the sheep deaths or determine the effects of the test on human health.[43] The 1953 test gave radiation doses to the population of Utah and Nevada estimated to be between 40 and 500 times the dose which prompted the evacuation of pregnant women and children at the Three Mile Island nuclear accident in March 1979.[44]

The radioactive particles released in these and other nuclear tests are not confined to the region near the test site, but pollute the air, water and land globally. The population of the entire world, in a very real sense, must pay for US, Soviet and other nations' gains through nuclear testing. The arms race is slowly depleting the human 'store' of health.

Soviet nuclear·testing since 1949 has been conducted north of Lake Balkhash in Siberia. Two towns, Rubtsovak and Semipalatinsk, each of about 100,000 inhabitants, are within 62 miles of the site. No published reports on the effects of this testing have been released by the Soviet government.

British tests on Christmas Island and at Monte Bello and Maralinga in Australia have also been shrouded in secrecy, although there was an admission of 'excessive fallout' during one test due to an unexpected violent storm which caused fallout radiation three times greater than predicted. At least three of the British Maralinga tests were in the megaton range, i.e. about 1,000 times more powerful than the Hiroshima bomb. Many Aboriginal people in Australia, unable to read the warning signs near the Maralinga test site, unknowingly hunted and fished in the area and were exposed to the radioactive fallout. An Adelaide newspaper reported in April 1979 that twenty servicemen, federal policemen and civilians who worked at Maralinga had died of cancer.[45] The Atomic Veterans Association in Queensland, Australia, has called for a full-scale investigation of eighty cases of premature death.

According to a London *Times* report from Tony Duboudin,

Melbourne, 30 April 1984, a former RAF technician dying of cancer in Adelaide revealed that he had discovered the bodies of four Aborigines in a bomb crater after one of the British nuclear tests in 1963. A veteran, Mr John Burke, aged 63, also reported finding a number of deformed animals. The Australian Department of Resources and Energy confirmed a series of five 'minor' atomic tests in May and June of that year.

The French tests in the Pacific have also left a toll and the numbers continue to mount.[46] About 100 nuclear blasts have been set off in Moruroa. Nuclear tests are still being conducted there beneath the lagoon and information on the experience of other communities has been denied to the French Polynesian people. It is not unusual for Polynesian critics of French nuclear testing to be imprisoned in France.

A US government fact-finding committee in 1980 concluded that the Atomic Energy Commission had not only been 'inadequate' in its protection of public health from radioactive fallout, but it had also disregarded and actually suppressed all evidence suggesting that radiation was having harmful effects. Although, for example, the AEC knew the danger of ingesting contaminated food, it did not prevent pregnant women, infants and children from drinking milk from cows grazing downwind from the Nevada nuclear test site. The milk contained radioactive iodine, known to be able to be absorbed into the blood, pass the placenta barrier and lodge in the foetal thyroid gland, emitting gamma radiation there. Damage to the foetal thyroid can cause hypothyroidism, mental retardation and other physiological problems.[47]

The AEC refused to monitor the milk until forced to do so six years after the weapon testing began. In 1965 Dr Edward S. Weiss documented an unusual increase in leukaemia deaths in south-western Utah. The federal government investigated the deaths and 'failed to draw any final conclusions on causation'. The report was suppressed. In 1967, Dr Weiss drew attention to the twofold increase in thyroiditis and fourfold increase in thyroid cancer in Utah between the years 1948 and 1962. Before researchers could compile conclusive evidence, the federal funding for the research was discontinued. A memorandum from Dwight Ink, General Manager of the Atomic Energy Commission, to AEC Chairman Seaborg, of 9 September 1965, stated:

Although we do not oppose developing further data in these areas, performances of the above US Public Health Service studies will pose potential problems to the Atomic Energy Commission. The problems are: (a) adverse public reaction; (b) law suits; (c) jeopardising the programs at the Nevada Test Site.[48]

The responsibility for health research was shifted to the US Environmental Protection Agency, an agency with no funds to do the work.

An earlier AEC internal report in 1963 had pointed out the same health problems. This report was also deliberately suppressed. A memo sent by Nathan Woodruf, Director of the Division of Operational Safety, on 11 January 1963, read:

We do not recommend any new radiation protection guides for nuclear weapon testing at this time. The present guides have, in general, been adequate to permit continuance of nuclear weapon testing and at the same time have been accepted by the public, principally because of an extensive public information program. To change the guides would require a re-education program that

---

Classified

8.6.1 The general fallout area for this shot (Upshot Knothole, Nevada Test Site 1953) was narrow, not greater than 20 miles wide at more than 200 miles from ground zero. The radiation intensity was very high even at the 250 mile limit. The infinity dose at this distance was 5 rad. The only locations where fallout hit populated areas were a service station west of Bunkerville, and Mesquite, Nevada. The 25 rad infinity iso-dose line extended for 60 miles from ground zero, and the 10 rad infinity iso-dose line extended for 110 miles.

8.6.2 The fallout was so great on the highways near Glendale that several vehicles were found highly contaminated. It was necessary to establish roadblocks and decontamination stations at Las Vegas, St George and Alamo. These stations were very effective in their decontamination processes.

8.6.3 A considerable number of monitor and project personnel were overexposed on this shot.

Partial document
Verified unclassified 21 June 1979
Verified unclassified 11 March 1980
Obtained through Freedom of Information request

could raise questions in the public mind as to the validity of the past guides. Lastly, the world situation today is not the best climate in which to raise the issue. Therefore, we recommend the continuation of the present criteria.[49]

The US military establishment has never found the international climate favourable for promulgating radiation protection standards which reflect the true health hazard posed by nuclear materials. The façade was even strengthened in 1978 when a strong call was raised to allow yet more liberal radiation standards. The nuclear establishment likes to refer to its long 'successful' history of handling radioactive material.[50]

Atoms for Peace

On 8 December 1953 President Eisenhower launched the 'Atoms for Peace' programme in his address to the United Nations. There followed a concerted United States effort to make nuclear technology, with all its support industries, acceptable to society in general. Military commitment to nuclear technology required expanded uranium mining, milling, enrichment facilities, transportation, research and development and waste disposal. Weapon production required large amounts of electrical energy. Enrichment, for example, is a very energy-intensive part of the process. There was also an increasing military need for trained civilian personnel to operate nuclear support industries and a need to involve the academic and medical communities in the growing problems of nuclear development. Jerome H. Kahn, researcher at the Brookings Institute, estimated that between 1953 and 1960 an average of $35 billion (in 1974 dollars) was spent *annually* in the USA on 'strategic capabilities'.[51] This type of spending on military research and development exceeded any previous spending, even in time of war.

Moreover, US legislation and the climate of opinion prohibited the sharing of nuclear weapon technology with non-nuclear nations, whether friend or foe. The necessary scientific and technological information could be shared through the 'peaceful atom' programme. In case of an international crisis the 'peaceful atom' technicians could quickly be engaged by the military, just as the military had engaged the radium dial industry in the Manhattan Project. It was an ingenious idea for winning broad-based co-operation, with everyone's conscience feeling 'clean' and everyone's intentions appearing idealistic.

As more money was poured into university and industrial research into nuclear technology, actual weapon production was also quietly increased. Most research had dual uses. A billion-dollar weapon industry – unprecedented in war or peacetime – was woven into the fibre of the United States economy without most Americans being aware of it. Politicians vied for military contracts for their constituents, and the 'man in the street' perceived growing US militarism only as a source of jobs and economic expansion.[52] The old historical wisdom which said that a strong military defence policy would assure both peace and prosperity was popularly accepted.

The continued fear of military nuclear retaliation seemed to fuel the US desire to be 'strong' militarily. Patriotism, pride and progress were also strongly connected in the US popular perception of superiority in nuclear technology.

Government and military leaders of the 1950s took courage from the favourable reports about the growing 'pentomic forces', i.e. men trained to obey orders under stressful nuclear war conditions. The relatively optimistic reports from the Atomic Bomb Casualty Commission calmed some of the fears about radiation, and the Japanese will to recover made some think that nuclear war could be survived and 'won'.

Foetuses had fared worst at Hiroshima and Nagasaki. Many were spontaneously aborted and those surviving to birth had underdeveloped brains. Researchers decided to measure only profound retardation as a 'health effect', i.e. children unable to respond to a greeting or to feed and clothe themselves. They took comfort in the fact that not all Japanese children exposed *in utero* were born profoundly retarded.

Few survivors were able to have children during the first three years after their exposure. 'Monsters' were spontaneously aborted, and those victims with poorer genetic composition prior to the bombing did not survive the catastrophic post-bombing conditions and have children themselves. Most early deaths were due to blast wounds, severe burns, pneumonia and other serious infections. There was irreparable damage to lungs, gastro-intestinal tracts and bone marrow, disfigurement by keloid scars and intense grieving.[53] But they *were* alive, they had 'survived', and this encouraged military planners.

The land was 'habitable' after cleansing by the late summer

monsoon winds and rains. Japanese people came to Hiroshima and Nagasaki to rebuild the cities and reclaim their human dignity. The victims were silent and an illusion of well-being was created. The mutated vegetable strains died out after a few seasons and were replaced.

In the 1950 Japanese census, 100,000 survivors were identified for inclusion in a Life Span Study at the Atomic Bomb Casualty Commission. This study population, of course, was seriously biased toward the hardiest residents of these cities, since a prerequisite for being in the study population was being alive five years after the event.[54,55]

Dr Shingetō, a busy physician who kept records in defiance of the occupation force's prohibition, discovered an excess of leukaemia cases among survivors. The well-funded research arm of the US government, the Atomic Bomb Casualty Commission, had not noticed. While non-findings and minimising of problems might have pleased the military and government planners, Dr Shingetō's records and findings condemn the incompetence of the US scientific staff. The Atomic Bomb Casualty Commission failed from the beginning to be an independent scientific research facility. Whether or not health effects are 'detectable' depends on the sensitivity of the measurement, the competence of the scientists and the integrity of reporting.[56]

An Accident at Chalk River, Canada

On 13 December 1950, the first major nuclear generator accident took place at the NRX reactor at Chalk River, about 150 miles north-west of Ottawa, Canada. A hydrogen explosion occurred, killing one man and seriously contaminating five others. The reactor core was largely destroyed and 1 million gallons of highly radioactive water flooded the structure. One of the young US Navy technicians sent to assist in the emergency was Jimmy Carter, later to become President of the United States.

This accident exposed the inadequacy of medical preparation for nuclear accidents. The on-site emergency facilities were oriented towards 'routine' decontaminations from minor spills, not towards handling individuals who were both seriously wounded and con-taminated. Some excerpts from a US Atomic Energy Commission document dated 9 February 1951 highlight the problems.[57]

- Doctors hesitated to make hypodermic injections through contaminated skin.
- The decontamination of wounds was complicated by the lack of a probe sufficiently directional to tell from what part of the wound the radiation was coming.
- A new kind of decision had to be made concerning the extent to which the patient should be decontaminated before transfer to the townsite hospital could be allowed.
- Unprecedented problems arose in connection with transferring the contaminated body to an undertaker.

All these problems seemed surmountable to governments determined to 'survive' in the nuclear age. The accident was shrouded in secrecy in the United States, Canada and Great Britain for the sake of national security. Understanding the implications of nuclear accidents might reduce civilian co-operation with the growing weapon industry. Decisions required by the expanding technology were made by an elite few capable of 'rational' decision-making on emotional issues. The policy of 'secrecy so as not to panic the public' became an accepted way to manage popular opinion in the nuclear age. It is strangely reminiscent of Nazi management of the Jewish ghettos.

As late as April 1979 it was apparent that US medical personnel responsible for hospital management were not well versed in the lessons learned twenty-nine years earlier at Chalk River, Canada. Most were completely unprepared for a possible major catastrophe such as was occurring at the Three Mile Island nuclear reactor.

A second major accident occurred at Chalk River in 1958. Overheated uranium fuel rods ruptured within the core, and in the removal process a piece of burning uranium was dropped into a shallow maintenance pit. It spewed fission products to accessible areas inside the reactor building and contaminated a sizeable area downwind. More than 600 men were required for the clean-up.

Official Atomic Energy of Canada Limited (AECL) reports stressed the small number of individuals 'overexposed' and implied that there would be no adverse health effects from the radiation doses received. Similar to US policy, the Canadian government failed to do a follow-up study on the men to find out what their actual experience was.

One of the workers, Bjarnie Paulson, had his first cancer surgery in 1964, six years after the accident. In 1966 part of his nose was removed because of cancer. He has had forty operations since, with cancer occurring in the rectum, scalp, chest, pubic and anal region. Paulson, a Royal Canadian Air Force corporal at the time of the Chalk River accident, kept silent about his exposure in keeping with his solemn promise to do so in view of military secrecy. Finally, in 1978, in an effort to obtain a military pension, he mentioned his fear that it was the clean-up operation which had so disrupted his life. Just as happened in similar cases in the United States, the Canadian pension board told him there was no record of his ever having been in Chalk River. Under pressure, AECL eventually released the record. They had recorded the external gamma exposure dose only. Paulson's cancers were probably due to unrecorded alpha particles which had adhered to body hair follicles.

AECL had never done a follow-up study of any of the exposed workers at Chalk River.

Another RCAF corporal named McCormand, who had been in the Chalk River clean-up with Paulson, has been found to have cancer of the throat. Efforts are now being made to contact other workers in a full-scale investigation launched by the Canadian veterans' pension board.[58] It remains to be seen whether or not the findings of the study will confirm the men's experience and whether they will be compensated for their years of suffering.

The International Commission on Radiological Protection

A brave new world dependent upon nuclear technology must systematically learn to accept radiation fission products as a 'fact of life'. Although medical uses of X-ray had been accepted for a long time, the Hiroshima and Nagasaki experience made necessary a whole new public 'education' to ensure acceptance of the Peaceful Atom programme. Public acceptance was necessary if the military programme was to become a reality and nuclear support services thrive. Only if large numbers of people willingly co-operated could the grandiose plans be accomplished.

Some countries had established radiation protection standards as early as the 1920s, but as yet in 1950, there were no uniformly accepted international recommendations for 'permissible' levels of exposure to humans. 'Permissible' literally means accepting whatever damaging health effects occur within the permitted exposure. Then, as now, it was realised by scientists (although not publicly admitted by many) that every exposu e to ionising radiation causes some biological damage.

In 1950, at a meeting in London, the International X-ray and Radiation Protection Commission (which had been in existence since 1928) reorganised and assumed the name International Commission on Radiological Protection (ICRP). Members also formed a companion organisation called the International Commission on Radiological Units (ICRU). Although ICRP and ICRU became independent organisations, they both continued to receive financial support for secretarial services and suggestions for new members from the International Congress of Radiology. Members of ICRP and ICRU ordinarily 'donated' their professional services, being financed by member governments or their nuclear industry employers.

In the spring of 1956 in Geneva, Switzerland, ICRP and ICRU became formally affiliated with the World Health Organisation (WHO) as a participating non-governmental organisation (NGO). They have provided information for, and received financial assistance from, the United Nations Scientific Committee on the Effects of Atomic Radiation (UNSCEAR). The ICRP has recommended the radiation standards and practices now generally accepted and implemented throughout the world.

Membership of the ICRP is highly selective and controlled. Prospective members must be recommended either by current ICRP members or by members of the International Congress of Radiology and then approved by the ICRP International Executive Committee. *Through this structure, participation in standard setting has been dominated by colleagues from the military, the civilian nuclear establishment and the medical radiological societies who nominate one another.* Participation of physicians in ICRP is limited to medical radiologists. People in all these categories *have a vested interest in the use of radiation and depreciation of the risks in its use.* There is an added problem of military secrecy in many countries, including the USA, about radiation health effects, since these are the results of a nuclear bomb. This again limits the pool of 'experts' available to ICRP. There is no independent body, even the World Health Organisation, which can place a person on the ICRP. *It is, in every sense of the term, a closed club and not a body of independent scientific experts.*[59]

ICRP would have benefitted from broader medical and scientific disciplinary representation (paediatricians, internists, cell biologists, and so on) and by including within its structure those physicians and scientists whose research and statements have challenged their philosophy and/or recommendations. Epidemiologists, biostatisticians and public health specialists, i.e. those who could provide an audit of ICRP predictions, are excluded from membership. The ICRP should be comprised of people elected from various other related organisations, rather than being as it now is, a self-perpetuating group of users of radiation.

In its functioning since 1950, the ICRP has never taken a public position in favour of protecting public health in any of the controversial radiation-related problems encountered: it has not taken a stand against above-ground nuclear weapon testing; it has not condemned radiation experiments on humans (prisoners, military personnel and terminally ill patients); it has not called for a reduction of exposure of uranium miners to radon gas by increasing mine ventilation; it has not called for a reduction of medical uses of radiation for diagnostic purposes; it has not called for a reduction of exposure levels for nuclear workers as experience and research showed that their danger had been underestimated; it has not taken a position against the nuclear industry practice of allowing transient workers in the high-risk radiation exposure category to move from job to job without adequate control of cumulative exposures.[60]

The hydrogen bomb which exploded on 1 March 1954 on Bikini Atoll escalated the nuclear weapon race and marked a definitive commitment in the Western world to 'peace' through military stength, regardless of the cost in personal suffering and destruction of the life-support system. It was the Western response to the Soviet detonation of its first nuclear bomb in Siberia in 1949 and was followed, predictably, by a Soviet explosion of a hydrogen bomb in 1955. Hydrogen bombs were also exploded in Australia, on Christmas Island, and the island of Novaya Zemlya in the Arctic Ocean. ICRP became part of the elaborate structure built to support the nuclear arms race, even if some of its members failed to realise this.

A Soviet Nuclear Accident

The Soviet Union also had its accidents and difficulties with nuclear development. A major nuclear accident occurred during the winter of 1957 in the Ural mountains. The first public mention of the accident came in 1976 when a former Soviet scientist, Dr Zhores Medvedev, referred to it in a British journal.[61] It was later described more fully in a book.[62]

The article, 'Two Decades of Dissidence', startled the Western world by its casual mention of a major disaster due to the explosion of stored nuclear waste near Kyshtym and Chelyabinsk in the Ural mountains around 1957. Medvedev himself was startled to find that this major disaster was unknown outside Russia. His story was generally received with incredulity in Great Britain, the United States and France. A member of the UK Atomic Energy Authority, Sir John Hill, declared it 'rubbish' and 'a figment of the imagination',[63] calling such an explosion impossible.

The Central Intelligence Agency in the United States admitted that there had been an 'incident' in late 1957 or early 1958, but denied that it was related to nuclear waste as stated by Dr Medvedev. They said a plutonium-production reactor (i.e. a nuclear reactor operated just for the sake of producing spent fuel rods) at a Soviet nuclear weapon complex went out of control. The American intelligence report said the reactor was 'only distantly related to present nuclear power plants and the accident's relevance to the safety of civilian nuclear power today is probably minor'.[64]

The anonymous American analyst added: 'there was no evidence that the accident put a crimp in their plutonium production or that it gave them [the Soviets] second thoughts about their weapons programmes. They just piled dirt over the reactor – buried it that way – and went about their business.' It is, however, well known that in early 1958 Khrushchev suddenly announced unilateral suspension of atmospheric atomic weapon tests in the Soviet Union. In October 1958, despite the protests of a number of Soviet atomic scientists, Khrushchev announced resumption of the nuclear tests.[69] He was at the same time negotiating with the United States and Great Britain for the above-ground nuclear test ban, and was perhaps trying to stop the testing but also to negotiate from a 'position of strength'. The above-ground test-ban treaty was finally signed in 1963.

After his revelation about the Soviet accident, Dr Medvedev experienced a character attack by anonymous scientists which was never substantiated. A newspaper article hinted at the connection between his revelation and British plans to build a large nuclear waste treatment plant at Windscale in Cumbria.[65]

The American intelligence report stating that the Ural disaster was a reactor accident, and not a waste storage facility explosion, reached the Israeli press and was read by Professor Lev Tumerman, a 1972 émigré from the USSR who had travelled through the radioactive zone in the Urals in 1960. In December 1976, Professor Tumerman, a nuclear power advocate, disturbed that the American report might jeopardise the Israeli nuclear reactor programme, wrote a letter to the Editor of the *Jerusalem Post* corroborating Dr Medvedev's statement that the disaster was caused by negligence and carelessness in the storage of nuclear waste.[66] He stressed the fact that scientists and ordinary people in the USSR all told him that the disaster was due to mismanaged nuclear waste. He also gave an eye-witness account of the vast territory contaminated and of the road-blocks and warnings to travellers. The latter were advised to drive quickly through the contaminated areas, not stopping or getting out of their automobiles because of the high level of radioactive chemicals in the land and water.

On 24 January 1977 the US Central Intelligence Agency released a report received from an unnamed 'source' and marked 'unevaluated information', as a response to the disclosures by Medvedev and Tumerman. The document refers to articles by two former USSR citizens describing a region of 'vast nothing' in the Ural mountains and suggests that the wasteland could be explained as a Soviet nuclear weapon test, reported in a 'top secret Soviet film':

According to the film the USSR constructed a completely new city in a valley in the Ural Mountains region for the test. A subway was constructed under the village, and one of the major purposes of the test was to see if the subway could withstand a nuclear attack. The inhabitants of the village were goats and sheep, and the post-explosion photography showed the effects of a nuclear blast upon animal life as well as building material. Military equipment was placed around the village and the effects of the explosion upon armaments of war were depicted in the film.

The bomb itself was described as a 20 megaton device which was dropped from an airplane. The flash of the explosion illuminated the mountains which surrounded the village. The city virtually was eliminated, but the subway survived the explosion.[67]

The world press quickly picked up the story without labelling it 'unevaluated' and without realising that it was the USSR nuclear test site on Novaya Zemlya in the Arctic – geographically a continuation of the Ural range but 1,250 miles north of Chelyabinsk – which was the designated area of 'vast nothing'. It is also unlikely that the USSR would explode a 20 megaton bomb in a populated area, and especially an area containing a large military reactor built in 1947 for the production of plutonium. The large production plant was completed in 1948, and generated enough plutonium for the first atomic bomb to be exploded in September 1949, a goal set to celebrate Stalin's seventieth birthday. The plutonium plant, built with such pride by the Soviet military establishment, was only ten years old at the time of the Kyshtym–Chelyabinsk explosion.

The bomb test described by the CIA was common to many tests done both in the USA and the USSR during 1958 and 1959. The account was apparently 'sanitised' so that an impression of safety in the subway ('bomb shelter') was conveyed. Actually the fire storm created by a 20 megaton bomb would consume all the oxygen in the subway and victims seeking shelter would be asphyxiated. The underground structure, if it survived the blast, would serve as a mausoleum.

These attempts to explain away the Chelyabinsk nuclear explosion while satisfying the superficial reader failed to be credible to the informed public. A British television company, Granada, managed to find two more Russian emigrants, now in Israel, willing to discuss the event. They gave their testimony in a radio broadcast in November 1977:

Around the end of 1957 we began to hear rumours that a terrible accident had occurred at Chelyabinsk 40 (the military designation of the site), that there had been a terrible nuclear explosion, an accident caused by the storage of the radioactive waste from the plant. Soon after, the routes between Sverdlovsk and Kapaesk were closed. I could not see my parents for about a year.

Also, during that year, I spoke with friends who were doctors. I went once to a hospital in Sverdlovsk for the removal of a wart, and one of my friends, a doctor, told me that the entire hospital was crammed with victims of the Kyshtym catastrophe. He said that all the hospitals in the entire area were crammed full, not only in Sverdlovsk but also in Chelyabinsk. These are huge hospitals with many hundreds of beds. The doctors all told me that the victims were suffering from radioactive contamination. It was a tremendous number of people. I believe thousands. I was told that most of them died.[68]

A nurse in Israel provided further testimony:

There were no signs of destruction [in Kyshtym in 1967] but everything which they bought at the market place or even if they went to the woods to gather mushrooms, they had to measure with radiometers, and they had little radiometers with them.

When I came there [1967] I got pregnant there and the doctors told me to get rid of the child because of the radiation. They were afraid that something might be irregular and so I had to make an abortion.[69]

The identity of these independent witnesses was protected because they had relatives still living in the Soviet Union. The information they gave was independently verified by the correctness of detail, such as Chelyabinsk 40, the post office box number of the military installation, and by later disclosures by CIA releases under the US Freedom of Information Act.

The TV commentary added that the Israeli witnesses described a large fenced-in area containing piles of contaminated topsoil. Common weeds with distorted shapes and sizes were growing on it. Local people referred to the area as 'the graveyards of the Earth'.[70] Apparently homes were also burned by the government so that people would not be tempted to retrieve contaminated possessions.

Censorship in the USSR and self-serving secrecy in the Western world which was preparing the public to accept the Peaceful Atom programme (first announced in 1954) kept this event of 1957 hidden for twenty years. The question of whose interest is being protected by this secrecy and manipulation of public opinion is a serious one.

Through the Freedom of Information Act in the United States, the Natural Defence Research Council and Ralph Nader's Critical Mass were able to obtain some documents held by the US Central Intelligence Agency describing the event in more detail.[71] One of the fourteen documents released a year after the Medvedev disclosure stated that 'hundreds of people perished'. A report dated 23 May 1958 revealed:

Various Soviet employees and visitors to the Brussels fair have stated independently but consistently that the occurrence of an accidental atomic explosion during the spring of 1958 was widely known throughout the USSR.

A woman who had been in the hospital at the time of the explosion reported that she saw many victims brought into the hospital. A Foreign Intelligence Report on the 'Nuclear Explosion at Chelyabinsk' in the Ural mountains, dated 20 September 1976, said that the local Russian press reported the 1957–58 event as 'an unusual occurrence of the northern lights', despite the fact that the hospitals were filled with people 'suffering from the effects of radiation'.

Shortly after the explosion a scientific research institute was established at Chelyabinsk to study the effects of radiation:[72]

All stores in the Komensk-Uralskiy which sold milk, meat and other footstuffs were closed as a precaution against radiation exposure, and new supplies were brought in two days later by train and truck. The food was sold directly from the vehicles, and the resulting queues were reminiscent of those during the worst shortages during World War II. The people of Komensk-Uralskiy [near Chelyabinsk in the Urals] grew hysterical with fear, with an incidence of unknown 'mysterious' disease breaking out. A few leading citizens aroused the public anger by wearing small radiation counters which were not available to everyone.[73]

In another probably related event, an American U-2 spy plane piloted by Francis Gary Powers was shot down during a surveillance flight over the Sverdlovsk region of the Urals shortly after the accident occurred. Khrushchev's memoirs record many other U-2 flights over this disaster area, so it may be presumed that US intelligence-gathering on the event was fairly comprehensive. However, total censorship in the USSR and secrecy in the USA prevented public understanding of the nuclear waste disposal problems, the superpower relationships and the significance of the U-2 incident.

After the Chelyabinsk disaster, Professor Vsevolod Klechkovsky, director of the Department of Agrochemistry and Biochemistry at the Timiriazev Agricultural Academy in Moscow, was asked to organise research into radiation effects on plants, animals and humans in the Chelyabinsk region of the Urals. Dr Medvedev had worked under him in Moscow.

Medvedev was invited to be part of the research team. Since such participation required agreement not to publish findings, not to meet or correspond with foreigners, a prohibition on travel abroad and acceptance of KGB surveillance of all contacts with other Soviet citizens, Medvedev refused. However, some of the young researchers in Medvedev's department accepted the attractive offer to participate in the radiation research. Later their names provided Medvedev with the key he needed to reconstruct from their research papers, published after 1966, some of the details of the accident. These research papers were not easily identifiable by Westerners who did not know the names of the Chelyabinsk scientists.

In 1947 the Soviet Union set up a secret science centre in the Ural mountains under the Ministry of State Security (MGB) where the main work was done by experts deported from Germany and by Soviet prisoner–scientists. There were also free workers at the secret centre, but their publications, communication and travel were restricted by agreement with the Soviet government. This first scientific research centre was headed by Timofeev-Resovsky, an internationally recognised Soviet geneticist who had migrated to Germany in 1926 and was returned to the Soviet Union as a prisoner in 1945. Khrushchev declared this centre an 'open laboratory' in 1956, making it a biophysics branch of the Soviet Academy of Science. Timofeev-Resovsky was not allowed to do genetic research (which had been declared illegal in the Soviet Union in 1949 and remained illegal until Khrushchev's retirement in 1964). However, it appears he was a first-hand witness of the biological genetic changes in the Urals after the Chelyabinsk accident. He was well qualified as an observer, having published more than 100 papers in radiation genetics between 1926 and 1945, whilst in Germany. Overall charge of the work on radiation at the secret science centre was under A. I. Burnazian, who held both a military title of lieutenant-general and a civilian title of deputy minister of health. Apparently in the Soviet Union there was recognition of the implications of radiation-related research for policy decisions.

After the 1957–58 accident, according to Medvedev, a second secret radiation research laboratory was opened near the prison science centre, perhaps because of tight security control of findings. There were apparently two radiation research laboratories near the Chelyabinsk disaster area between 1956 and 1965, an open one and a secret one.

Timofeev-Resovsky was allowed to publish papers in his own

name after 1956, and in 1965 he was given Soviet academic degrees, freed from his prison status, and appointed director of the genetics department at the Institute of Medical Radiology at Obninsk, about sixty miles from Moscow. This new institute sprang up as genetics, cytogenetics, radiobiology, population genetics and other special-ised disciplines became legal in the Soviet Union. At the same time, the system of censorship was decentralised away from the State Committee for Atomic Energy to local commissions. This allowed some of the science, sanitised of specific details, to be published in research journals.

Although Lieutenant-General Burnazian received the rewards, titles and decorations for the science done at the prison–science centre, it was eventually Timofeev-Resovsky who was recognised as the capable scientist. Medvedev, who worked under Timofeev-Resovsky in Obninsk, credits him as the real founder of radio-ecology in the Soviet Union.[74]

It is necessary to recognise which Soviet research comes from open and which from secret laboratories in order to understand the omission of information on geographical location, research design and purpose of the 'experiment' in the latter. Obviously, a catastrophic accident was not designed to fit usual research criteria.

In the *Milwaukee Journal* of 23 August 1978 a leading Soviet geneticist, Nikolai P. Dubinin, head of the Soviet Institute of General Genetics, was quoted as saying:

> Any further advancing along the path of uncontrolled damage to the biological basis of mankind's [humankind's] existence could bring about great losses in the biological quality of the human population.[75]

Dubinin was addressing 2,000 geneticists from all over the world attending the first International Congress of Genetics ever held in the Soviet Union. In his opening-day remarks, Dubinin noted that the percentage of children in industrialised countries born with congenital defects had more than doubled between 1956 and 1977. In 1956 about 4 per cent of children born in industrial countries had mutations which could be classified as serious congenital diseases. Congenital disease occurs during embryonic or foetal development, producing diseases in children which were not present in their parents. Some of the congenital diseases can be passed on to the children of the affected offspring, some are not inheritable.

In 1968–69 the World Health Organisation estimated the global mutation rate (sometimes called genetic load) at 6 percent, with the highest rates in developing countries. The UN Council for Radiation reported in 1977 that the genetic load had reached 10.8 percent. For each visible or expressed genetic disease there are thousands of persons carrying the genetic mistake but, because its effects can be masked by a normal gene, it is not observable as a disease. The rare person who receives a defective gene from both father and mother will have an observable genetic disease. The increase in global genetic load was blamed on environmental pollution.

Dubinin is cited by Dr Medvedev as one of the scientists who did research at Chelyabinsk. In Dubinin's autobiography, *Eternal Motion*,[76] he mentions a 1970 report to the Presidium of the Soviet Academy of Sciences on the results of eleven years of experiments in forest and meadow areas

> which had been contaminated by high doses of radioactive substances . . . In these circumstances some species died out, some continued to suffer for a long time, their populations reduced in size, and some evolved toward a higher resistance.[77]

Reports by Dubinin and fellow workers give evidence that, in this contaminated area, all the pine trees died and about 30 percent of the birch trees were severely damaged by the radiation. The amount of biomass which was common grass (*Calamagrostis epigeios*) increased to three to five times what it had been before the accident. Higher plants and trees were replaced by radio-resistant grass.[78]

Sir John Hill, the vociferous British scientist who had called the early report of Dr Medvedev 'rubbish' in 1976, was one of the chairpeople of the radioecology section of the IAEA meeting in Vienna in 1971 where the Soviet papers on the Ural disaster were first presented. Not a single question was asked of these Soviet scientists following their presentation or, if asked, the questions and answers were not published in the proceedings.[79]

In a 1972 paper by Dubinin and his colleagues, a type of field mouse inhabiting a radioactive environment for a long period of time was examined for radiosensitivity.[80] They reported an increase in both somatic and genetic mutations in the mice. The cells of these defective mice showed less chromosomal rearrangements than non-chronically irradiated mice when strontium 90 was injected into them. This was interpreted as 'increased resistance to radiation'. He did not discuss the lifespan or the general vigour of the mouse population.

In another population study of 200th to 400th generations of a single cell algae (*Chlorella*), some strains of which survived in the radiation-polluted environment, a new strain of algae resistant to some doses of radiation emerged.[81] The area studied must have been near the centre of the Chelyabinsk accident since the contamination described approached one curie per square metre of soil, a level lethal for all animals and higher plants.

In terms of humans, those most complex biological organisms which are dependent on lower biological forms for food, oxygen production, and recycling of all organic waste material, the futile hope of rapidly evolving a radio-resistant body is a cruel hoax proposed to lay to rest fears about the nuclear age. If grass, algae and mice are only partially 'successful', the human species has no possibility of quickly, or in large enough numbers, evolving a 'strain' able to survive widespread radiation contamination. It appears that all the superpower military and political decision-makers are aware of this fact. The general public, however, encouraged in an attitude of false hope by misleading interpretations of scientific data and reluctant to face the truth until it is forced upon it, still clings somewhat to the false hope.

According to Tumerman's personal witness, in 1960, of the results of the explosion which spewed radioactive waste across dozens of miles in the Urals:

> On both sides of the road as far as one could see the land was 'dead': no villages, no towns, only chimneys of destroyed houses, no cultivated fields or pastures, no herds, no people . . . nothing . . . I was later told that this was the site of the famous Kyshtym catastrophe in which hundreds of people had been killed or disabled.[66]

The US study of the accident, done at Oak Ridge National Laboratory, estimates that most of the industrial city of Kasli was contaminated and fourteen lakes and about 625 square miles of land were poisoned (see pp. 77–78).

The Soviet Union has never released an estimate of the number of immediate or delayed fatalities attributable to the accident. The Russian people who survived this accident have never had the human comfort of talking with the survivors of Hiroshima and Nagasaki, Micronesia, French Polynesia, Utah or the other regions of the world victimised by the nuclear age. The voices of these nuclear experts, made so by their personal experiences, are not heard in the Kremlin, Pentagon, corporate boardrooms, or other sterile environments where visions of short-term futures are concocted and long-term tragedies suppressed.

On 23 April 1979, the Soviet Union's Power Minister admitted to visiting US congressmen that serious problems resulting from 'minor' nuclear accidents had caused them to make significant improvements in the safety technology of 'peaceful' nuclear installations. Details on these other minor accidents are not reported in the general scientific literature of the free world.

It is believed that the CIA has at least fifteen other classified documents which are needed by the global scientific community in order to assess unknown nuclear dangers revealed in the 1957 Soviet accident. Implications of other Soviet accident experiences are equally inaccessible to concerned scientists.

In the light of a major attempt to begin civilian atomic involvement, the need to classify reports such as the Soviet accident for 'national security' is understandable but by no means commendable. The commercial industry would have begun much more slowly and cautiously had the full truth been told. Perhaps it would not have been begun at all, and that was the underlying fear.

A US Mistake in Early Nuclear Waste Management

The weapon installations at Chelyabinsk and Hanford have provided the USSR and the USA with first-hand experience of chronic radiation exposure, either from an explosion–contamination accident or through routine occupational handling of radioactive material. However, the learning advantages of these experiences were limited because of information control in both societies. Only selected people had full access to the facts.

It was not known generally until 1973 that Hanford came close to a Chelyabinsk-type explosion in one of its burial trenches:

Due to the quantity of plutonium contained in the soil of Z-9 [a low level waste burial trench] it is possible to conceive of conditions which could result in a nuclear chain reaction. These conditions would be the rearrangement of contaminated soil, flooding of the enclosed trench following a record snowfall and rapid melting, and failure to implement planned emergency actions (pumping of flood water from adjacent terrain and addition of neutron absorbing materials to the enclosed trench).[82]

Hanford's arid climate and a $2 million clean-up project saved it from this potential plutonium disaster. However, the quantity of nuclear waste accumulated there and elsewhere since 1943, and the problem of millions of years of storage for this waste, unveils another facet of the nuclear option. The nuclear waste problem is by no means unique to Chelyabinsk and Hanford, but these are undoubtedly among the most contaminated spots on the globe.

As the Western and Eastern superpowers deepened their military and economic dependencies on nuclear armament, both realised the need for the committed involvement of the general public in terms of active and passive co-operation with the expansion of nuclear industries. It was within this 'climate' that the military–industrial complex was formed and the 'peaceful atom' programmes were developed. The chart entitled 'Nuclear Fuel Cycle and Nuclear Weapon Cycle in the United States' (see pp 186–187)) clearly illustrates the mutual components of both industries in the American plan. The dotted lines indicate components desired but not in commercial operation in the USA. Nuclear generators are a necessary part of the nuclear weapon cycle and the central dilemma of a 'pure' peaceful atom programme in the USA consists in whether to force the military to operate their own 'weapon reactors' or to let them derive plutonium from the spent commercial fuel rods. President Reagan opts for the latter.

Individuals involved in the seven shared operations are encouraged to think of themselves as working to solve the 'energy crisis' or to better the standard of living of the poor. This enables nuclear research and development to be undertaken in colleges and universities and nuclear physicists and engineers to be trained for the nuclear age with untroubled consciences. 'Electricity will be too cheap to meter', and the third world will be enabled to share in the wealth of the economically 'developed world'. These goals serve to justify work which is really needed by a massive weapon industry. Most Americans are unaware of the necessary link between the commercial and military atom.

From the early days of the Manhattan Project to the present time anything related to nuclear technology has been seen as requiring strict information control for the sake of national security. In the USA, special agencies and powers were created to deal with this new and unique danger. The National Security Act of 1947 created

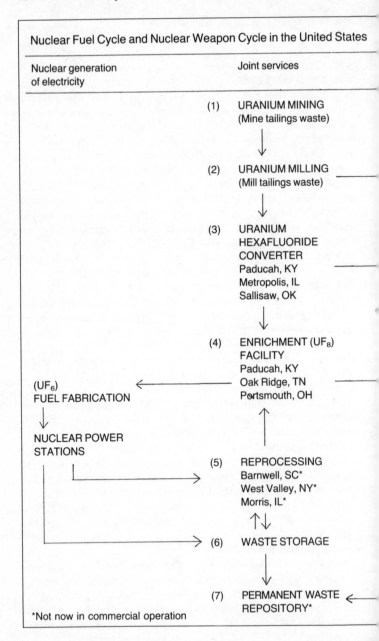

Nuclear Fuel Cycle and Nuclear Weapon Cycle in the United States

Nuclear generation of electricity	Joint services

(1) URANIUM MINING
(Mine tailings waste)

(2) URANIUM MILLING
(Mill tailings waste)

(3) URANIUM HEXAFLUORIDE CONVERTER
Paducah, KY
Metropolis, IL
Sallisaw, OK

(4) ENRICHMENT (UF_8) FACILITY
Paducah, KY
Oak Ridge, TN
Portsmouth, OH

(UF_6) FUEL FABRICATION

NUCLEAR POWER STATIONS

(5) REPROCESSING
Barnwell, SC*
West Valley, NY*
Morris, IL*

(6) WASTE STORAGE

(7) PERMANENT WASTE REPOSITORY*

*Not now in commercial operation

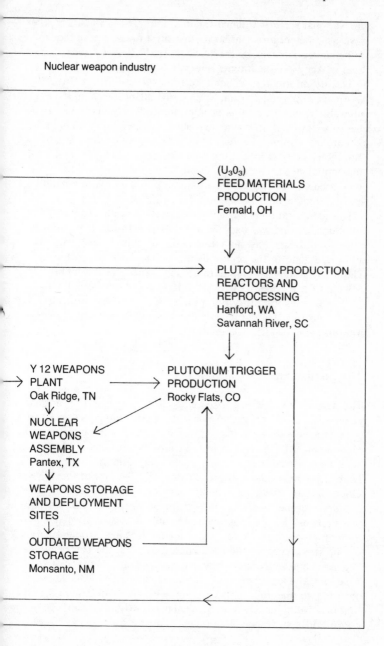

Nuclear weapon industry

(U_3O_3)
FEED MATERIALS
PRODUCTION
Fernald, OH

PLUTONIUM PRODUCTION
REACTORS AND
REPROCESSING
Hanford, WA
Savannah River, SC

Y 12 WEAPONS
PLANT
Oak Ridge, TN

PLUTONIUM TRIGGER
PRODUCTION
Rocky Flats, CO

NUCLEAR
WEAPONS
ASSEMBLY
Pantex, TX

WEAPONS STORAGE
AND DEPLOYMENT
SITES

OUTDATED WEAPONS
STORAGE
Monsanto, NM

the National Security Council and the Internal Security Act of 1950 gave government broad authority to protect the secrets of the atom and manage survival in the nuclear age.[83] In 1954 the Atomic Energy Act gave the Atomic Energy Commission extensive power to classify information under a security system derived from the above acts. Wiretapping and other surveillance techniques were allowable if nuclear secrets or interference with nuclear programmes were involved. In the popular mind, co-operation with both the military and commercial nuclear industries was linked with patriotism. Nuclear decisions were considered outside the realm of democratic decision-making. Dissenters from this popular view were accused of being communists, bent on undermining national security and economic progress.

This identification of the 'anti-nuclear' stance with communism or anti-Americanism was reinforced by Soviet calls for 'An atom-free zone in the Far East and the entire Pacific Basin[84] and 'An atom-free zone in the region of the Balkans and the Adriatic'.[85] These Soviet moves were interpreted by some as an attempt by the Soviets to leave the West vulnerable to their aggression, since a nuclear-free 'world' was *not* proposed. It became difficult for Americans to speak out against the nuclear pollution of the earth without being labelled communist.

The Commercial Nuclear Industry

The late 1950s saw the beginning of the commercial nuclear generation of electricity. Among the first commercial reactors operating were six reactors near Troitski in the Soviet Union, one in Shippingport, Pennsylvania, in the United States; one in Marcoule near Gard, France; and four reactors at Cumbria and four at Dumfriesshire, in the United Kingdom. By the end of the 1950s, Great Britain, the United States and the Soviet Union all had military nuclear reactors and well-developed nuclear weapon programmes. France was about to embark on its own nuclear weapon programme. According to a United States Atomic Energy Commission report to the US Congress in 1959, the United States had given nineteen foreign research reactor grants between 30 May 1956 and 31 December 1959. Grants, totalling $47,270,000, were awarded to the following countries: Brazil, Sao Paulo; Spain, Madrid; Netherlands, Petten; Denmark, Risoe; Japan, Tokai-mura; Portugal, Lisbon; Venezuela, Caracas; Italy, Ispra; Greece,

Aghia Paraskeri; Sweden, Studsvik; Israel, Rehovoth; West Germany, Munich; Belgium, Mol; China, Hsinchius; Austria, Vienna; Norway, Kjeller; Korea, Ahyang; Thailand, Bangkok; Vietnam, Dalat.[86]

The 1950s ended with the three superpowers (the United States, the USSR and Great Britain) seriously planning to end the above-ground nuclear explosions, France deciding to begin such a weapon-testing phase and all four countries beginning to generate electricity through 'harnessed' nuclear fission.

The United States, the USSR and Great Britain had chosen to test weapons on their own territories, colonial possessions or trusts. France, however, proposed to test in the Sahara and Morocco, which had achieved independence and joined the United Nations on 12 November 1956. The Moroccans took the issue to the UN General Assembly on 4 November 1959. The French Representative, Moch, stated in response:

So long as there remains the agonising insecurity of a world dedicated, as it is, and despite ourselves, to the arms race, each State has the right – and each Government the duty – to ensure the protection of its country, France as well as all others.[87]

On 17 November 1959, the UN General Assembly adopted Resolution 1376 (XIV) calling for additional studies on the effects of ionising radiation. This resolution was sponsored by Argentina, Canada, Czechoslovakia, Ghana, Ireland, Italy, Japan, Mexico, New Zealand and Norway. On 20 November 1959, the General Assembly adopted Resolution 1376 (XIV), expressing 'its grave concern over the intention of the Government of France to conduct nuclear tests' and requesting 'France to abstain from such tests'. The phrase, 'in the Sahara', which appeared in the original draft after 'nuclear tests' was deleted in the final version.[88] The resolution was sponsored by twenty-two member nations and passed by a vote of 51 to 16, with fifteen absentions.

The Arms Race

At 7.00 a.m. on 18 February 1960 the President of the French Republic, General de Gaulle, gave orders to explode a plutonium device in the Tanezrougt area of the Sahara Desert. General de Gaulle expressed gratitude to 'ministers and scientists, officers and engineers, industrialists and technicians'[88] for placing France in a position to defend herself against communism. Having nuclear weapons placed France again in the 'higher' diplomatic echelon, entitling her to attend the international nuclear arms talks and enter into agreements with atomic powers on disarmament. The prominent role given to scientists, engineers and technicians in the rebuilding of France's role as a world power should be self-evident. France's desire to 'enter into agreements between atomic powers' was an unabashed bid to regain former international status.

At a news conference on the same day, President Eisenhower expressed US understanding of the French action.

> We must realise that this spirit of nationalism of which we hear so much is not felt just by the under-developed nations, the ones that the people want suddenly to be independent; it is felt by all of us. The matters of pride and national prestige impel people to do things, I think at times, that would not be necessary.[89]

Eisenhower fell short of saying that this nuclear bomb testing was not only unnecessary but also destructive of a fragile desert life-support system on which many humans and animals were depending.

World opinion was quickly mobilised against the French weapon testing in the Sahara. However, a request by twenty-two nations failed by six votes to gain the needed majority (forty-two votes) required to convene a special session of the UN to consider the problem. All NATO countries either voted against the session or abstained from voting.

France persisted in her determination to test in the Sahara until forcibly evicted by the Moroccans. Testing was then continued near the island of Moruroa* in French Polynesia, to the distress of the indigenous leaders who were aware of the consequences. The sad story of French nuclear testing in the Pacific is well documented in a book by Bengt and Maria Thérèse Danielsson, *Moruroa Mon Amour*.[90]

On 19 December 1959, in Geneva, the three other nuclear powers, the United States, the USSR and Great Britain, agreed on a draft treaty for discontinuance of nuclear testing above ground. The actual cessation of above-ground tests by the United States and

---

* The French frequently spell the name of this atoll Mururoa, but French Polynesian spelling is Moruroa.

the Soviet Union did not take place until 1963, although there were intervening periods of voluntary cessation. Great Britain undertook seven 'legal' nuclear tests and five 'illegal' nuclear tests (after the ban) in 1962–63 at Maralinga near Adelaide, Australia.

The Australian desert is still contaminated with fission products and Australians continue to demand clean-up. One complaint states that the British buried some radioactive debris on an Aboriginal reserve, and covered it with 2 to 3 inches of soil. The British insist that there was no *agreement* to remove the waste, only an 'understanding'.

There have been no known attempts to decontaminate the Nevada or Siberian test sites. They are still being used for underground detonations. These test sites will probably never be usable by humans in the future.

Many nuclear war-related issues, such as the question of protection against surprise attack, were left unsettled by the Nuclear Test Ban Treaty. The question of total disarmament was not mentioned. In the 'free world', the democratic process was an encumbrance to the desired secrecy of nuclear policy, and was seen as increasing vulnerability to surprise attack. Then, as now, nuclear attack meant possible destruction of the entire nation in a matter of days or hours. The Soviet Union proposed reconvening the conference of experts which had drafted the treaty to ban above-ground weapon testing so that they might recommend practicable measures to prevent surprise attack. In response, President Eisenhower made the following statement at a news conference in Gettysburg, Pennsylvania, on 12 August 1959:

As I have pointed out so often, how can a democracy make a surprise attack – for the simple reason that we have to engage in war by the will of the Congress . . . A free country, in my opinion, is absolutely helpless when it comes to launching a surprise attack.

In view of later developments, such as the Vietnam war, which was waged by the USA without congressional approval, the gradual erosion of democratic processes in the USA with respect to managing international affairs is apparent.

Other facets of international politics became distorted by the nuclear arms race. Faced with the possibility of an unthinkable nuclear holocaust, the three nuclear powers took to manipulating other countries, extending there by their 'spheres of influence', presumably to increase their own security. Free countries began covert international operations without open debate of policy and without parliamentary or congressional approval. Even 'first strike' or surprise attack eventually became acceptable to democratic governments.[91] In his 'Fiscal Year 1980 Arms Control Statement to the US Congress', President Jimmy Carter stated:

> The addition of highly accurate Trident II missiles with higher yield warheads would give US-SLBM [US Submarine Launched Ballistic Missile] forces a substantial time-urgent-hard-target-kill capability for a first strike . . . The countersilo capability of a (deleted) KT Trident II missile would exceed that of all currently deployed US ballistic missiles.

The UK is also building Trident submarines in Cumbria for its own use.[92]

Since 1945 there have been 125 violent conflicts between nations. About 95 percent occurred in the developing world and many have been considered surrogate wars between the superpowers. The USA or NATO countries have intervened in 75 percent of these wars; 6 percent involved the USSR or Warsaw Pact Treaty nations; 15 percent involved intervention by other third-world countries.[93] These wars have caused almost 9 million civilian casualties and 6 million military deaths.[94]

Although economists agreed that the arms race was causing great hardships in all countries, bringing about inflation and unemployment, draining off the best minds, the most promising human potential and the nations' wealth for military purposes, no one seemed to know how to stop the spiral. The arms race has obviously been drawing other nations into its frenzy and seems to be building up to a cataclysmic crisis.

Nuclear 'experts' were expected constantly to defuse the political crises which threatened to touch off nuclear war. This gave birth to the Pugwash Commission which took on the role of advising a world armed with enough nuclear weapons to destroy itself. The Pugwash Commission has helped, but it cannot constitute a lasting form of relief valve for global tensions.

The unmanageability of democracy probably prompted much of the feverish effort of the US Atomic Energy Commission to promote understanding and acceptance of commercial nuclear technology. This agency spent millions of US taxpayers' dollars on 'peaceful atom' booklets which were distributed free of charge to schools, on summer institutes for science teachers, on educational films and teacher-training sessions held at national laboratories. The 'promotional' aspect of the Atomic Energy Commission in the USA so overshadowed its regulatory functions and constituted such a blatant conflict of interest that the US Congress dissolved the Commission in 1972, creating in its place the Nuclear Regulatory Commission (NRC) for regulation and the Energy Research and Development Administration (ERDA) to manage research and promotion. The latter agency was raised to Cabinet level in the form of the US Department of Energy (DOE) in 1978, demonstrating a clear government priority for nuclear research and promotion over regulation. The regulatory agency, the NRC, on the other hand, has had little power truly to regulate and is usually the scapegoat blamed for nuclear problems in the public mind.

The US Department of Energy has a mandate from the US Congress to develop nuclear weapons. In 1980, 39.5 percent of its budget was devoted directly to weapon production and deployment. This does not include research. The proportion increased to 41.4 percent in 1981.[95] The 'energy crisis' has always seemed to be a secondary objective of DOE policy. This department has developed out of the crisis management of a world become nuclear. Part of its management objective is to elicit popular support for nuclear options. Its behaviour reflects its origin, the US Atomic Energy Commission. Its promotional role consists in 'assisting' the public to accept the inevitable nuclear pollution connected with national dependence on nuclear weapons for military and economic 'security'. Many Americans are unaware that the Department of Energy budget includes the Nevada nuclear testing, intercontinental ballistic missile silos and nuclear weapon laboratories. Citizens therefore do not complain about the increased budget for this agency as they do for the Department of Defence.

Bilderbergers

The business segment of society, very dependent on international security for its economic health, was the chief stabilising and unifying force in the post-Second World War era. In Europe especially, the divisive effects of national rivalries were recognised as detrimental to European recovery. Participation in a nuclear world dominated by two superpowers with large territorial holdings and unified political stances also caused Europe to examine its structure. The in-fighting between traditional European enemy-nations was directly counterproductive for survival.

Dr Joseph Retinger, former top aide to General Sikorski, head of the Polish government in exile in London, was the first head of the European Movement, designed to restore power to Europe. The United States, involved in the Korean War and threatened by Soviet testing of nuclear weapons in Siberia, generously funded the European Movement. It was anxious to re-arm Europe (including Germany) in order to enhance its military position vis-à-vis the Soviet Union.

This organisation proved to be unworkable because of growing European discontent about re-arming Germany, US policy in South-east Asia, the Cold War, McCarthyism and the process of European integration. Finally, Retinger resigned from his position and the organisation collapsed.

Retinger, sensing that an informal relationship between key 'actors' in Europe and North America would be more fruitful than the European Movement had been, set about drawing up a European critique of US policy designed to initiate constructive dialogue. He engaged leading Europeans to contribute to the critique and then confronted William Averell Harriman, the US Marshall Plan administrator in Europe, with a carefully worded document.

Harriman counselled Retinger to wait until after the 1950 US elections, elections at which General Eisenhower was elected President. In the interim, Harriman submitted the European critique to the US Committee for a National Trade Policy, which drafted a reply. Finally, after many informal preparations and small meetings, the recognised 'charismatic' leaders of Europe and North America met in May 1954 at the Hôtel de Bilderberg, Oosterbeck, Holland. According to Retinger, invitations were sent to

> only important and generally respected people who through their special knowledge or experience, their personal contacts and their influence in national and international circles can help to further the aims of Bilderberg – defending Western ethical and cultural [life].[96]

The name 'Bilderberg' for the organisation, and 'Bilderbergers' for people invited to attend one or more of the meetings, has persisted to this day.

The Bilderbergers were concerned with communism and the

Soviet Union, with 'dependent' nations and peoples, with economic policies and problems, European integration and European defence. The European Economic Community and the European Atomic Energy Commission (Euratom) are believed to have originated at their 1955 meeting.

One of the most difficult confrontations between Europeans and Americans revolved around the US Senate Hearings on Unamerican Activity being conducted by Senator Eugene McCarthy. Although Bilderbergers were anti-communist, they also had an only too real remembrance of fascism. McCarthyism in the USA was too close to dictator thought-control methods for their taste. It is said that C. D. Jackson, President Eisenhower's 'assistant for psychological warfare', commented to European Bilderbergers:

> Whether McCarthy dies by an assassin's bullet or is eliminated in the normal American way of getting rid of boils on the body politic . . . by our next meeting he will be gone from the American scene.[97]

Prior to his political appointment, Jackson had been President of the Committee for a Free Europe (which broadcast the anti-communist Radio Free Europe), the publisher of *Fortune* magazine and the managing editor of *Time-Life*. Bilderbergers were careful to include the media moulders of national opinion in their elite circle. For whatever reason, McCarthy was silenced as Jackson had predicted.

Prince Bernhard, the German Prince of Lippe-Biesterfeld, husband of Princess Juliana of Holland, remained in charge of convening Bilderberg meetings and centralising its activities until his resignation in 1976. Retinger was permanent secretary until his death in 1960. Bilderbergers have never had elections because of their informal, changing membership. They exert a significant influence on Western national policies when they achieve consensus, and also, through their policy of not acting divisively, achieve a cohesive effect even when not acting on a consensus. After the 1973 OPEC crisis it was the Bilderberg Conference which 'managed, in effect, to set the world's monetary system working again'.[98]

Because of its stable core of personnel, its flexibility in inviting people to its meetings, and its commitment to seek consensus and provide a climate of dialogue, the Bilderberg has reduced national tensions in Europe and among NATO countries. It has close connections with the British Institute for Strategic Studies in

London, and the Centre for Strategic and International Studies at Georgetown University, Washington, DC, in the USA.

> Bilderberg does not make policy. Its aim is to reduce differences of opinion and resolve conflicting trends and to further understanding, if not agreement, by hearing and considering various points of view and trying to find a common approach to major problems. Direct action has therefore never been contemplated, the object being to draw the attention of people in responsible positions to Bilderberg's findings.[99]

The institutions on which the informal consensus forged by the Bilderberg meetings has had an impact include the United Nations, the International Monetary Fund (IMF), the World Bank, the Organisation for Economic Co-operation and Development (OCED), the North American Treaty Organisation (NATO) and the European Economic Community (EEC).

Bilderberg has brought about other more informal relationships, such as the nuclear technology transfer between West Germany and South Africa. At the September 1955 Bilderberg meeting in West Germany, the agenda included the political, strategic and industrial aspects of atomic energy. The agenda was designed as a response to the Eisenhower 'Atoms for Peace' proposal in 1954 and the establishment of the International Atomic Energy Agency at the Geneva meeting of international scientists, on 8–20 August 1955. West German Bilderbergers invited South African Bilderbergers to send scientists to the Karlesruhe Nuclear Centre to learn about uranium enrichment technology.[100]

These informal Bilderberg meetings occur outside the scrutiny of people directly and indirectly affected by their decisions. Bilderberg holds no legitimate mandate empowering it, it cannot be held responsible for actions emanating from its meetings, and it is beyond the control of any national or international bodies. It represents business, financial and communications interests, while neglecting to include spokespeople for people lacking food, shelter, clothing, medical care and recognition of their personal dignity, that is, the majority of the world's citizens. These impoverished and disenfranchised planetary citizens are neither participants nor observers of the global dealings of Bilderberg.

While efforts towards consensus and unified action have had undeniable benefits, Bilderberg's elite and secretive meetings cannot be a long-term solution for global unification problems. It is a powerful global agent for change which is beyond the scrutiny of the global community.

Beginnings of Dissent

In the 1950s, anti-nuclear protest was totally focussed on the bomb. Peace groups actively protested against escalation of the arms race, and scientists protested against above-ground weapon testing.

A little-known protest against a nuclear reactor occurred, however, as early as 1959. The United Auto Workers' president, Walter Reuther, led the AFL-CIO, the principal US trades union organisation, to file suit in protest against the Fermi 1 Breeder Reactor being built near Detroit, Michigan. Three million people, including 500,000 UAW members, lived within a 30-mile radius of the plant. Leo Goodman, Reuther's adviser on health and safety, was one of the most knowledgeable and vocal spokespersons against nuclear pollution from the earliest days of the nuclear age. He had fought. courageously against military control and for congressional control of the Atomic Energy Commission. He had persuaded Reuther and the whole union to protest against this nuclear reactor which even the Atomic Energy Commission would not declare safe. The union lost the suit in the Supreme Court and the reactor was built.

A major accident occurred at the Fermi breeder reactor in 1966. The breeder came close to spewing plutonium over the city of Detroit – which would have made it uninhabitable.

It is generally admitted by nuclear physicists that for a nuclear reactor to achieve criticality, i.e. the ability to explode, the uranium must be 'enriched' to at least 20 percent. This means that the fissionable uranium, U235, or plutonium, needs to be concentrated from less than 1 percent of the mixture, as it occurs in nature or spent fuel rods, to about 20 percent of the mixture. US commercial light water reactors use fuel enriched to about 3 percent and are not expected to explode through nuclear fissioning. However, breeder reactors use fuel enriched to 25 percent or more, and are fully capable of uncontrolled fissioning at start-up. While the configuration of fuel in a breeder does not match that of an atomic bomb, if a breeder reactor explodes, it could release great quantities of plutonium in aerosol form.

The accident at Detroit was of much greater potential harm and magnitude than that at Three Mile Island, although the publicity given to it was much less.[101]

The commercial nuclear industry expanded and grew with government favour and financial assistance during the 1960s. Many people dedicated to this commercial industry had no idea of its origin or the many 'national security secrets' which had to be kept to ensure its popular acceptance. Many onlookers secretly wished the industrial technological phenomenon success so that some shred of hope and healing could blossom in a world shattered by the dropping of two unforgettable bombs in August 1945. Scientists working on the commercial nuclear enterprise quickly divorced themselves from the nuclear weapons industry. The weapons industry slipped into the background and everyone hoped for the best from the new commercial venture.

Disillusion:

The Sixties and Since

A Full-blown Commercial Nuclear Technology

> Summing up, a uniform irradiation of the whole human species, sufficient to reduce the actual rate of reproduction, might now be regarded, if it were practicable, as not by any means disadvantageous.
>
> Uneven irradiation, on the other hand (arising from atomic explosions rather than atomic energy), is serious for individuals or groups who receive heavy doses, especially heavier doses than their neighbors, and especially sudden doses rather than accumulations.[1]
>
> C. D. Darlington, 1950

What did not seem 'practicable' in the 1950s became a reality in the 1960s: the basic research, mining, milling, enrichment, transportation and decommissioning involved in the nuclear process could serve two nuclear industries. As we saw in the chart outlining the connections between parts of the US nuclear commercial and nuclear weapon industries, enriched and natural uranium is normally diverted (before reaching weapons grade) into fuel for nuclear power generators. Nuclear reactors, whether or not they are used for the generation of electricity, are a necessary part of the nuclear weapon industry. After six to eight months of use in a reactor, the 'spent' rods can be reprocessed for removal of fissionable uranium and plutonium, needed for weapons. Further, governments can freely subsidise the shared portion of the nuclear industry in the name of energy. In this way the public pays for its energy in taxes rather than electricity bills, making it appear 'cheap'. Citizen co-operation and public financing overtly seem to be helping to ease the 'energy crisis'; weapon industry needs are provided for indirectly, and everyone is enabled to work in good conscience to promote 'peaceful' uses of nuclear technology.

199

Prior to the signing of the Limited Test Ban Treaty by the US, UK and USSR on 5 August 1963, there were 329 US tests,[2] 19 UK tests and two joint US–UK tests, and 142 USSR tests.[3] In 1960–61, France also set off four nuclear bombs in the atmosphere. The temporal and geographic distribution of the tests and fallout was uneven. The true number of tests is not publicly known. The US announced 331 tests during this period, but the United Nations' estimate credits the United States with only 193, perhaps not counting tests of small yield. Since there were no indications of which tests were counted or not by the UN, all reported US tests are noted here. The UN estimates for the Soviet Union, Britain and France may also be too low. The table on p. 284 gives the yearly numbers of tests. Between October 1958 and September 1961 there was a voluntary moratorium on testing by the superpowers.[4]

Most of the information wanted by the US and USSR military was gained by the weapon testing done prior to 1960.

US Pacific tests prior to 1958 took place in the Marshall Islands. In 1958 the US conducted two tests at Johnston Island, one over the Pacific Ocean, and the remaining 32 in the Marshalls. The 1962 Pacific tests were at Johnston Island or the British-ruled Christmas Island. Other US nuclear tests were conducted at the Nellis Air Force Base Bombing Range (near the Nevada test site); in the South Atlantic and at Carlsbad, New Mexico. The tests at the bombing range were plutonium dispersal shots; those in the South Atlantic were rockets set off about 300 miles above the earth; and the Carlsbad test involved a 3 kiloton blast in a salt cavern.[2]

Scientists from all over the world were by then exclaiming against the serious genetic effects and cancers being caused by above-ground nuclear tests. However, the International Commission for Radiological Protection remained silent, never taking a stand against the atom bomb tests, the reckless exposure of civilian and military personnel, or the building up of a layer of radioactive particles in the upper atmosphere which will slowly drift to earth over the next decades. Further, nuclear testing produces nitric oxides in the stratosphere, where they also deplete the ozone layer. They later return to earth as acid rain. The beta particles react with atmospheric oxygen and nitrogen in the same way as lightning.

There was only scattered citizen opposition to the nuclear tests in the United States and Canada, most people being unaware of either the events or the danger. The US Atomic Energy Commission had sponsored a heavy public relations campaign near the site to convince the public that low level radiation posed no threat to health. Opposition to nuclear testing was made to appear unpatriotic.

The following is a direct quotation from an Atomic Energy pamphlet dated 1 January 1955:

Each Nevada test has successfully added scientific knowledge needed for development and for the use of atomic weapons and needed to strengthen our defense against enemy weapons. An unusual safety record has been set. No one inside the Nevada test site has been injured as a result of 31 test detonations. No one outside the test site in the nearby region of potential exposure has been hurt. There were instances of property damage from blasts such as broken windows. Some cattle and horses grazing within a few miles of the detonations suffered skin radiation burns, but the damage had no effect on their breeding value nor the beef quality of the cattle.[5]

In spite of such optimistic reports, many serious accounts of radiation-related illness were reported by the off-site population. Even a death from beta burns sustained by a young man who drove unknowingly through an atomic fallout area in his convertible was suppressed. A US Public Health Service report on handling civilian reports of illness after the 'Plumbob' test in 1957 stated that 'in most cases it was sufficient for the medical officer to discuss the possible biological effects of radiation with them and they would realise that the symptoms which they had could not be due to fallout.'[6] Even after leukaemias began to occur, citizens complaining of the tests to the Atomic Energy Commission received letters telling them there was no evidence of a link between nuclear weapons and leukaemia. This was, of course, well after the dramatic increase in leukaemia at Hiroshima and Nagasaki.

The US Congressional Committee charged with investigation of the health complaints of citizens downwind of the test site concluded in August 1980 'that the AEC's desire to secure the nuclear weapon testing programme took precedence over the Commission's responsibility to protect the American public's health and welfare'.[7] Significant scientific discoveries were disregarded by the AEC if they brought into question the safety of the tests. Evidence of damage to human health was suppressed or ignored. It was within this era of secrecy and reckless military expediency that nuclear commercial ventures began. The commercial companies did not question the standards set for human radiation exposure; they merely began to design an industry which could stay within the permissible pollution levels and which would be cost-competitive with other commercial forms of generating electricity. This latter requirement was, of course, assisted by extensive government subsidy. The consumer paid the bill as part of his or her taxes rather than by direct electricity rates.[8]

As universities began training nuclear physicists and engineers for work in a commercial field, all the shared aspects of the nuclear weapon industry took on a 'peaceful' purpose. Pure scientific research, the development of health and safety practices, the packaging and transporting of nuclear materials, nuclear waste disposal, all were perceived as civilian problems inviting creative solution by the intelligent young scientists joining the ranks. These scientists were unaware of their unwitting benefit to the military nuclear establishment. Governments developed commercial licensing processes, set legal limits to liability for nuclear accidents and gave research fund allocations and other subsidies, hidden and overt, to make this commercial nuclear venture a success. The crucial steps of enrichment of nuclear fuel, atomic research, insurance, decommissioning and waste management have been non-commercial governmentally supported links in the nuclear fuel cycle in the USA since the beginning of the commercial industry. A free people could not be expected to devote unlimited money and personal resources to producing weapons of mass destruction during peacetime. They *could* be expected to rally behind efforts to promote economic prosperity and jobs – the reward offered for solving the energy crisis. 'Free' is here used in the sense of 'having a vote in public elections'. Democracy, of course, cannot exist without an informed electorate.

The US Federal Radiation Council

In the USA, a Federal Radiation Council (FRC) was appointed to parallel the International Commission on Radiological Protection. President Eisenhower set up the FRC on 14 August 1959, commissioning it to set national radiation standards for health and safety.[9] The move seemed to be a political response to head off legislation meant to remove the setting of radiation health standards from the US Atomic Energy Commission and place it with the US Public Health Service. The concern about health and safety in the USA was increasing, of course, because of continued military testing of nuclear weapons in Nevada.

On 26 August 1959 John A. McCone, then Chairman of the AEC, appeared before the congressional Joint Committee on Atomic Energy, and requested military exemption from any national health or safety standards which might be set by the FRC.[10] McCone admitted that:

> as atomic weapons, nuclear propelled vessels and other military reactors in the custody of military departments became more numerous, the requirements of military readiness in training and manoeuverability may exert more and more influence toward less restrictive safety procedures.

The proposed bill to exempt the military did not go beyond the congressional committee.

The FRC, in consultation with the Secretary of Defence, the Chairman of the AEC, the Secretary of Health, Education and Welfare and the Secretary of Commerce, developed a 'compromise' on health and safety.

The executive order signed by Eisenhower on 13 May 1960 read:

> The Guides may be exceeded only after the Federal Agency having jurisdiction over the matter has carefully considered the reason for doing so in the light of the recommendations in this paper.

The Defence Department had a much more sweeping permission to exceed standards than it would have had from the defeated congressional bill. The defeated congressional bill would have deferred the power to the President, who could give permission 'to the extent he deemed necessary in the national interest'.

The FRC was little more than a façade, with a budget of $48,000 per year and no full-time staff for the first four years. The first chairman was Secretary Calabrese, from the Department of Health, Education and Welfare, whose philosophy was stated at a congressional hearing of 6 June 1963: 'The health implications associated with fallout and the possible influence of protective measures on the US economy and the national security are major considerations to which proposed actions would need to be related.'

The executive director of the FRC was Dr Paul C. Tompkins, whose previous positions included Head of the US Naval Radiological Laboratory and Atomic Energy Commission. Whilst serving as executive director of the FRC he received a salary from the AEC. He and Lieutenant-Colonel James B. Hartgering, radiologist with the US Army, apparently dominated the FRC. The latter served as assistant in charge of radiation matters to Dr Wiesner, Science Adviser to the President. The FRC provided most of the congressional testimony and information on radiation used as a basis of national US decision-making in launching the commercial nuclear sector.

At the national meeting of the American Public Health Association on 12 December 1962, Professor Russel H. Morgan of John Hopkins University condemned the FRC policy of exempting the military from meeting standards for radioactive material in the environment.[11] The military–scientific answer to the 'problem' was to propose raising the standards by a factor of 10 or more. Secretary Calabrese referred to the fact that 'national security' was a major consideration, along with concern for disruption of the food industry and agriculture (by admitting to contamination). Little or nothing was done to alert the public to the increased cancer and genetic damage it would experience for the sake of national security, the food industries and agriculture.

Windscale (Sellafield)

One of the most serious nuclear accidents occurred on 9 October 1957 at Windscale, England (now called Sellafield). A uranium fuel element burst, causing a uranium fire and a massive release of radiation into the atmosphere. Sellafield is in Cumbria, on the Irish Sea, about 300 miles north-west of London. Fallout from the accident was measured in Ireland,[12] London, Mol (Belgium) and Frankfurt (FRG).[13] The radiation levels were so high directly downwind of the plant that cow's milk was confiscated from an area covering 200 square miles and dumped into the Irish Sea. It is said that government officials tried to minimise the accident to the public, stating that all harmful radiation had been blown out to sea.[14]

In addition to the radioactive fallout from the fire, for 20 years the Windscale nuclear plant has been regularly discharging radioactive waste through a pipe extending more than a mile into the Irish Sea. With recent reports of increased levels of thyroid cancer and leukaemia in the vicinity of the plant, the fire of 1957, the quarter-ton of plutonium routinely discharged into the sea and various other hazardous radioactive discharges are coming under suspicion as probable causal agents.

In April 1984, a Yorkshire Television documentary reported elevated pollution levels and abnormally high cancer rates in the Sellafield area.

They also found plutonium 239, americium 241, ruthenium 106 and cesium 137 in house dust at Ravenglass. None of these radioactive chemicals occur in the normal earth environment.

Seascale, south of Sellafield, has ten times the national average rate of leukaemia. The general beach region in the vicinity of the nuclear plant has five times the national average rate. Yet despite expert opinion supporting the Yorkshire Television findings, British Nuclear Fuels, who own the Sellafield plant, have denied being responsible for the effects Yorkshire Television claimed to have discovered.[15]

In a survey of the Sellafield area conducted by academics from Leeds University after the Yorkshire Television programme, it was found that 64 percent of the 400 people interviewed were more frightened by unemployment than the radiation risks, despite realizing the reality of the dangers (*Sunday Times*, London, 26 August 1984).

In a research paper published in November 1983, Dr Patricia M. E. Sheehan and Professor Irene B. Hillary reported that young women living in an Irish boarding school across from Windscale (Sellafield) at the time of the 1957 fire had given birth to six Down's Syndrome children.[16] Forty-seven married fertile women participated in the study. Their average age at time of birth was 26.8 years, and since it is unusual for mothers of this age to give birth to Down's Syndrome babies, this was a highly significant statistical result.[17]

Eight other cases of Down's Syndrome have been reported in Maryport, 16 miles from the Sellafield plant. While one would expect only one Down's Syndrome child per 1,600 live births to women in their twenties, Maryport has reported about ten per 1,600. All were born to young women (average age 25 years) who were alive or *in utero* at the time of the 1957 Windscale (Sellafield) fire.[18]

Following the nuclear reactor accident at Three Mile Island in 1979, there was also an increase in Down's Syndrome births in Tompkins County – 200 miles north of the reactor site – and Broome and Chemung counties for 1980 and 1981.[19]

Will anyone keep track of the later reproductive experience of women exposed to Three Mile Island radioactive releases? Who will piece together the puzzles and determine precise causes and biological mechanisms? It will probably not be an overworked and underfunded local department of health. Given the high level of mobility of the US public, if there is an abnormal increase in Down's Syndrome babies among these young women the occurrences will be lost to history in national statistics collected in a non-meaningful way.

The SL1 Reactor, Idaho Falls

On 3 January 1961 another major accident occurred in one of seventeen experimental nuclear reactors at the US Atomic Energy Commission's testing grounds near Idaho Falls, Idaho. Three men were killed in this accident, one impaled on the ceiling from the force of the accident. The bodies had to be dismembered because parts were highly radioactive. The highly contaminated parts were buried in lead coffins while less contaminated parts were taken to Arlington cemetery (the national military cemetery) for burial. These were unsung heroes of the hidden atomic war for 'national security'.

Big Rock Nuclear Plant

The Big Rock Nuclear Plant, which began operating in December 1962, is owned by Consumers Power Company. It is located on a 600-acre site on the shore of Lake Michigan, three miles north-west of Charlevoix, Michigan, USA. It has a thermal (heat) capacity of 240 MW (megawatts) and produces 63 MWe (megawatts of electricity). This is small compared with the 1,100 to 1,200 MWe plants currently being constructed. It is one of the few plutonium-fuelled plants operating in the USA.

Northern Michigan is primarily a resort area where a larger proportion of the population is engaged in recreational services and in farming than in the state of Michigan as a whole. It has correspondingly less occupational involvement with manufacturing than does the state. For a long time it has been a retirement haven for the elderly. One would expect Charlevoix to be a 'healthy' place to live, remote from industrial and urban problems.

Dr Gerald Drake, an internist in Petoskey, Michigan, and long-time resident of the area where the Big Rock nuclear generator is located, became concerned about changes in public health trends in the vicinity of the Big Rock Nuclear Plant, and reported this concern to the US Atomic Energy Commission (AEC) in 1974. Dr Drake's principal findings relating to the first ten years of operation of the plant were as follows:

1. Infant deaths per 1,000 live births per year were 22.4 in Charlevoix County and 23.0 in the State of Michigan for 1962–66. In 1967–71 the death rate rose to 22.6 in the county and dropped to 20.7 in the state. The difference was not statistically significant, but disturbing.
2. There was an average increase of 21 percent in low-birth-weight infants (under 5½ lbs or 2,500 grams) in the county while state increases were only 1 percent for the same time periods. This was statistically significant with the probability of it happening by chance less than 0.005. This increase around the nuclear reactor cannot be explained by factors affecting the whole state, such as nuclear weapon fallout or food additives, or by long-standing local health problems. Change must be initiated by changed circumstances.
3. Cancer deaths increased by about seven people in the 16,000 county population per year, and increased by only one person per 16,000 in the state when the same two time intervals were compared. This was statistically significant, with a probability of its happening by chance of less than 0.001.
4. Although the leukaemia death rate stayed stable in the state, it rose from 1 per 16,000 to 2 per 16,000 per year in the county. This was not a statistically significant change but a disturbing trend.
5. Congenital defects per 100 live births decreased in both the county and the state, but the county decreased by 8.5 percent while the state figures decreased by 12.5 percent. This difference was not statistically significant.

In response to these indicators for concern, the US Atomic Energy Commission hired the Argonne National Laboratory to conduct a study relating to the allegation that these negative public health trends might be due to radioactive releases from the Big Rock Nuclear Plant.[20] The Argonne report clearly states the philosophy underlying the nuclear industry's claims that radiation exposure poses no harm to human health:

When dealing with extremely low levels of radiation, therefore, the issue becomes one of satisfactorily eliminating from consideration, where possible, every other factor known to influence the birth or death trait that is in contention. The need for statistical significance is obvious, and this demands adequate sampling statistics. Unfortunately, the latter is not always attainable in human population studies.

Not only is the industry assumed 'innocent' until proven guilty beyond a shadow of a doubt (every other possible cause must be proven not responsible), but it is admitted that the proof required is 'not always attainable'. In fact, using ordinary vital statistics as collected in the USA and most other places, it can *never* be proven because such records give only gross numbers of occurrences and no background information on the persons who have died. Since we do not know the details about occupation, illnesses, ethnic and socio-economic backgrounds of those who have died, it is impossible to 'satisfactorily eliminate every other factor known to influence the birth or death trait in contention'. Inadequate government record-keeping, plus the AEC philosophy, produces a vicious circle in which the victim is unable to 'prove' damage by the AEC rules.

For example, although we can prove that there are extra leukaemia cases in the vicinity of a nuclear power plant, we have no information on how long these people have lived in the area, their medical or occupational histories, or other exposures, personal lifestyle habits or hereditary vulnerability. The 'cloud of unknowing' is, of course, not necessary, and could easily be dispelled by better record-keeping, but it serves very well to quieten those who try to dissent from the 'expert opinion' of government and industry that 'low level radiation poses no threat to human health' unless the public proves otherwise.

Dr Drake is still seriously questioning nuclear pollution and still devoted to protecting his patients and others from its damaging effects, but his observations were rejected by the 'experts'. The tactics used to diffuse his warnings illustrate well the operation of the US government's nuclear laboratory system. As we have seen, the health problems in the vicinity of this reactor worsened with time in spite of 'expert' assurances. The health study done by Martha Drake was discussed on pp. 129–130.

The response prepared by the Argonne National Laboratory was worded carefully. It did more to confuse than to resolve the issue. For example, in estimating the radiation exposure of the population these experts used 'industry-reported releases, industry-estimated population and meteorology, as modified by AEC staff, and radiation dose models'. No testing of air, soil, water or food was done.

The Argonne experts included in the population they studied the eight northern counties in Michigan, not just Charlevoix County. These added counties were neither downwind nor downstream of the Big Rock Plant hence including them served to mask and dilute the health effects. Although the experts admitted an infant mortality problem in the eight counties they stated, without proof, that it was not due to radiation exposure, and then without follow-up theorised that 'it may reflect maternal age or socio-economic factors known to influence human survival'. They saw their job as exonerating the power plant rather than as helping in a public health problem. They merely raised hypothetical problems: 'illegitimacy may be a problem', 'the median age of reproductive women in upper Michigan may be greater' and then found the ultimate excuse, 'available data are incomplete'. The experts dismissed the increase in cancer rate by noting that the eight northern counties had a lower cancer rate than that of the state as a whole, 'but the differences are narrowing slightly'. Rural rates are always lower than the rates in highly industrialised areas. Drake had pointed out the disturbingly higher rate of increase in cancer but this problem was avoided by the experts, although they rather obscurely noted it as fact.

The Argonne experts omitted to deal with the combined negative trends in five important public health indicators but attempted to explain or dilute away each problem in isolation from the others. Their final clinching argument was that 'present day knowledge of radiation effects would have to be in error by a factor of at least 5,000', if one were to admit that radiation pollution contributed to the health problems. This may seem like a big 'margin of safety', but given Argonne reliance on the industry's own measurement of its pollution and mistakes which the AEC had made in the rate of biological uptake of radioactive cobalt in milk (they underestimated by a factor of 2,300), an error of 5,000 is easily credible.[21] Errors are normally multiplied, so if the pollution level was twice the industry-reported level, the dose from radioactive cobalt ingested was at least 4,600 times higher than the Argonne estimate.[22]

There is evidence that radioactive technicium emanating from a power plant is taken up by humans at 100 to 1,000 times the AEC official estimate; thorium, and presumably plutonium, concentrate in the human liver at a rate 10 times greater than predicted. Current research indicates that the official prediction of radiation-induced cancers is 10 to 20 times too low.[23] Combinations of two or three of these errors could easily lead to the underestimation of health effects by a factor of 5,000 or more.

Hearings and Licensings

Commercial nuclear power posed new and singular problems to society. An elaborate licensing mechanism, safety inspectors, insurance against accidents or sabotage, transportation regulations for nuclear materials passing through densely populated areas, evacuation plans, crisis management, and nuclear waste disposal are only a few dimensions of the problem. Compounding it was the great need felt by the military decision-makers to gain public acceptance of the nuclear industry and all its support services. They made great efforts to divorce the concept of nuclear power from nuclear weapons in the public consciousness. This divorce was also promoted by the commercial nuclear sector, many members of which totally rejected nuclear weaponry. However, they found it difficult to promote a 'safe, clean, efficient' energy source and at the same time explain to the public that evacuation plans are necessary in case of an accident. The unwillingness of private insurance companies to assume liability for catastrophic accidents was a further public relations difficulty. It became necessary to demonstrate 'faith in' nuclear power rather than deal realistically with safety problems, leading to such excesses as the Indian Point and Zion reactors being located in close proximity to the densely populated New York City and Chicago areas. The Diablo Canyon and North Anna plants are located in close proximity to earthquake faults in California and Virginia showing the rash planning tolerated at public expense and risk. Even ordinary common sense which opposes these plants is labelled anti-nuclear bias or irrational fear.

Many citizens in the USA come to confront the reality about nuclear power plants through reactor licensing hearings. A fairly typical example of this awakening process is that of Ilene Youngheim and Carrie Dickerson, both long-term residents of Oklahoma in their sixties, who became citizen intervenors in the licensing process of two nuclear power plants. Mrs Dickerson's husband is a dairy farmer and she has developed her own health food store and nursing home in Claremore, Oklahoma. Mrs

Youngheim's husband is an engineer and she first became aware of the nuclear threat when she learned about pollution of the Oklahoma City drinking water by the Kerr McGee uranium operations. Oklahoma is 'oil company territory' and it has been almost impossible for the citizens there to oppose oil company policies. The oil companies virtually own the nuclear industry, controlling, amongst other aspects, the uranium mines which provide the fuel. In Oklahoma they also control much of the communication media and are a major employer in the Tulsa area.[24,25]

Thirteen of the top twenty corporate holdings of domestic uranium are oil and gas companies. Twelve of these firms control 51 percent of the market. Five of the eight firms controlling US uranium milling capacity are oil companies. These five oil companies control 88 percent of the milling.[26]

The Public Service Commission of Oklahoma proposed to build two large nuclear power plants, Black Fox 1 and 2 in Inola, just outside Tulsa. In this dry, dust-bowl area, they wanted a guaranteed 62.5 million gallons of water per day from the Oologah reservoir and the city of Tulsa sewage plant effluence, for fifty years. Of that amount, 44 million gallons a day would be required to operate the two nuclear plants. The rest of the water was to be used for expansion of another generating station at Oologah. Because of the intense heat generated by the fissioning fuel rods once a nuclear plant is operating, it cannot be denied water. Loss of cooling water results in a major accident. Therefore, in a time of short water supply, it is the local population who would have to do without water or move elsewhere. If the water level in the reactor fell below plant needs, water would have to be brought into the area regardless of distance or cost. The electricity company planned to discharge water effluence from the two large nuclear generators into the Verdigris River, drinking water source of the city of Broken Arrow which has a predominantly American Indian population. This pollution of the water was protested against by Umesh Mathur, former water-quality chief for the Indian Nations Council of Governments. He had little popular support, partly because of general ignorance of the proposal, but also because of general disbelief that any industry would do this to the city's water supply.

The water problems in Oklahoma are compounded by the fact that because of uranium mining and milling, the Arkansas River, west of Tulsa, is already contaminated with radium in excess of federal standards. During the Black Fox hearing the lawyer for the citizen intervenors tried to plead that the Verdigris River should be kept unpolluted since it was the only water supply available to the people of Tulsa other than the already polluted Arkansas River. The three-member Atomic Safety and Licensing Board ruled that this was not a problem since the Arkansas and Verdigris Rivers never 'merge'. The pollution of the Verdigris would not worsen the pollution of the Arkansas. The Board rejected the 'mixing' of the two rivers in the drinking water of the people of Tulsa as being outside the terms of reference of the discussion.

In the *Oklahoma Energy Reporter* of 31 August 1977, published by the Oklahoma Business News Company, the public information official for the Public Service Commission, Joe Bevis, commented with respect to the hearings: 'Frankly, we expected the [public] reaction to be worse. It has been much worse in other parts of the country.' The people of Tulsa apparently had not understood the implications of the hearings and had willingly and passively accepted the planning of the experts.

During the hearing, citizens were allowed, if they wished, to make five-minute presentations of their opinion before each day's proceedings. These comments do not affect the licensing process and need not be considered by the Licensing Board. A Tulsa Catholic priest, Father Bill Sheehan, remarked, 'The little people have been given a hearing, but not heard. The democratic niceties have been fulfilled.' Joe Bevis of the Public Service Commission explained that 'people do not understand the purpose of the public hearing.' He said the hearing was not to determine whether or not Black Fox would be built, but 'if the structure will meet the requirements established by the NRC'.

Mrs Dickerson and Mrs Youngheim took money from their small savings accounts to try to pay legal fees and travel expenses for scientists willing to testify on their behalf to 'humanise' and, if possible, improve the nuclear happening in their neighbourhood. They were affected by a popular caricature which attempts to discredit all citizen intervenors, blaming them for the delays and mounting costs of nuclear generator construction. Citizens are given little recognition for drawing attention to the real problems connected with commercial nuclear technology. The US Nuclear Regulatory Commission has admitted to 123 safety-related generic issues (i.e. affecting all or a large number of nuclear plants) which are 'unresolved'. They also report that about 85 percent of power plant delays are due to manufacturing and construction problems (as determined by the US experience up to May 1978). These are the same problems plaguing reactor reliability and safety which the NRC lists as 'unresolved'. Nevertheless, citizen concern is often treated in the media as deliberate delaying tactics or irrational attempts to harrass those sincerely seeking a solution to the energy crisis. Energy is seen by politicians as needed for military supremacy and economic advantage in a lawless international arena. The water supply problem of Tulsa is, in the minds of those 'managing' national survival, a risk well worth the benefit. Citizen intervenors are therefore portrayed as hysterical women holding up progress in the real (i.e. economic) world.

Another caricature of citizen intervenors portrays them as irrationally concerned about the eradication of some exotic rare species of animal or plant. This situation stems directly from the US Nuclear Licensing Board's declaration that all health and safety problems are 'generic', i.e. to be settled once and for all on the federal level. These generic issues cannot be challenged at a local hearing. Only local differences can be brought up and only modifications of the regulations to accommodate local differences are considered. Hence, citizens unable to address serious health and safety problems sometimes seek to use the legally permitted intervention issues such as endangered species in order to get a hearing.

Oddly, the local differences which may be legally addressed in reactor licensing hearings do not include localised differences in the health of individuals at risk from the plant. Only the 'standard man' is legally considered exposed to the radiation emissions. The usual US Environmental Impact Statements give only the number of people living near the plant. Not even their age distribution or present cancer rates are considered. The Calvert Cliff nuclear power plant near Baltimore, Maryland, is located in close proximity to a school for mentally retarded and handicapped children. Their presence did not have to be considered in the licensing of the plant because they did not reside at the school. Their special vulnerability to radiation apparently made no difference to the planners.[27]

Formation of a 'Generic' Argument

The US Nuclear Regulatory Commission has a generic estimate of the number of deaths risked for the benefit of one year's nuclear generation of electricity, including the entire nuclear fuel cycle required to provide one year's supply of fuel. A citizen intervenor, Dr Chauncey Kepford, noticed that the NRC staff had estimated deaths due to uranium mill and mine tailings needed for one year's fuel only until the closing of the uranium mines, normally worked for twenty-five to thirty years. This is not realistic, of course, because the sealing of the mine has no effect on the mine tailings left outside in the environment. The deaths from radon continue. When the 250,000 years of persistence of radioactive decay products were considered, the number of deaths to future generations caused by one year's supply of nuclear reactor fuel runs into hundreds of thousands. This made nuclear a very undesirable option relative to other technologies. There was no way to 'increase' the number of deaths caused by some alternative technology, such as coal, to make it appear to be a 'worse' choice. However, in the US licensing fiction, all alternative methods of solving the energy need *must* be considered and the 'best' method chosen. The attempt to produce a new yardstick, more favourable to the nuclear option, would be, of course, a corruption of science, ethics and human decency. Yet this is the reality. It serves to underline government's strong preference for the nuclear option on the basis of reasons other than human well-being, or even the energy crisis.

Some proposals for reducing the number of deaths logically attributable to this long-lived toxic radon gas pollution have been: to assume an ice age in which everyone would die anyway; to assume short-term containment of the waste, expecting each successive future generation to restabilise it; and finally, to dismiss the longevity of the problem with a simple insertion of 'deaths per year' without indication of how many years were involved. The latter solution came from Dr Walter H. Jordon of the US NRC in a memo to James R. Yore, Chairman of the Atomic Safety and Licensing Board Panel, on 21 September 1977.

To quote from page 6 of this memo, now available in the NRC public document room:[28]

Let us now consider how to treat properly the radon from tailings piles associated with the uranium mills. Although the mill recovers most of the very long lived uranium from the ore, the thorium 230 which was in radioactive equilibrium with the uranium is returned to the tailings piles. Consequently, the radon 222, a daughter of thorium 230, is continually generated in the tailings pile and will diffuse to the surface of the pile and escape into the atmosphere. Since thorium 230 has a half-life of about 80,000 years, the tailings pile becomes a long lived source of radon. Therefore, the total amount of radon 222 that is emitted by 2.7×10^5 MT of tailings becomes a very large number when integrated over the radioactive life of thorium 230. NUREG-0002 does not include that number, however it does estimate that the amount of radon 222 that would be emitted *each year* from the 1.6×10^9 MT of tailings in piles at the end of this century would be about 420,000 Ci, assuming a 2-foot thick earth cover over the piles. If this number is divided by the 5,875 annual fuel requirements (AFR) which produced the piles, one arrives at a figure of 71 Ci/year. This is numerically near the 74.5 Ci figure of Table 5–3 . . . hence changing that table from 'Ci per AFR' to Ci/year per AFR might be the easiest way out.

The correct estimate of radon gas pollution for one year's fuelling requirement was estimated to be 8,000,000 Ci rather than 74.5 Ci given in the original generic statement – a rather incredible error. Correcting the error by simply inserting the phrase '74.5 Ci per year', with no indication of the number of years involved, is even more incredible.

It should be noted here that only the uranium used in commercial nuclear electrical generation is included in this estimate. Nuclear weapons reactors and other technological and industrial uses of uranium, thorium or radium are also major causes of radon gas pollution.

National Security and Control of Workers

Uranium mine milling and refinery problems, with their long-lived waste, have occurred at Elliott Lake and Port Hope in Canada, and appear to be destined for repetition in the western part of Canada, where rich uranium finds have been made at Cluff and Key Lakes and very recently at Wollaston. Because of communication barriers, little is known about the uranium mining being carried on in Namibia, a country suffering since the Second World War from political domination and exploitation by South Africa. The Australian trade unions have strongly resisted uranium mining and exporting in their country, but have experienced growing governmental retaliation for their stand through direct penalties, threats to their pension funds and other suppressive action. Mr Jim

Donaldson, Assistant National Secretary of the Australian Railways Union, reported at a meeting in Osaka, Japan, on 3 August 1978:

> In Australia, the conservative government has passed legislation which imposes the greatest threat to civil liberties for those who oppose the mining and export of Australian uranium. The Atomic Energy Act and the Defense Projects Protection Act were both passed in the hysteria which flowed from the Cold War. Both were aimed at military situations. Both are now capable of being used against workers who oppose the government's uranium policies. Workers could be liable to 12 months jail or $10,000 fine if they hinder a uranium mining project. The legislation can be used against any Trade Unionist who interferes in any way with the mining, transportation or export of uranium. Hard won industrial rights will be cast aside to let uranium go through.

Jim Donaldson then made an eloquent plea for a new understanding of human rights, one which would allow the worker to protect individual survival needs when threatened by national priorities:

> Human rights should mean the right to a decent job, adequate shelter and care, food for one's children, the right to share in a democratic control of production, the right to determine the character of labor and the nature and disposal of its products.[29]

Australia, like Canada, has refused to allow overseas reprocessing of its uranium, to prevent diversion of the plutonium into weapons. However, in November 1980, Australian policy 'softened', apparently to enhance its marketing position vis-à-vis Canada. There are indications that Canada may soon follow suit, changing its tough nuclear safeguard policies in order to maintain economic competition.

The so-called 'softening' of uranium export policy is in direct opposition to the recommendations of the Ranger Uranium Environmental Inquiry to Australia, in 1976.[30] Given the worldwide decline in nuclear reactor sales since 1974, the efforts to increase uranium sales may indicate a desire to mine and sell uranium before its use becomes obsolete for energy production.

In Canada, the Atomic Energy Control Act was passed in 1946, invoking Section 92–10(c) of the British North American Act, which allowed Parliament to declare a 'local work and undertaking' to be 'for the general advantage of Canada'. The Atomic Energy Act authorises the Canadian Atomic Energy Control Board to control atomic energy materials, equipment and information in the interest of national and international security, to award grants in aid of atomic energy research, and to administer certain aspects of the Nuclear Liability Act. (Statutes of Canada 1946, Chapter 37, now the Atomic Energy Control Act, Revised Statutes of Canada 1970, Chapter A–19.)[31, 32]

The implications for civil liberties of these typical national acts governing nuclear information and material is well documented in the international best seller *The New Tyranny*, by Robert Jungk.[33] Robert Jungk is well known for his earlier popular book on the atomic bomb, *Brighter Than a Thousand Suns*.[34]

The French Experience

COGEMA is a French private energy consortium which manages uranium mining and milling in France. While it is a private industry, its government connections are undeniable. For example, André Gagnadre, a geological engineer with COGEMA for twenty-six years, is the Socialist mayor of Limousin, site of some of France's richest uranium deposits.

Permissible levels for radon gas emission from uranium mining and milling processes have been set 10 times higher in France than was recommended by the International Commission on Radiological Protection. Those ICRP recommendations already favoured the nuclear industries. The French estimate that if ICRP recommendations were followed only 20 percent of the world's uranium resources would be exploitable, both for technical reasons involving containment of the radon gas and economic reasons due to the high cost of health protection that would be necessary.

Uranium ore emits penetrating gamma radiation as well as radioactive radon gas. In October 1978 the Monts d'Ambazac defence committee contracted with experts to measure gamma rays in the uranium mines. The average rate was 6.0 mrad per hour, 12 times higher than the maximum rate of gamma radiation exposure permitted to miners in France. At the observed average rate, a miner would receive 48 mrad per day, 12,000 mrad a year, in addition to the internal exposure from breathing in radon gas. ICRP recommendations are that radiation exposure be kept 'as low as reasonably achievable', and that under no conditions should they exceed 5,000 mrad per year, on average, after age eighteen.[35]

The abandoned mine pit, Brugeaud, in Haute Vienne, has been used as a dumping place for solid radioactive waste from a uranium refinery since 1974. The radioactive debris and contaminated metal are trucked 250 miles from Paris for burial at Haute Vienne, while trucks move in the other direction with ore. Signboards near Haute Vienne read: 'Uranium Service Station' and 'Uranium Restaurant'.

The mining companies promised jobs and prosperity in areas traditionally dependent on grazing and wine, a promise attractive to the young. They also claimed they would benefit the local community, bringing prosperity to a depressed regional economic situation. However, in June 1979 the mayor of Haute Vienne declared before the General City Council: 'Uranium contributes nothing to local communities or the department. On the contrary, it means bigger outlay on roads.' While uranium sold for 400 francs per kilo to Electricité de France at that time, the mining companies were contributing less than 1 franc per kilo to the local community.[35]

The tension between those French farmers who choose to continue their family farms and resist the mining and those who profit from the uranium mining has grown over the years. It is now so great that the subject is seldom discussed even in the home. Many families are divided on the issue. The impact of uranium mining on the air, water and land is gradually becoming known,[36] further aggravating the problem and depriving the farmers of a true choice between farming and mining.

On 24 March 1980, in the village of Grandmont, the tension erupted into a public protest and citizens set up a road block. French police moved in forceably and dismantled the barricade while the villagers stood by helplessly with little or no legal or political power. After such spasmodic outbreaks, the protest again sinks below the surface, a hidden bleeding wound.

Some French farmers compare their lot under the 'New Seigneurs', engineers of the mining company, to the resistance days under German occupation. The 'New Seigneurs', being caught up with the desire for immediate profits and dreaming of a high technology superworld, consider the farmers an impediment to progress. They take for granted the earth, the water and the air, and do not mourn their degradation or worry about their loss.

The concern for national prestige and military strength is in deep conflict with the concern for food and preservation of the life support system in the pysche of the French people.

The Trilateral Commission

With student uprisings in France and the USA in the late 1960s and early 1970s and with disagreements in the West over the Russian invasion of Czechoslovakia in 1968, Bilderbergers turned much of their effort inwards, towards managing dissent within and between Western nations. Their preoccupation with military/security matters annoyed some participants who wished to discuss economic and business strategies. Moreover, the secrecy and lack of written policy statements so characteristic of European diplomacy went counter to American taste. These factors plus the growing influence of Japan on world economy sparked the beginning of a new elitist group: the Trilateral Commission. Some persons belong to both Bilderberg and the Trilateral Commission but the styles of the two groups differ markedly. Bilderbergers include heads of state, top government officials and royalty. They meet on a regular basis maintaining strict secrecy and confidentiality. Personal relations between influential elites account for their effectiveness. The Trilateral Commission is explicitly organised as a pressure group. It produces public documents and lobbies publicly for its point of view. Members of governments are excluded from the Trilateral Commission during their term in office.

Zbigniew Brzezinski's book *Between Two Ages: America's Role in the Technetronic Era*,[37] called for the formation of a community of developed nations composed of the United States, Europe and Japan. He characterised these nations as the 'most vital regions of the globe', in the vanguard of 'scientific and technological' progress. He envisioned this unity eventually spreading to include Eastern bloc nations. This coming together of developed nations was deliberately designed to avoid the United Nations, which had grown to include the seventy nations which gained independence after Gandhi's dramatic non-violent restoration of home rule in India. Brzezinski found the UN a clumsy tool for developing 'effective programmes' because of the Cold War and North–South tensions.

As the global situation deteriorated, Brzezinski began the Tripartite Studies at the Washington-based Brookings Institute in the USA. Scholars from the Tripartite Studies met with those from the Japanese Economic Research Centre and the European

Community Institute of University Studies to brain-storm solutions for common economic problems.

David Rockefeller joined with Brzezinski to present the idea of a Trilateral Commission at the 1972 Bilderberger meeting. The idea was favourably received and in July 1972 a planning group of seventeen people met at the Rockefeller home in Tarrytown, New York, to form the Trilateral Commission. They included political, financial and educational leaders in the United States, Europe and Japan.

The founding session of the Trilateral Commission was held in July 1973 and the first Executive Committee meeting was held in Tokyo on 22–23 October 1973. Zbigniew Brzezinski was the Commission Director.

In the midst of this attempt of the so-called 'First World' to co-ordinate its plans for a global economy several crises intruded on to the global stage: the OPEC oil crisis, the defeat of the USA in Vietnam and the 'Third World' call for a New International Economic Order. Both Kissinger, who served under President Ford, and Jimmy Carter, were Trilateral Commission members, although they were also members of the Republican and Democratic parties respectively – traditional rivals in US politics. Trilateral foreign policy blurred political lines within nations.

The influence of the Trilateral Commission peaked in 1976–77 with many members becoming political leaders in their own governments. The most notable Trilateral 'success' was Jimmy Carter's presidency in the United States.

In 1975 the Trilateral Commission produced its controversial book *The Crisis of Democracy*[38] and in 1977 its taskforce report: *Toward a Renovated International System.*[39] The first world elitist group clarified its goals and legitimised operating outside the United Nations. It deliberately excluded third world countries and the working classes and minorities within the member countries. Members described themselves as an unofficial grouping of concerned citizens trying to suggest ways of bettering life on planet earth.

The OPEC Crisis

In 1960 five countries – Venezuela, Saudi Arabia, Iraq, Iran and Kuwait – formed a producers' association to protect their oil export earnings. They were soon joined by Libya, the United Arab

Empire, Algeria, Nigeria, Qatar, Indonesia, Equador and Gabon. Together these nations produce 28 million barrels of crude oil a day, about 54 percent of world consumption. Saudi Arabia and the other five Arab countries produce 75 percent of the OPEC oil.[40]

In 1970 Saudi Arabian crude oil sold for $1.80 a barrel. In 1973 the price was abruptly raised to $5.12 a barrel. Though most business people admitted that this was a fair price, the suddenness of the change caused severe economic dislocation and distress.

It has been suggested by CBS Television in the USA, on its investigative news programme *60-Minutes*, that the OPEC cartel was originally the suggestion of Henry Kissinger. The US Congress was averse to giving weapons to Iran even though the reigning Shah was friendly towards the USA at the time. From a US military point of view, intelligence-gathering from Iranian bases bordering on the Soviet Union was important. By increasing oil prices Iran was placed in the position of economic ability to purchase US arms. The US public paid for these arms at the petrol pump rather than through the democratic processes by which Congress would control such 'aid'. Officially this scenario has been neither confirmed nor denied. It cannot be denied, however, that in fact much of the 'oil money' has been spent on US arms, high technology transfer and elegant living for a few elite decision-makers. OPEC oil revenue was $67 billion (US) for the ten years prior to 1973, and $1,600 billion (US) for the ten years after 1973.

The non-oil-producing developing countries have been seriously hurt by the increased oil prices which by 1981 had risen to $34(US) a barrel. Non-OPEC oil producers quickly moved to 'world prices' so as not to be left out of the profit. Some developed countries, such as Canada, maintained a lower price for domestic oil purchases but increased their price on the export market. No country extended special prices to the non-oil-producing developing countries.

As the non-oil-producing countries were forced to cut back on economic development programmes and reduce domestic programmes for meeting the basic food, shelter and clothing needs of their people, their foreign debt soared. The cumulative debt since 1973 is more than $220 billion (US). The exorbitant oil bill for these countries reached $18 billion (US) in 1978 and $28 billion (US) in 1979.

The plight of these countries is acute, with much of their food production and economic strategy based on oil which provides transportation, fertiliser and other petrochemical products. Iraq has proposed selling oil to these countries at a lower rate but this has been continuously voted down by the other OPEC countries.

The oil crisis was also used by some developed countries as a strategy to 'sell' nuclear power. Nuclear power was the 'keystone' of energy independence schemes although it never had the ability to displace dependence on oil. Nuclear power produces electricity, not petrol, fertilisers and the myriad petrochemical products on which the developed world depends.

The Environmental Liaison Centre in Nairobi, Kenya, formed by 100 Non-governmental Organisations at the United Nations to co-ordinate programmes with the UN Environment Programme and the UN Centre for Human Settlements, has estimated that the single greatest waste of energy in the United States and most other developed countries is the automobile. In the USA, 154 million automobiles, about one for every two Americans, consume 60 percent of the world's petrol. The synthetic rubber for tyres and floormats, plastic for dashboards and steering wheels, polyvinyl for seat covers and roof-linings, and other petroleum products, plus highway construction, make up 20 percent of US energy use.[41]

The energy consumption of the military sector is seldom mentioned but must also contribute heavily to oil consumption for transportation and petrochemical products in the developed world.

Nuclear generation of electricity in no way lessens these dependencies on oil.

Most energy and economic planners admit that small-sized fuel-efficient cars would alleviate the oil crisis more quickly and effectively than capital and energy-intensive nuclear power plants. However, the media generally confuses 'energy' and 'electricity' and the public is often made to think that nuclear generation of electricity would be the only way to prevent 'freezing in the dark' when the oil runs out. However, even if the public believes, however fallaciously, that nuclear power is the only substitute for oil, it could still choose to reduce its dependency on either.

Contrary to the expectations of major oil companies, which also control the sale of nuclear technology, the 1973 oil crisis did not cause popular acceptance of nuclear power in developed countries. Instead, deliberate popular efforts to reduce electrical consumption began and a rapid decline in nuclear reactor purchases followed.

In the USA, the electricity companies, faced with unsolved nuclear safety issues, a growing nuclear waste problem and the onset of decommissioning costs and realities, found the drop in electrical demand a fine persuasive reason for cancelling reactor orders. In addition to the engineering and economic problems, the

American public, initially supportive of the nuclear industry, was growing disenchanted with the reality. Citizen groups becoming involved in the licensing process learned more and more about the danger and unsolved technological problems. They also learned to use the licensing system to gain added safety features for their local nuclear plant and to delay construction of unwanted plants. There were a large number of plant cancellations in 1974 and a sharp reduction in new plant sales.[42] New domestic sales in the USA reached zero in 1978, one year before the Three Mile Island accident. Although the nuclear reactor companies in the USA are still filling a back-log of reactor orders, domestic sales have been zero for four years. Unless these companies are able to obtain sales in the developing countries, they will be forced to declare bankruptcy or phase out their nuclear divisions.

The grass-roots anti-nuclear forces in France, West Germany and Sweden managed to stall the ambitious European plans to meet the 1973 OPEC crisis. With dwindling domestic markets, France's Framatome and West Germany's Kraftwerk Union stepped up competition with the United States' Westinghouse and General Electric corporations for third world nuclear reactor sales. Even international monetary policies were changed to aid the failing nuclear industries as domestic markets declined. For example, between 1974 and 1976 the Export-Import Bank authorised $2.4 billion (US) for nuclear power in the developing world, in sharp contrast to the $0.17 billion (US) average per year spent for this purpose prior to 1973. As of 30 June 1978 nuclear power plants became the largest single item financed by the Export-Import Bank. Developing countries, unable to use a conservation strategy and limited by the exporting countries' financial lending policies, were forced into purchasing the technologies unwanted in Europe and North America.

In a similar move, the EEC increased the Euratom loan scheme from £335 millions to £670 millions to boost investment in nuclear projects.[43] Unlike the Export-Import Bank, the EEC can override financial or other advice for political reasons.

In October 1978 Shah Mohammed Reza Pahlavi declared that his plan to purchase twenty nuclear power plants for Iran might have been too ambitious. On 27 January 1979, Prime Minister Shahpour Bakhtiar announced Iran's pull-back from nuclear reactor orders from France and West Germany, as well as Iran's probable withdrawal from participation with the Eurodif uranium enrichment plant. The plant construction also involved France, Italy, Belgium,

Switzerland, Sweden and Spain. In addition to these nuclear projects, the Iranians had agreed to assume 25 percent of the cost, together with France, Italy and Niger, to explore for uranium in Niger. With the fall of the Shah, all this changed, and Iranian oil money was lost to the nuclear projects.

Just as the Export-Import Bank uses US money to purchase US nuclear reactors for the developing world, so the French nuclear industry relies on governmental financial assistance. The Iran contract was officially insured at a 90 percent rate by the government. The French government is responsible for most of the country's nuclear spending, using public revenues to shore up the failing nuclear industry that it perceives as necessary for national economic and military survival.[44] Although Western countries extol the free market, government policies protect nuclear technologies from such a market in extraordinary financial, informational and legal ways.[45]

The United States was the chief exporter of nuclear technology in the late 1960s. Westinghouse and General Electric had a complete monopoly of the non-US Western bloc market in 1966.[46] But, by 1971, the US share was reduced to 40 percent of the market, due to aggressive marketing by West German and French subsidiaries of Westinghouse. By 1976, US firms were talking about retaining 35 percent of the international market but this proved to be too optimistic a goal. Between 1976 and 1978 General Electric and Westinghouse were unable to win any foreign orders for nuclear reactors, and Germany's Kraftwerk Union and France's Framatome became the leading exporters of nuclear technology.[47]

In June 1975 West Germany and Brazil sealed a $5 billion (US) contract for the construction of two to eight large nuclear reactors. The contract included joint participation in the entire nuclear fuel cycle: uranium mining, enrichment, reactor manufacturing and the building of a reprocessing facility.[48]

However, it seems that Brazil's first nuclear reactor, Angra I, will never achieve more than half its capacity. Angra 2, the first of eight projected nuclear plants, is behind schedule. Most observers believe that the programme would be dead were it not for military interest.[49]

The inclusion of enrichment and reprocessing facilities, which makes large-scale production of nuclear weapons possible, is called a 'sweetener' in nuclear industry jargon. Both Framatome and Kraftwerk Union used sweeteners to attract developing countries away from US sales and US influence.

Under President Carter's policy of non-proliferation of nuclear weapon capability, the USA refused to provide enriched fuel for the Koeberg nuclear reactor which Framatome built for South Africa. The French reactor company was unable to keep its promise to supply nuclear fuel until 1984 when South Africa is expected to be able to operate its own large-scale uranium enrichment installation, also built with the help of Framatome. South Africa, like France, is not a party to the Nuclear Non-Proliferation Treaty, and it probably already has a store of nuclear weapons. The *New York Times* of 14 November 1981 reported that South Africa had been successful in obtaining enriched fuel in spite of the US refusal.

Some experts said the source could have been Italy or West Germany, both of which have scaled back ambitious civil nuclear power programs because of political opposition. But there was also speculation here [Paris] that the fuel might have originated in China, which has long been suspected of having secret nuclear contacts with South Africa, and that the material had been channeled through Swiss or other European brokers.[50]

As the number of 'players' on the international scene increases, the amount of national control of nuclear proliferation decreases. There is now no way to curb nuclear weapon production globally without an effective international security system capable of satisfying legitimate national security needs and dealing effectively with aggressive national governments.

The Canadian Experience

Canada has, in general, been a less aggressive nuclear competitor than Europe and the United States. Canada has stricter policies for control of weapon-grade plutonium, stemming from its negative experience with India. In 1974 India exploded a plutonium bomb, joining the 'nuclear weapon club' nations. Its nuclear physicists had used the small reprocessing plant at Bhabha Atomic Research Centre outside Bombay to extract plutonium from the used fuel rods from its CANDU research reactor near Delhi. The CANDU reactor, which can be refuelled while operating, is especially vulnerable to theft or diversion of weapon materials. The Canadian public reacted strongly against Canada's involvement in nuclear weapon proliferation.

Canada was the only nation that was a partner to the Manhattan Project (which produced the Hiroshima and Nagasaki bombs) and had not produced nuclear weapons after the Second World War. Although Canada has allowed four nuclear weapon bases on its soil, the weapons have been handled by American rather than Canadian personnel.

Canada devised its own stricter safeguards for nuclear materials produced in CANDU reactors sold abroad after the India incident, thus reducing Canada's economic competitiveness with US, West German and French firms. A second problem with selling the CANDU is that its initial cost is about 25 percent higher than the cost of a light water reactor. This price difference is offset by lower fuel costs and greater efficiency after the reactor is in operation, but the initial capital outlay can be a deterrent.

Following the tightening up of CANDU safeguards sales plummeted. There were no sales between 1978 and 1980[51] and only a few countries – Korea, Mexico, Egypt and Romania – have shown an interest in future purchases. While the CANDU industry is capable of building fifty to sixty reactors during the 1980s, the most optimistic forecast indicates a maximum of seven sales.

The Canadian government, pressed with inflation and unemployment, appears to be regretting its $3 billion subsidy of the nuclear industry. It has purchased private heavy water plants in Nova Scotia, engaged in 'cost-sharing' on the construction of the first nuclear plants in Ontario, Quebec and New Brunswick, cancelled a $850 million debt of Atomic Energy of Canada Ltd when Ontario Hydro reduced its over-ambitious nuclear programme, and provided other industry 'aids'.[52] These government subsidies artificially reduce Canadian nuclear electricity costs, with the rate payers paying in taxes rather than domestic bills, but many politicians are finding this give-away programme unpalatable with the increasing pressures of inflation.

On 25 April 1981 a report prepared by a Montreal consulting firm and sponsored by four Canadian private-sector nuclear suppliers – CAE Electronics Ltd, Canatom Inc, Dominion Bridge-Sulzer Inc, Valan Inc – and by the Royal Bank of Canada, called for transfer of the control of CANDU sales to private manufacturers. The firm called for 'diffusion of CANDU technology by all feasible means' and 'reorientation of Canada's nuclear safeguards policy so it is more in line with its major competitors in the nuclear industry'. The proposal suggests raising CANDU sales from 5 percent of the world market to 20 percent by the year 2000.

Most speculators see this as a last attempt to save a dying industry. Between 1964 and 1978 about twenty-four CANDU reactors were sold in Canada and abroad. Sales stopped in 1978 and in June 1981 the Ottawa meeting of the Canadian Nuclear Society clearly confirmed the serious crisis. An industry which employed 31,000 workers in 1977 was reduced to 16,000 in 1981, with forecasts of a further reduction to 8,000 by 1983.[53]

Unlike the United States, France and West Germany, the Canadian reactor failure could not be assigned to anti-nuclear protests. Canadian nuclear power projects have been cautious. The domestic Canadian nuclear power plants were built quite recently with deliberate 'retirement' of non-nuclear-operating generators in order to use nuclear technology. The Canadian electricity surplus is expected to continue into the mid 1990s. It is still possible for Canada to meet its energy needs without using the nuclear plants already built. A few individuals and organisations such as the Canadian Coalition for Nuclear Responsibility, Energy Probe and Voice of Women have tried to make Canadians aware of the complexity and 'dangers concomitant with nuclear technology, but generally the public is poorly informed and uninvolved. There is also a strong element of Canadian disassociation with nuclear weapons, a course deliberately pursued after the Second World War, and a pride in the Canadian technology which developed the CANDU reactor. These attitudes lead the public to reject arguments against nuclear power based on the desirability of military nuclear potential or disenchantment with non-Canadian nuclear technology. Canadians have no desire to develop nuclear weapons. Canadian technology is different from the technology which has failed in Europe and the USA.

The Canadian government fully co-operates with the nuclear military policies of NATO (the North Atlantic Treaty Organisation) and NORAD (the North American Air Defence Command), and was willing to allow three or four nuclear weapon bases (under the control of US military personnel) on its territory. Only one Canadian military base, in Comox, British Columbia, has nuclear warheads under Canadian control and the government has promised to relinquish control to the USA at this base also in the near future. Although Canada sits strategically between the United States and the Soviet Union, and although Canadian scientists helped to develop the atomic bomb, Canadians generally exclude themselves from the picture of global nuclear war as if it were not their business but a US problem.

The shift of CANDU sales to aggressive private-sector marketing in the developing world may well arouse Canadian public opinion. Yet the nuclear industry sees this strategy as a solution to its economic woes. Atomic Energy of Canada Ltd (AECL), a Crown corporation, believes it lost a reactor sale in Argentina in 1980 to Kraftwerk in West Germany and Sulzer Bros Ltd in Switzerland because of the stringent Canadian nuclear safeguards. Imitation of or collaboration with West Germany may provide a way to avoid government safeguards.

Prior to the economic summit meeting held in Ottawa in July 1981, West German Chancellor Helmut Schmidt and Prime Minister Pierre Trudeau of Canada announced a decision to work together to sell nuclear reactors in 'a different international market'. Gordon Osbaldeston, secretary of the Canadian Ministry of State for Economic Development, and special projects director Harry Swain, were sent to tour West German nuclear centres to explore opportunities for co-operation. Mr Osbaldeston and Mr Swain called on the top management of Kraftwerk Union. The West German government has claimed that it is not responsible for the reactor deal with Argentina since it was an agreement reached by private commercial interests.

Canadian officials do not speculate publicly on the chances of softening Canada's nuclear safeguards policy. The pressure for sales of CANDU are great with the purchase in July 1981 of a reactor by Romania being the first CANDU sale in four years. Marketing is now focussed on Mexico and on the Province of Alberta. Mexico is searching for a provider of nuclear equipment for the generation of about 2,400 MWe power. Alberta has been told by AECL that nuclear power is needed to separate oil from oil sands.[54] Both projects seem unnecessary since Mexico and Alberta contain sufficient local natural gas and oil to produce either the electricity or the steam necessary for the oil separation process. Vern Larson of ESSO Resources Canada Ltd claims that nuclear power 'probably wouldn't be used in Alberta until the year 2000'. Floating these plans is an obvious 'testing of the water' of public opinion in Canada. If there is no serious outcry from the public, nuclear reactor marketing in Canada will probably move out of the public domain and into the boardrooms shrouded in commercial corporate secrecy.

Canada offers an excellent example of a country which deliberately renounced its nuclear weapon 'advantage' after the Second World War and which just as deliberately developed its own nuclear technology for the generation of electricity. The nuclear experts, some of whom participated in the Second World War Manhattan Project, are well supported and honoured in Canada in terms of both financial rewards and academic prestige. The main Canadian research centres are located at Chalk River, Ontario and Whiteshell, Manitoba. They maintain relationships with the university system providing research grants and jobs for graduates.

Nuclear-related decisions in Canada are relegated to the Atomic Energy Control Board (AECB), a small group of elite decision-makers whose task is to promote the peaceful uses of nuclear power. This board has the power to set limits on information flow to the public, thus in a real sense controlling public understanding and participation in nuclear decisions. For example, the Pickering nuclear power plant, completed in 1970, predates the Canadian safety requirement for new nuclear power reactors. New regulations require two back-up emergency systems. About 2 million residents of metropolitan Toronto are at risk from this plant. When the Pickering town council voted to request the AECB to conduct a public hearing into the safety of the Pickering reactor, the AECB refused this request. The AECB has also refused to order the Pickering plant to restrict operation to 70 percent of capacity, a move which would partially compensate for the lack of back-up safety protection and reduce radiation contamination in case of an accident.

Public hearings up to this point had been one of the main vehicles for informing the Canadian public about nuclear decisions but the process has been costly for the nuclear industry.

The Royal Commission on uranium mining in British Columbia was halted in 1979 before completing its hearings and a seven-year moratorium on uranium mining in the province was declared.[55] The moratorium declaration prevented presentations on the ethics of uranium mining with respect to rights of future generations and testimony on the effects of chronic exposure to low level radiation. The failure of Eldorado Nuclear to convince the people of Port Granby, Ontario or Warman, Saskatchewan at public hearings to allow it to build a uranium refinery resulted in a decision to locate two new refineries at Blind River, Ontario and Port Hope, Ontario each incorporating a part of the original plan for one refinery complex, without permitting further public inquiry. We might wonder how long the Canadian nuclear industry can maintain its control of public opinion through secrecy and non-debate within Canada; and how credible Canadian technology would be internationally if public debate was allowed and it lost domestic support.

Other Canadian public hearings have been criticised by the Canadian people as shams. The Porter Commission on nuclear power in Ontario was supposedly deliberating on whether or not the province should depend on uranium as a source of electricity. While the public inquiry was in progress the provincial government was closing a deal with the mining companies for uranium fuel deliveries well into the twenty-first century. The Bayda Commission of Inquiry into Uranium Mining at Cluff Lake, Saskatchewan was perceived by the people as so biased in favour of mining that they boycotted the Key Lake Inquiry which followed it. The elaborate radiation protection plan presented at the Cluff Lake Inquiry by the mining company was never completely implemented. Its designer, Margaret Swanson, eventually resigned from the company in frustration.[56] The testimony was apparently a hoax put on by the company to impress the Hearing Board.[57]

The French influence on Canadian nuclear policy can be seen in the Amok management of the rich uranium deposits in Saskatchewan. The managing director at Cluff Lake is Marcel Tabouret, who also serves as France's vice-consul in the Saskatoon Region and is adviser to the French government on trade with western Canada. Mr Tabouret has been a vocal critic of Canadian concern for participation in nuclear decision-making and Canadian efforts to prevent uranium diversion for use for nuclear weapons. At a meeting of the Canadian Gas Association in Saskatchewan he was reported to have said: 'Let the government go back and make a few decisions, popular or unpopular, on its own behalf, and not continue to duck the issues by forming boards of inquiry which only provide forums for professional biomass disturbers.'[58]

The Canadian nuclear industry has the record of being strictly 'peaceful' and yet it has adopted many of the elitist decision-making and secrecy strategies used by the military nuclear industry to survive public scrutiny. Even with strong government financial and political backing, the Canadian nuclear establishment has not been able to develop a viable commercial domestic or foreign market for CANDU reactors. The existence of CANDU tends to lend legitimacy to Canada's role in providing uranium mining and refining and nuclear research for global nuclear endeavours. Its future may involve further concessions to global economic, social and military pressures to 'conform' and contribute more openly to the nuclear military establishment.

Dissenting Voices

In spite of the handicap posed by the inadequate information routinely gathered on public health in the USA, several attempts have been made by scientists to demonstrate an increase in birth defects, neonatal deaths or cancers due to nuclear weapon testing or contamination near nuclear installations, both commercial and military.

An outspoken critic of above-ground nuclear weapon tests was Dr Ernest Sternglass, a physicist with Westinghouse in the USA. He estimated that 400,000 infants had died because of the testing.[59] For Sternglass the death of a child or the causing of a genetic disease was not an abstract happening. His own daughter died aged three from Tay Sacs disease in 1947. The Atomic Energy Commission asked the scientists at the Lawrence Livermore Laboratory to review Sternglass's charge, hoping they would discredit the esti-mate. Dr Arthur Tamplin did reduce the estimate of infant deaths (on paper) to 4,000. His department Chairperson, Dr John Gofman, defended him against the AEC which found the number still too high for its public image.[60]

Dr Sternglass commented to me that even though Westinghouse was in the nuclear reactor business at that time, he was not pressured at work to be silent about the nuclear weapon testing. On the other hand, Drs Tamplin and Gofman eventually were forced out of the Lawrence Livermore Laboratory apparently for their refusal to change their estimates of infant deaths due to the weapon testing, but as defence workers they were able more freely to criticise the commercial nuclear industry. Dr Gofman has recently declared (1983) that he thinks now that Sternglass's number, 400,000 infant deaths, was more nearly correct than Tamplin's lower estimate.

After Dr Sternglass had accepted an invitation to become head of the Laboratory for Radiological Physics at the University of Pittsburgh School of Public Health, he discovered through the writings of Gofman that nuclear power plants were also emitting radioactive material. He began studying the vital statistics for the area around the Shippingport nuclear reactor, the first US commercial plant, and found an increase in infant mortality. His allegation that babies were dying in the vicinity of nuclear power plants provoked the first official fact-finding commission on the health effects of nuclear power in the USA, called by the Governor of Pennsylvania.[61]

The Shippingport Fact Finding Commission reported in 1974 that neither the vital statistic information on health effects, nor the monitoring information on radioactive releases from the nuclear plant, were adequate for assessing Dr Sternglass's claims. They could not prove him wrong. Neither could they prove him right. Their recommendations for improvement in the quality of information have yet to be fully implemented in Pennsylvania or elsewhere in the USA. In view of the Three Mile Island (TMI) accident, the Pennsylvania failure to keep adequate health records, pointed out by the Governor's Commission in 1974 and never corrected, is tragic. With poor records on pre-accident health parameters for people living near TMI, it will be difficult to document any change since the accident.

In 1960 the Department of Health in Washington County, Utah reported alarming increases in leukaemia and other cancers among people exposed to nuclear test fallout. These findings were suppressed and 'experts' assured the people that radiation levels were much too low to cause health effects. The population was too small to yield statistically significant results using standard methodology designed for large populations. Few public health officials or county health departments had the mathematical background, the record co-ordination capacity, the funds or the staff to attempt more sophisticated analyses. There was, moreover, a constant subtle pressure in the USA not to interfere with the military nuclear programme. To do so was to be labelled a communist. The general public assumed the problem was scientific uncertainty, failing to understand the political underpinnings.

The health problem was not eliminated as easily as were the written reports of health problems. In February 1979 the *New England Journal of Medicine* published a research paper on leukaemia mortality among Utah children in the high fallout area: 32 deaths occurred when 13.1 would have been expected, and even in the low fallout area, 152 deaths occurred when 118.9 would have been expected.[62] Because of population mobility and occupational exposures to other carcinogens, adult health effects are more difficult to document. An excellent analysis using Mormon Church records on its members, even those who had moved from Utah, was finally published by Carl Johnson in 1984.[63]

The Governor of Utah has filed a suit against the US government for the excess cancer deaths in Utah. Nearly 1,200 individual lawsuits have also been filed. These suits are primarily for cancer deaths. Many other human tragedies such as spontaneous abortions, thyroid disfunction, aplastic anaemia and hormonal diseases related to radionuclide pollution of the living space have yet to be recognised and documented, much less brought to litigation. It takes a long time for the public to discover what is happening to themselves and to their living space.

Nuclear Proliferation

When observed only from its public face, the growth of nuclear power has been phenomenal. Many futurists were enchanted by the dream of unlimited energy 'too cheap to meter' and by extravagant schemes to use such energy in the third world. Horror at the Hiroshima and Nagasaki bombing gave way to hope and the nuclear technicians were seen as those experts who could guide humankind into a safe future.

This hope for scientists to provide technological solutions to human problems prevailed at the highest level of government. In an address before the UN General Assembly on 29 September 1960, the British Prime Minister, Harold Macmillan, stated that the Geneva Conference on disarmament:

> started by reaching agreement between the three Powers concerned that, in the initial stage, our representatives who were meeting were not to play a political role. They were to study the problem from the scientific and objective point of view. They were scientists, not diplomats or politicians, and they were to report whether, in their view, effective measures could technically be devised by which, if an agreement were made to stop nuclear tests, the agreement could be enforced.[64]

This expectation was reinforced by both the new intelligence-gathering capability, technologically made possible by advances in communication, and the experience of commonality between Eastern and Western bloc scientists meeting on committees of the UN or attending Pugwash meetings. Scientists often found themselves in greater accord with one another than with their respective governments.

Macmillan proposed that scientists devise a surveillance system against surprise attack, so that fear and distrust between nations could be put to rest forever.

Science was seen as intervening to protect nation from nation.

There seemed to be no thought of scientists actually causing national destruction through ecological disaster. Undermining the relationship between a nation and its life-supporting food, water and air through undisciplined and distorted scientific development was not even considered a problem. If an adverse environmental or human health problem was perceived, science was invoked to 'cure' it rather than being asked to cease the activities which were causing the problems. Prime examples of unsuccessful attempts at technological solutions are the 'war on cancer' and genetic engineering.

The USA, forbidden by the US Atomic Energy Act to transfer nuclear weapons to any other country, or to transfer any information which might assist any other country not already having substantial nuclear capacity to design or manufacture a nuclear weapon, began to assist friendly nations to acquire 'peaceful atom' technology. The USA also established weapon bases in friendly countries where Americans would themselves 'manage' the nuclear bombs. Local scientists trained in commercial nuclear technology would provide a trained back-up personnel in case of nuclear war.

The co-operation of the United States and Israel in a small research reactor project near Rehoboth in Israel attracted international attention, requiring reassuring comments by the US Department of State that only peaceful uses were intended. The US entered into special agreements on sharing nuclear power technology with the European Atomic Energy Community (EURATOM), the International Atomic Energy Agency (IAEA), and with thirty-nine separate nations and West Germany. Besides the controversial reactor project in Israel, the USA gave reactor grants to Rawalpindi, Pakistan; Istanbul, Turkey; and Ljubljana, Yugoslavia.

The chart on pp. 236–237 shows the growth of the commercial nuclear industry.[65] Reactors smaller than 30 MWe (megawatt electric) output or retired from service by January 1984 were omitted and no military reactors or nuclear installations for support services were included. Since the health impact of radiation would be expected to become noticeable twenty to thirty years after exposure, the dramatic increase in generators may well precede a dramatic increase in cancer and birth defects in the 1980s and 1990s. Unfortunately such an epidemic of radiation-related tragedies would continue for decades after all nuclear activities cease.

As was evident through the French experience and later through happenings in India, South Africa and Israel, the sharing of 'peaceful' nuclear technology is sufficient to provide the scientific know-how and fissionable material needed to produce a nuclear bomb. In November 1979, the Washington-based Council of Hemispheric Affairs said that Brazil and Argentina had been working on nuclear bombs for the previous two years, with the aid of South Africa. South Africa is believed to have received its weapon capability from Israel. The Brazilian source of weapons-grade plutonium is nuclear power generators built for them under a $10 billion contract with Kraftwerk Union, West Germany. Kraftwerk Union began as a foreign subsidiary of US Westinghouse. Argentina has recently signed a reactor contract with the same West German company. Both Argentina and Brazil deny that they are developing nuclear weapons, but their failure to sign the Treaty of Tlatelolco, prohibiting nuclear weapons in Latin America, and the scale of their technology continue to arouse international concern.

In order, apparently, to offset and minimise weapon proliferation, the USA

> found it necessary and desirable to conclude with NATO in the interests of collective self-defense . . . arrangements under which the defense forces of the North Atlantic Treaty Organisation have atomic weapons for their protection. American, as well as Allied forces, [have] in their possession vehicles capable of carrying such weapons. The weapons themselves are maintained in a stockpile system under the custody of the United States in accordance with existing United States policy and law.[66]

Thus the USA became inextricably involved in foreign weapon bases and in promoting global reliance on nuclear power so that allies could become familiar with nuclear technology and assist military action if necessary.

The US government, which had provided all the money for the basic research needed to launch the American nuclear reactor industry, also operated the uranium enrichment installations and provided the Price-Anderson Act to assume public liability in case of a nuclear reactor accident. The US public is not covered financially for loss in a nuclear accident, receiving only a few cents on a dollar lost, but the nuclear industry is protected to some extent from its losses in an accident. The Organisation of European Economic Cooperation (OEEC) and EURATOM assumed public liability for the EEC countries. Later, after the waste problem was realised, governments even assumed responsibility for radioactive waste disposal.

Global Distribution of Commercial Nuclear Generators*

Number of units and capacity in megawatts of electricity generated

Country	1958–64	1965–69	1970–74	1975–79	1980–84	1985–
Argentina			1		1	1
MWe			335		600	692
Belgium				3	2	2
MWe				1,650	800	2,000
Brazil					1	2
MWe					626	2,490
Canada		1	5	4	6	8
MWe		206	2,310	3,100	3,576	6,290
Czechoslovakia				1	3	6
MWe				440	1,320	2,640
Egypt						2
MWe						1,800
Finland				2	2	
MWe				1,105	1,105	
France	2	5	3*	5	28	19†
MWe	80	1,450	1,288	4,500	23,040	23,070
Germany (DR)		1	1	2	1	2
MWe		70	440	880	440	880
Germany (FR)	1	1	2	6	4	15*
MWe	52	328	1,270	6,067	4,593	16,972
Hungary					1	3
MWe					440	1,320
India		2	1		2	5
MWe		400	202		422	1,100
Iran						2
MWe						2,400
Iraq						1
MWe						900

* This chart is derived from information published in *Nuclear News*, February 1984.

† One is a liquid metal fast breeder reactor.

Italy	1	1			1	3
MWe	150	260			875	2,004
Japan		1	7	14	6	14
MWe		159	3,548	10,759	4,547	12,350
Korea				1	2	6
MWe				556	1,234	5,476
Libya						1
MWe						300
Luxembourg						1
MWe						1,250
Mexico						2
MWe						1,308
Netherlands		1	1			
MWe		50	445			
Pakistan			1			
MWe			125			
Philippines						1
MWe						620
Poland						2
MWe						880
Romania						6
MWe						4,840
South Africa					2	
MWe					1,844	
Spain		1	2		5	10
MWe		153	920		4,564	9,731
Sweden			2	4	4	2
MWe			1,035	2,690	3,600	2,100
Switzerland		1	2	1	1	2
MWe		350	670	920	942	2,065
Taiwan				2	3	1
MWe				1,208	2,809	907
Turkey						1
MWe						440
United Kingdom	4	11	2	5	6	4
MWe	1,456	2,812	1,180	2,750	3,700	2,640

Yugoslavia					1	
MWe					615	
TOTAL	29	34	72	80	131	182
MWe	3,093	9,637	42,627	64,181	110,079	173,079
United States	3	6	35	24	27	44
MWe	445	3,149	25,384	22,156	29,127	48,614
USSR	8	2	7	6*	22	14
MWe	910	250	3,475	5,400	19,260	15,000

* One is a liquid metal fast breeder reactor.

Summary:

USA	units	139	26%
	MWe	128,507	32%
USSR	units	59	11%
	MWe	44,375	11%
UK	units	42	8%
	MWe	14,538	11%
Japan	units	42	8%
	MWe	31,363	8%
France	units	62	12%
	MWe	53,428	13%
West Germany	units	29	6%
	MWe	29,282	7%
Canada	units	24	5%
	MWe	15,482	4%

World units   528
World MWe   403,510

With the concept of the peaceful atom accepted by the public and financial liability for the industry somewhat guaranteed, the commercial nuclear industry began to expand in the late 1960s. However, government involvement was not generally understood, and most people assumed they were financially protected. Many nuclear physicists and engineers were trained in and oriented towards commercial technology only and they mentally dissociated themselves from the nuclear bomb, never having co-operated with that aspect of the nuclear industry. Dissociation from the weapon industry was possible for those involved in research, radiation protection activities, mining and milling,, transportation, enrichment, fabrication, reprocessing, waste disposal, training nuclear engineers or physicists and producing educational or public relations material. Individuals were able to shut out of their consciousness the atomic weapon world and wholeheartedly promote the 'peaceful atom'. The work of their minds and hands, however, could and would be used for both purposes either directly or indirectly. For example, much of the extravagant 'need' for electricity in the USA is for the production of aluminium for bombers and submarines or for other weapon-related industries. Much of the theoretical work in the nuclear commercial industry is directly transferable to nuclear-powered ships and submarines. Theoretical physics research serves both weapon and commercial technology. Uranium mining, milling, transportation, enrichment and other shared aspects of nuclear technology support both civilian and military programmes. Many other examples could be given, but the overall curtain of respectability appears to be the main advantage of the 'peaceful atom' programme when viewed from a military perspective.

By 1984, the USA was operating 95 (27.5 percent) of the world's commercial nuclear power plants, with a generating capacity of 80,261 MWe power (35.0 percent). The USA together with her four closest allies – Great Britain, West Germany, Japan and Canada – had a total of 191 nuclear generators (55.2 percent) with a generating capacity of 132,674 MWe power (57.8 percent). This was more than half of the global nuclear energy production.

The USSR had 45 (13.0 percent) of the global nuclear power plants by 1984, with a generating capacity of 29,295 MWe power (12.8 percent). The Eastern bloc nations – Czechoslovakia, East Germany, Hungary and Yugoslavia – account for an additional 11 reactors (3.2 percent) with a capacity of 4,645 MWe (2.0 percent), bringing the Warsaw Pact nations' total to 56 (16.2 percent) nuclear generators with a generating capacity of 33,940 MWe power (14.8 percent). In spite of this reality, many Americans feel pressured to build even more nuclear generators for fear the Russians will get 'ahead'.

The Treaty of Non-Proliferation, which was negotiated in the 1960s and entered into force on 5 March 1970, obliged non-nuclear-weapon states to conclude safeguards agreements with the International Atomic Energy Agency (IAEA) on their peaceful atom activities and their efforts to prevent diversion of nuclear materials to bombs. The signatories agreed to assist other nations to acquire and develop peaceful nuclear technology and to help in formulating a general nuclear disarmament treaty.[67] The health and safety problems of nuclear technology and the indirect shield of respectability it provides for military nuclear activities, making the weapon industry more possible, were not addressed by the treaty.

Countries already possessing nuclear weapons can continue their programmes under the Treaty of Non-Proliferation. Non-nuclear nations can come as close to weapon production as they wish. It is also now possible to build weapons and assume their reliability without testing, hence nations can stockpile parts of weapons (which can be quickly assembled) without violating the treaty. The NPT has ten signatories and involves 106 parties (as of 1 February 1979). The United States, the USSR and Great Britain are parties to the treaty. France, Brazil and Argentina are not parties to the treaty, although there are indications that they recognise and comply with the provisions.

In 1980, with the major nuclear weapon powers failing in their responsibility to bring about disarmament and de-escalation of armed conflict after thirty-five years of negotiation, popular confidence in the NPT is waning. Since the treaty is also non-enforceable it has little significance as a basis of international confidence. The 1985 review of the treaty may provide an opportunity to revive and strengthen it.

The First Committee of the UN General Assembly meeting on 12 October 1978 began formal consideration of matters of disarmament and international security. One motion proposed that:

> the use of nuclear weapons will be a violation of the Charter of the United Nations and a crime against humanity . . . The use of nuclear weapons should therefore be prohibited, pending nuclear disarmament.[68]

The motion was presented by India with the support of Algeria, Argentina, Cyprus, Ethiopia, Indonesia, Malaysia, Nigeria and Yugoslavia. The resolution passed with 84 'yes' votes. There were 16 'no' votes, including the United States and the European Economic Community. The Soviet Union, Japan and fourteen other countries abstained. China was absent.[69] Unfortunately, the UN vote had no effect on either the NATO build-up of nuclear weapons or on the US decision to implement the MX missile plans. The Soviet Union responded by moving the SS-20 into the European 'theatre'.

In this continued climate of superpower failure to achieve or co-operate with disarmament, it appears logical for other nations to seek at least some access to the feared nuclear technology. Tragically, this path moves the world inevitably towards catastrophic nuclear war.

Living with Titan II Since 1963

Verne Woner, who has lived 2,065 feet from one of the 54 Titan II missiles since 1963, was fascinated by the Air Force project. Each Titan II is 10 feet in diameter, 103 feet tall, has about 33×10^4 pounds of launch weight, a flight range exceeding 6,000 miles and more than 5,000 kilotons of blast power (about 225 times the size of the Hiroshima bomb). The depth and size of the hole, the giant spring, the elaborate communication system, the realisation that the 700-ton lid over each silo can be opened in about 20 seconds, and that once the Titan II is fired it cannot be recalled or destroyed – all these technological feats can distract attention from the grim purpose of the Titan II system.[70]

It was somewhat alarming on 24 August 1978 when clouds of dark orange gas spewed from the underground silo at Rock, Kansas, in the 43rd recorded 'incident' in the US nuclear weapon programme since 1950. The cloud was not radioactive but was caused by a valve failure which released an oxidiser (nitrogen tetroxide) which had just been pumped into one of the two propellant tanks. It formed a toxic gas, killing Staff Sergeant Robert Thomas immediately and another crew member nine days later. Two other men were seriously injured and hundreds of people from nearby homes were evacuated. In a few hours the Air Force had the incident under control, everything returned to 'normal' and there was little further media attention on this small Kansas community.

Verne describes himself as 'your conservative old farmer'. He is a veteran of the Second World War and believes it is his patriotic duty to co-operate with the Air Force nuclear defence system. He is typical of the patriotic Americans who still depend on 'defence' as a peace strategy.

There is a Launch Enable System (LES) consisting of two electronic signals originating from outside the Strategic Air Command (SAC) base where the Titan launch crew is located. Only in the event of a nuclear war would the signals be broken off enabling the missile to be fired. Sometimes bored crew members have ovecome the LES and the system has had to be corrected. Two crew members, 'key-turners', are needed to detonate a system. The local people consider it 'safe'.

Two former US Pentagon officials have admitted to a nearly catastrophic accident which occurred on 24 January 1961 when a B-52 bomber pilot, knowing his plane was crashing, jettisoned two nuclear bombs over Goldsboro, North Carolina. One bomb broke apart on impact, scattering its radioactive content. The other bomb came close to detonating after five of six safety switches were released when its parachute became entwined in a tree. That bomb was a 24 megaton size – 1,800 times the size of the Hiroshima bomb.

In September 1980, in Damascus, Arkansas, a Titan II missile silo exploded, blasting off the 700-ton cap and expelling the nuclear warhead. One US Air Force maintenance man was killed and several others were seriously injured. Fortunately, the warhead, most likely a hydrogen bomb, did not explode.

On 15 September 1980, in addition to the Titan II accident, there was a fire in a B-52 (carrying about 30 nuclear weapons) at Grand Forks Air Base, North Dakota. There was also a crash of a nuclear-armed FB-III off the New England coast in October of that year.

Reuter News Agency acquired a document from the US Department of Defence in December 1980 acknowledging 27 'broken arrows', the military code for a serious nuclear weapon accident. Military sources estimate an additional 10 serious accidents 'too politically sensitive to discuss'.[71]

The Stockholm International Peace Research Institute (SIPRI) has data on 125 nuclear weapon accidents between 1947 and 1977 – an average of about one every three months. In 32 of those accidents, US nuclear weapons were believed to have been destroyed or seriously damaged. SIPRI reported 22 Soviet, 8 British and 4 French nuclear weapon accidents.[72] None of these accidents involved a nuclear explosion, but several have involved extensive radioactive contamination of land and water.

Military policy dictates attracting as little attention as possible to such 'incidents' or 'accidents', even when the public is at risk. The conflict of interest between national security and individual harm in the name of 'defence' is usually resolved in favour of what are perceived as security needs. This means growing military control over the flow of information.

Rocky Flats, a Nuclear Trigger Factory

Dr Carl Johnson of Jefferson County, Colorado has reported health effects due to radioactive contamination from the weapons industry at Rocky Flats, 16 miles north-west of Denver. The plant makes plutonium triggers for all the nuclear bombs and missile warheads produced in the USA. Large residential areas downwind have been contaminated with plutonium for a period of over thirty years.[73]

Large offsite emissions of plutonium occurred during an explosion and major fire in 1957. All 600 filters in the main stack at the plant blew out. The filters had not been changed in the previous four years of the plant's operations. There was a second major fire at the plant in 1969. Other major releases of radioactive material to the air occurred in 1968 and 1974, and from spills of cutting oil containing metal millings of plutonium and uranium leaking from several thousand corroded barrels stored outside the plant.

The most important contaminant is plutonium 239, but other plutonium isotopes, americium 241 and cesium 137 are also present. The concentration of plutonium on the loose surface dust of the soil (easily able to be airborne and inhaled) ranges as high as 3,390 times the 'background' levels from fallout due to weapon testing. Edward Martell and S. E. Poet were the first to discover and report the massive contamination to the public in 1969.[74]

Dr Johnson found that the rate of congenital malformation of children (reported at birth) in Aravada (downwind from the plant) was 14.5 per 1,000 live births. The rate for the rest of the country was 10.4 per 1,000 live births. He also found significant increases in the incidence of all cancers, and especially cancers of the more radiosensitive organs, with increasing pollution levels.[73]

Each nuclear accident is in itself a story which requires a book. Each is accompanied by carefully managed public information, followed by increased cancers in the population and official opinions that there can be no proof of connection between the two events.[75] If the cancer rate increases on the east coast of the USA, in Great Britain or the Netherlands, it is blamed on industrialisation. If it increases in rural Ireland or Utah, USA it is blamed on lack of medical care. Meanwhile, there is no sensible record-keeping which would enable us to assign fractions of the total number of cases to industrialisation and medical care and fractions of the total number to the weapon testing or nuclear accident. Were it money and not health being lost, there would be a public audit and strict accountability.[76]

Insuring the Uninsurable

While the uninformed citizen was slowly and trustingly learning to live with the 'peaceful atom', the more realistic insurance industry, lacking actuarial data, was refusing to insure it. In the USA the clause 'not covered in the event of radioactive contamination' was written into all property insurance policies. In order to protect the desired new industry, the US Senate enacted the Price-Anderson Act in 1957 to provide insurance for nuclear industries for ten years. The hope was that ordinary insurance mechanisms would be able to take responsibility for insurance at that point, as is the custom in all high-risk ventures.[77]

However, the Price-Anderson Act had to be extended and amended in 1965, 1966 and 1975. The present nuclear insurance policy in the USA, at tax-payer expense, extends to 1 August 1987, assuring thirty years of federal insurance for the commercial industry.

Under the law, public recovery of damage from nuclear electricity companies is limited to $560 million and recovery from the nuclear manufacturing industry is altogether prohibited. The US Nuclear Regulatory Commission estimates that a major nuclear accident would cost around $15 billion or more. Others have estimated damage at $17 billion to $280 billion from the 'maximum credible accident'. Assuming a low-cost $14 billion accident, the victims would receive 4 cents on each actual dollar lost. Besides its financial inadequacy, the philosophy behind the Price-Anderson subsidy is seen by many as directly opposed to the free enterprise system. Price-Anderson is a good indicator of how much the commercial nuclear industry is desired by the US government, and how much it is protected from the usual market-place demands.

Electricity companies operating nuclear power plants can purchase insurance from 'insurance pools': Mutual Atomic Energy Liability Underwriters for liability, and American Nuclear Insurers for property coverage. No other home, automobile, property or business owner can be insured against nuclear accidents. Government-guaranteed liability coverage for the public has risen from $60 million in 1957 to $160 million in 1977. However, since for the same period the consumer index has risen by 281 percent, there is an actual decrease of 15 percent in liability coverage when measured in 'real dollars'.

On the other hand, property insurance available to electricity companies has increased from $60 million in 1957 to $300 million in

1977, a 219 percent increase in 'real dollars'. It is the consumer, of course, who pays for nuclear insurance through taxes. Most consumers are unaware of either the tax which increases hidden nuclear energy costs to the consumer, or the decrease in nuclear liability coverage for the public.

The EURATOM countries enacted a similar insurance scheme to protect the European fledgling nuclear industry and the European public also assumed this burden in taxes rather than in domestic electricity rates.

Meanwhile, the myth of cheap atomic generation of electricity was perpetuated. The hidden subsidies provided by governments were never included in the cost.

Dr Alice Stewart

Dr Alice Stewart began publishing her findings on the effects on infants of maternal X-ray exposures during pregnancy in 1958.[78] She later became the central figure in the dispute in the political–scientific–military world over nuclear industry liability for injury caused by low level radiation. Dr Alice Stewart carefully collected information on children born in England and Wales and through rigorous analysis showed damage from radiation at levels of 1/50 to 1/100, the levels being assumed as 'safe' by the nuclear military–commercial establishment. By 1970 Dr Stewart had studied some 16 million children, and through her painstaking documentation and dogged persistence had moved the medical world to recognise the harmful effects of the medical X-ray of women during pregnancy. Affecting the political world was not so easy, as was seen in the Hanford worker episode in the context of the conflict between national security and human health.

In spite of attempts to cut her research funds, attack her scientific reputation and weary her spirit, Alice Stewart continued methodically to prove her points about the seriousness of human exposure to radiation. She also just as methodically unravelled the carefully constructed 'proofs' and non-findings of the official radiation health experts.

By the end of the 1960s the nuclear military exploitation of human health was beginning to become visible to the general public in parts of the world most directly affected by nuclear weapon testing. Since the testing had been done in 'remote' areas, and since the victims did not have status in the scientific community or access to the media, information has been slow to disseminate into economic planning circles. Commercial industries relying on the military experience should have been less trusting on the health issue.

At the same time, the so-called scientific dispute over the health effects of low level radiation began to penetrate political conscious-ness through the carefully documented Hanford worker research of Alice Stewart, Thomas Mancuso, and George Kneale, recounted earlier. The unmasking of the official suppression of information on health which took place in the 1970s in the USA has caused much of the international civil unrest and anti-nuclear crises of the 1980s. To misunderstand or ignore this dimension of nuclear problems is to underestimate the depth and strength of the current anti-nuclear and anti-war movements.

As the public slowly awakened to their personal survival problem vis-à-vis nuclear war or the slow poisoning of the environment caused by preparation for war, the military fear for national survival in conflict and the political fear for economic and social survival of the nation-states heightened. The 1963 Cuban missile crisis brought the United States and the USSR to the brink of nuclear war and the prospect was terrifying. It is now known that even if the Soviet Union had installed missiles in Cuba their range would not have included all of Florida, much less posed a threat to the rest of mainland United States.[79] Submarines offshore pose a greater threat. However, most Americans supported confrontation over Cuban missiles and silence about Soviet submarines. The pollution health questions were ignored whenever the thought of a military confrontation was proposed.

Is Peace Desirable?

'Think tanks' trying to imagine a world with 'permanent peace' decided that while this state of world peace might be possible theoretically, and while many people would see it as desirable, the transition to a state of peace was *not* desirable and the state of peace would not be 'in the best interest of a stable society'. Think-tank members found poverty and unemployment necessary to maintain discipline and a reserve labour pool, military training to be a social service programme for undisciplined young men and war to be a means to reduce 'overpopulation'. Obviously, these opinions were not for common consumption because 'ordinary people' would not understand. The rift between governing and governed in democratic societies grew larger, the realm of secrecy widened, and tensions within nations rivalled tensions between nations in intensity.

The 1963 missile crisis brought the United States and the USSR into their closest confrontation since the beginning of the arms race in 1945. The Soviet Union, preparing to move nuclear missiles into Cuba, backed down after President Kennedy's threat of all-out war. Both countries were shaken by the prospect of a nuclear exchange with unprecedented loss of civilians and land and massive pollution of the earth.

The Ackley Committee was set up in the USA to study the possibility of a society based on peace rather than war. This was a 'show case' committee, providing a report meant for public consumption. A second committee, meeting in secret, was a probable counterpart, and the book *Report From Iron Mountain*,[80] a fictionalised account, no doubt comes close to revealing the politics of the secret meeting. The Russian response to the Cuban missile crisis is shrouded in silence.

The gist of these US national peace studies was that peace was not really desirable, and that if there was a transition to peace, preserving national sovereignty was not possible. War was not seen as something a nation does; it was, rather, the motivating reason for which the nation was formed. If one removed war the reason for the sovereign nation-state was gone. These 'think tanks' projected that there would be social chaos if nations actually decided that there could be no more war. Social chaos meant, primarily, loss of some sovereignty for the nation-state.

War and threats of war were thought to guarantee the social co-operation of citizens with their national leaders. Military service was thought to provide a social welfare programme to discipline unwieldy, rebellious youths. These youths could be killed if need be to prevent unwanted national 'over-population', or competition. War production could be stepped up or reduced at the whim of the national leader, providing an effective social control over jobs and taxes. In a word, war-making ensured the absolute sovereignty of the nation-state and its ruler, regardless of the democratic or totalitarian nature of the regime. It was necessary to *create* threats of war and periodically necessary to. *wage* war to 'relieve the tensions' within the nation caused by the constant readiness for war. The leader of the nation has life and death power over subjects, just as in early societies the father used to have life and death power over his wife and children.

There was little or no motivation on the part of the national leaders or military strategists to do more than make token efforts towards peace. However, they also could not afford to lose everything through all-out war. Walking on the sharp edge of nuclear terror became the norm in a world growing more and more out of touch with reality. This elite self-serving decision to avoid genuine peacemaking prevented constructive work towards either security or stability in the early nuclear era.

As we will see later this outmoded historical wisdom based on past war-making experience was invalid for the totally changed situation posed by a nuclear holocaust. Instead of fostering political and social cohesion, the threat of nuclear war is causing deep splits within nations. Even the promise to create jobs is failing to convince workers. In a meeting of aerospace workers in London in 1982, called by the International Metalworkers Federation, the argument that military production secured jobs was scorned. These labour leaders, as well as other trade unionists, have concluded that the money and resources committed to building armaments could, if directed to almost any other end product, produce more jobs.

The recovery of West Germany and Japan after the Second World War, and the strength of their economies, bodes well for predicting national productivity to be possible when there is no brain and resource drain into military production. As their consumer products gain ever larger shares of world markets, the astute observer notices a corresponding increase in pressure from the United States and other Western bloc nations for West German and Japanese rearmament. It has become increasingly evident that war-making, even in developed countries, drains precious human energies and scarce material resources, leaving the global human sector poor and needy. Continued reliance on war-making for social and economic vitality, even if a shooting war is avoided, appears suicidal.

Breaking through the Secrecy

It is essential, therefore, that ordinary people should have sufficient information about the effects of atomic energy to be able to form a judgment and to exercise a watching brief over the expert. Unfortunately, in the particular field [of radiation biology], secrecy prevents the exercise of this very aspect of citizenship. This is wrong . . . If military or industrial secrecy prevents this exercise of citizenship then it undermines a much more general principle to which I hold very strongly, which is that secrecy should be anathema to any civilized community, and particularly in a scientific community, no matter what its alleged justification.[81]

As early as 1950, Kathleen Lonsdale, FRS, of the Chemistry Department, University College, London, pointed out the impossibility of human culture surviving with military secrecy pervading atomic science. This problem has, however, persisted. The most recent disclosure of its serious effect on understanding radiation biology was made in *Science* of 22 May 1981.[82] During the 1950s and 1960s Oak Ridge National Laboratory, a US government-supported military nuclear research and production installation (managed by Union Carbide), established a method of calculating the dose from neutrons and gamma rays received by the Hiroshima A-bomb survivors. These calculations are now being challenged. The methodology supporting the calculations was classified as secret and the record of it was 'accidentally' fed to a shredder. Since the number of cancer cases in the survivor population remains the same, the question of radiation dose becomes crucial for estimating the number of cancers per unit dose. If the Oak Ridge scientists reported a higher neutron dose to the population than was the actual case, then the excess cancer deaths attributed to neutrons really resulted from the lower energy gamma radiation which has been officially considered less dangerous.

It is important to note that this atomic bomb research error and subterfuge need not be ascribed to some massive conspiracy among professionals to deceive the world community. One or two individuals 'slightly' falsifying the estimates of radiation dose at a weapon laboratory could provide this misinformation to the epidemiologists at the Japanese A-bomb research centres, causing every subsequent report to incorporate a systematic error of underestimation of the health impact of radiation. This error would be perpetuated regardless of the high professional quality of the research and impeccable integrity of the researcher.

Incredibly, the same in-house US government research system which made the error is now completely in charge of setting it right.

The dose estimates for A-bomb survivors have been a jealously guarded possession of the US government since 1945. When I was in Japan in 1978, Dr Issei Nishimori of the University of Nagasaki was attempting to obtain original A-bomb dose estimates for the survivors in his study on long-term radiation-related health problems. He had copious data on the health parameters and medical experiences of the victims, but no estimate of their original radiation dose which would enable him to determine which ailments and diseases showed increased severity or higher incidence rate with increased radiation dose.[83]

In an interview with Stuart Finch from the USA, then director of research at the Radiation Effects Research Foundation in Hiroshima, and the only person able to authorise access to US A-bomb dose calculations, the question of release of information to qualified Japanese researchers was raised. Stuart Finch's initial casual reply was that the Japanese researchers might be biased in their approach to the radiation health problems. When it was pointed out that such a possible bias would be subjected to the traditional scientific peer review before any findings were published and should not be prejudged when requests come from internationally recognised experts, a different line of reasoning emerged. Finch declared that the work of the Japanese experts was to care for the medical needs of the survivors and 'they did not need to know the radiation dose to do that'. The possibility of US military bias in reporting health effects was allowed to remain undiscussed.

The interview ended with Finch agreeing to review a grant proposal submitted by the Japanese scientist requesting information on individual doses. There are, of course, many ways of criticising any grant proposal and of rejecting even the most excellent proposal, for example, by saying that the research has already been done and showed 'no effect'. The request can be lost in a jungle of bureaucratic reviews and eventually rejected by a usually anonymous panel of 'experts'. In the waiting, more victims die and researchers give up hope of being able to obtain the information which they need.

Official rejections of research requests are probably justified by the desire for a unified and consistent analysis of survivor data so as to create a scientific consensus rather than chaos. However, most scientists find that truth is the best basis for a consensus, not control or management of information. Military and economic national security strategies demand that the public be willing to handle and be exposed to radioactive material. Unfortunately papers on A-bomb effects are perceived as strategic instruments in forming scientific and general public attitudes of acceptance.

The problem of data-sharing is also aggravated by the traditional jealous control of scientific research data bases. These data bases are constantly 'mined' for pieces of information which researchers can publish to justify their employment. It is the source of researchers' bread and butter and also their professional image. To be an expert on atomic bomb health effects means one has had access to information not available to others.

During the 1960s and 1970s several data bases of information on people exposed to ionising radiation were identified and analysed independently of the A-bomb survivor studies. The findings based on independent research in Great Britain, West Germany, Japan, and the United States began to emerge as a consistent testimony to the systematic underestimation of radiation health effects by official nuclear-related agencies internationally. The gradual build-up of independent scientific information has accompanied the gradual popular disillusionment with nuclear technology and the rise of the anti-nuclear movement. Wherever there is nuclear technology, members of the scientific community have been carefully and slowly documenting the situation and validating the basic survival sense of the population, thus providing a factual basis for activating a tremendous human energy for change.

For those whose scientific expertise is in commercial applications of nuclear physics, chemistry or engineering, and who are still unaware of the seriousness of the scientific challenge to accepted radiation biology claims, the popular anti-nuclear unrest appears to be due to ignorance, unsubstantiated fear and the sensation-seeking of incompetent anti-nuclear scientists. Those pro-nuclear scientists who have been fed 'official' assurances continually recommend new educational programmes and/or public relations efforts to educate the public on radiation.

Because of their strong but opposite positions, the pro-nuclear and anti-nuclear movements have the potential for frustrating meaningful political action to promote the human agenda globally. The pro-nuclear scientists can prevent action to stop the careless radiation pollution resulting from military miscalculation of the health effects of exposure to radiation. The anti-nuclear scientists are sometimes naive about political and military tensions affected by their protests, tensions which govern choices regardless of scientific findings. With two vocal lobbying groups holding opposing positions on 'development' and 'security', national politicians are most likely to rely on the 'experts' from UN or national government-supported agencies, to revert to further secrecy to prevent public protest, which they view as 'uninformed', or to make use of the general confusion to promote economic or political goals.

Nation-state behaviour is also affected by the lack of a clearly identified global public sector. Nuclear-related information cannot be entrusted to an existing international group of people who could be trusted to act consistently on behalf of all the people of the globe.

The logical agency for non-nuclear nations to turn to for information is the International Atomic Energy Agency in Vienna. However, with a United Nations mandate to *promote* the 'peaceful uses of atomic energy', and with limited money and personnel to research information independently, this agency has become little more than a promotional arm for nuclear technology. There is no comparable UN agency to promote any other energy options, providing balance. Nor is there a UN agency entrusted with the task of distributing serious anti-nuclear research. Even attendance at IAEA meetings is by designation of the member nation to which the scientists belong. Dissenters from national policy are rarely if ever sent.

The second most relevant agency to which nations turn for information is the International Commission on Radiological Protection. This agency, by its closed self-perpetuating structure, does not serve as an independent scientific body providing valuable and valid contributions to understanding the impact on humans of political nuclear decisions. Its record shows no action or public stand on behalf of public health since its founding in 1952, and its recent documents, ICRP 26 (1977) and 30 (1979), recommending an increase in radiation exposure standards for many radiosensitive organs, such as thyroid and bone marrow, given the 'economic and social benefits of the activities', leave little room for hope of future reorientation.

The final section of this book contains recommendations for structural change in international and national agencies to alleviate the social–political–biological crisis of the 1980s. Prior to these recommendations, however, some of the scientific advances made outside the established nuclear science world are mentioned, and the information gained during the 1970s presented. Some important events in the rise of the anti-nuclear movement are also highlighted.

The privately controlled informational bases developed at considerable financial and personal cost now supply independent information on radiation health effects in humans to the public. These include the Oxford data on childhood cancers in Great Britain, the Tri-State data on leukaemia incidence in three American states, and the nuclear worker data from the Hanford installation in the United States. Data on victims of nuclear weapon testing in Nevada and in the Marshall Islands and the results of vast population contamination from nuclear weapon production accidents near Chelyabinsk, USSR, Sellafield (formerly Windscale), England, and Denver, Colorado, USA have provided further factual information. Studies of US Navy nuclear submarine workers are still under way.

Related research done in Heidelberg, West Germany, has furthered understanding of the movement of radioactive chemicals from air and water into plants and animals and finally into the human body. This research has called into question the mathematical calculations officially used by the nuclear industry to estimate radiation doses to humans from measurements of curies of radioactive chemicals released from a nuclear plant. Several major studies now challenge the estimates of cancers presumed to be related to a unit dose of radiation received.

New sensitive biological organisms, such as the spiderwort plant, have been introduced to measure genetic mutations resulting from exposure to nuclear radiation. In the late 1970s, citizens in the United States, Great Britain, Japan and West Germany began to buy their own equipment and develop their own methods of detecting radiation in the environment.

New information released in the 1970s on the health effects of chronic low level radiation exposure differs significantly from effects which had been predicted on the basis of acute exposures to high doses of radiation among A-bomb survivors and X-ray therapy patients. Normally a direct audit would replace such predictions and analysis of appropriate populations would be preferred to optimistic forecasts based on extrapolation from inappropriate populations and conditions. However, as the scientific controversy increased during the 1970s, dialogue frequently degenerated into personal attacks on the reputations of researchers, economic penalties were imposed on scientists dissenting from the nuclear industry projection of 'no harm to human health', and attempts were made to increase the prestige of those scientists who tried to hold ranks and support the findings on health which were most favourable to political, military and economic national goals.

Kalkar: A Lesson in Upright Walking

The West German resistance to nuclear energy was clearly

demonstrated in the Alsatian–Badisch citizen-action group opposition to Wyhl in 1975. It was a true 'country-folk' revolt with young and old from all walks of life occupying the construction site for nine months, enduring the winter weather in makeshift tents.[84]

Some German citizens, trained in and committed to non-violent action, have photographs of police infiltrators who attempted to incite the peaceful anti-nuclear demonstrators to violence so as to discredit the German citizen protest movement. For the most part, however, the citizens remained non-violent throughout the whole ordeal.

This German peasant revolt had expected to meet police violence on 13 November 1976 at demonstrations near Brokdorf and Gröhnde. They were not wrong.

Because of previous violent citizen and police clashes, the German media created a civil-war type of atmosphere. In 1977 the German Trade Union Confederation and Mr Kuhn (Prime Minister of Nordrhein-Westfalen) cautioned against participation in a demonstration against the Kalkar breeder reactor, located between Wyhl and Brokdorf.

On 4 September 1977, we were driving in the early morning fog, in order to find 2 hours of sleep between the little town of Kalkar and the even smaller village Honnepel.

Thus begins the story of Roland Vital, a West German citizen who was driving within West Germany to attend a citizen protest rally against the breeder reactor under construction at Kalkar.

Suddenly, in true police manner with a machine gun, our family transport is searched and we are asked to produce identification. A dozen policemen surround our car. Police spotlights and damp night air awake the children. Sleepily, our oldest recognizes the barrel of a gun aimed at us. Panic-like screams: 'Will they shoot us dead?' ('Quiet my child – in dry leaves sings the wind . . .', Goethe's Erlkönig.) It is cold, but the police controllers take their time. Two children's boat paddle ends (they have been left unintentionally since the last summer vacation trip in the car) and the two longer ignition keys are criticized as 'weapon-like' tools and are triumphantly presented to the chief of the search action. The taking-in of the 'weapon-like' tools is avoided after much strong protest. The police team is disppointed and refuses us the declaration of having searched us. Nearly three minutes later:

another police team, same pattern (including machine guns) and five minutes after that the attempt of a third search after the same pattern. Before the town of Kalkar, there await us two more identification controls. The cynical commentary after our complaint against these multi-controls: 'You will be searched and controlled even more.'

Tens of thousands of citizens poured into Kalkar for the demonstration in spite of impediments placed on highways and government hold-ups of trains and buses. The people listened to speeches between 11 a.m. and 4 p.m. gaining a sense of hope and solidarity as the crowd swelled to 50,000. All had been 'controlled and searched'.

At 4 o'clock a 4-kilometer-long, disciplined march forms itself spontaneously and moves while singing and clapping toward the little village Hennepel.

At the road junction to the Nuclear Power Plant Kalkar (so read the signposts and after that the words 'No Entry'), the prohibition on the demonstration is to become effective. Here we choose whether or not we will break the prohibition and will move to the rally terrain. Over 50,000 demonstrators (except for 400–500) decide *with common will to move onward to the farming land of farmer Joseph Maas*, the rally terrain and the construction site of the fast breeder. Thousands of armed policemen have been pulled into position by their chiefs in front and deep in the area of the fortified construction site.

For the state and its appendages who instinctively know or feel that they are risking their credibility and their power base, this enormous police-force manoeuvre is a very necessary act of securing their dominion: they are under the compulsion and coercion to secure their dominion in the service of capital and plutonium through humiliation of the Alternative Movement. If we would reply to the terror with direct and personal violence, we would be caught in a trap. But for war, one always needs *two* parties. We refuse, as in Wyhl, to participate in this war-game. Our strength is non-violence. Was it responsible for us to take four children along? I believe: *Yes* and even more so, because they are those who struggle tomorrow and Kalkar was a lesson in upright-walking![84]

The European citizen protests followed long and frustrating

attempts to deal with government nuclear policy through political and legal action.

Creys-Malville

In July 1977 an anti-nuclear demonstration was held at Creys-Malville, a village in the Rhône Valley, between Lyons, France and Geneva, Switzerland. It was in protest against the Super Phoenix breeder reactor being constructed by the French government in partnership with West Germany, Italy and Great Britain. Switzerland, not a party to the project, lies downwind of the site. The Super Phoenix, a 1,300 MWe plutonium-fuelled reactor costing upwards of $2 billion (US), is expected to begin commercial operation in 1985. A prototype 250 MWe breeder reactor, the Phoenix, has been operating not altogether successfully on the site since 1975.

A small group of demonstrators at Creys-Malville had threatened to use violence and organisers of the large group, determined to keep the demonstration non-violent, isolated this violent group and prevented their being amongst the first to arrive at the site. Tens of thousands of unarmed peasants and professionals began to walk towards Creys-Malville to confront the armed French military police who were forbidding the demonstration.

When the crowd refused to stop walking, police fired into the front line of people and released more bullets above their heads. Vital Michalon, a mathematics schoolteacher in his early thirties, was killed. Many others were injured, either by the police or by the panicked crowd.

The police also tossed hand grenades which explosively released tear gas and small sharp fragments. Those who tried to pick up the grenades and throw them out of the crowd had small pellets imbedded in their hands. Some required amputation. One man sitting on the ground lost a leg when a grenade exploded near him. Needless to say, the demonstrators quickly disbanded. The more violent component of the protest group never reached the scene. There had been no provocation of the police by the demonstrators according to an eye-witness report given to me. The French have prided themselves on this aggressive treatment of protest, effectively causing the anti-nuclear movement to go underground.

Ian Smart, a British nuclear expert who has studied the French policy, has said: 'The decision-making process at the top levels of the French government is swifter and less easy to challenge than in other European governments.' Power is in the hands of a small elite who went to the same schools, know one another, and are personally convinced of the rightness of what they are doing. It is much harder to use the French court system or nuclear plant licensing system to modify or delay reactor production than it is in West Germany or the United States. The ordinary citizen lacks basic information on the consequences of government planning and cannot enter into the decisions which drastically affect his or her life, and that of future generations. Nowhere is it more obvious that democracy built upon informed consent is incompatible with the secretive centralised planning required for nuclear strategies. The secrecy surrounding long-term environmental and health effects and the deliberate policy of sacrificing some areas of the country to provide economic and military advantages to the nation-state, inevitably erodes human and civil rights.

The present French nuclear programme was formulated by President Georges Pompidou in 1969 when he decided to abandon the graphite-gas nuclear reactor developed by Britain and France during the 1960s. He chose the Westinghouse pressurised reactor even though importing American reactors was politically distasteful. The Framatome company formed an agreement with Westinghouse, imported the pressurised water reactor and made extensive changes in its design. The contract expired in 1981 leaving France to export nuclear technology to other countries without US approval after that date.

In South Korea in 1976 and Pakistan in 1978, the United States used its leverage to prevent French sales of reprocessing technology. South Korea, although it has ratified the Nuclear Non-Proliferation Treaty, is pursuing a nuclear policy which will allow it to produce nuclear weapons within a few years. A threat from China, a nuclear weapon state, could conceivably accelerate that development. Pakistan, which has refused to sign the Non-Proliferation Treaty and which has expressed strong interest in nuclear weaponry, is the traditional enemy of India, which exploded its first nuclear device in 1974. A reprocessing plant is necessary if South Korea and Pakistan are to recover the plutonium from the nuclear reactor fuel for weapon use.

The implications of nuclear reactor export policy are not unknown in France. Walter Schütz, foreign affairs analyst at the Centre d'Études des Politiques Etrangères, Paris, commented in May 1978:

It seems therefore that the timetable of actual [nuclear] weapons proliferation will be determined primarily by what is happening on the world scene. A breakdown of east-west detente on a large scale and a complete collapse of arms-control policy would inevitably have a triggering effect. And even without such extreme scenarios, the outbreak of local wars – a renewed Israeli–Arab conflict or a dramatic heightening of tensions in southern Africa – could have somewhat similar consequences at least on a regional scale. What one cannot take for granted is that within the next decade or so the nuclear status quo will be maintained. Nor can it be assumed that by the year 2000 there will not be around 35 states possessing their own nuclear arsenals as a consequence of varying combinations of concern for national security, international status, and internal political and bureaucratic pressure to assert the national interest through reliance on nuclear weapons.[85]

The French nuclear energy programme was accelerated by President Giscard d'Estaing after the OPEC oil embargo in 1973. Under his presidency the nuclear programme at Cap de la Hague and the breeder reactor programme at Creys-Malville were well financed and politically supported.

Giscard d'Estaing also began a 'nuclear park' at Cattenom in the north-eastern part of France, bordering on Luxembourg and West Germany. Cattenom, with 15,000 MWe generating capacity, will be the largest nuclear complex in the world. Downwind and downstream from this 'nuclear park' is the historic West German city of Trier. Its residents, as well as the West German government, are politically powerless to influence French government policy.

A major accident at Cattenom could render the entire country of Luxembourg uninhabitable, yet this nation-state with an antinuclear energy policy cannot affect French decisions.

Cattenom also raises a new climatic question. It is speculated by some professionals that a concentrated source of heat such as at Cattenom can cause a small tornado to form. It is a 'point' source of intense heat. The experiment will be tried in one of the most densely populated areas of the world.

It is obvious to most Europeans that the concepts of national sovereignty, impenetrable boundaries and security of land possession have become obsolete in the interdependent nuclear age. In response to this reality a European Parliament was formed in the summer of 1979, but it was given very little power. It has already censured France for locating major nuclear projects near the borders of other countries but the French government has apparently ignored the censure. It is unlikely that Europe, where national boundaries are frequently defined by waterways, will solve the problem of site selection for nuclear projects (requiring location on waterways) by parliamentary action. While a concept of national sovereignty and national sacrifice areas might be tolerated temporarily on the international level as beyond extra-national control, the inescapable fact that all things are connected must eventually force action. Newly structured decision-making modes are needed which preserve personal rights, national cultural identity and international security against national decisions destructive of air, water, land and food which are the common human heritage. Such shared resources are not national possessions and their pollution cannot be limited within national boundaries. This concept of the common heritage of humankind has in fact been given formal expression in the UN Law of the Seas. This is a document for the twenty-first century which contains an exceptional awareness of humanity.

The French Green Party, which arose in response to grass-roots concern over the aggressive military and civilian nuclear policy, managed to unseat Giscard d'Estaing and elect François Mitterand in spring 1981. Mitterand, in his election campaign, promised general public debate on nuclear policy.

Immediately after the election, Mitterand cancelled the Plogoff nuclear reactor in Brittany, where opposition to the project by citizens threatened to reach full-scale revolt.[86] He 'froze' two nuclear projects – reactors at Chooz and Golfech – promising that local electorates would be consulted, and he reduced by 50 percent the planned electrical generation at Cattenom.

In September 1981 the Groupe de Bellerive, Geneva-based energy policy consultants, challenged French nuclear proponents to a debate on the breeder reactor strategy. The debate followed the action of the WCC calling for public discussion of nuclear energy policy. It also followed a presentation to France of the Geneva Appeal in February 1981, which was signed by 50,000 people and endorsed by four Nobel Prize winners: Konrad Lorenz, Heinrich Böll, Jan Tinbergen and George Wald, and by scientists in all disciplines. This Appeal demanded:

> interdisciplinary hearings open to all views on the plutonium vs. soft technology alternatives . . . the people of Europe thus informed be asked to vote on the aforesaid alternative. . .[87]

The Geneva initiation cited the 'fait accompli' policy of the French government with regard to the Super Phoenix as an example of 'corrosion of democracy for reasons of State'.

Some of the probable if not certain consequences of the plutonium society include the concentration and expansion of power – in all senses of the word – the spread of the military practice of secrecy to civil affairs which will be justified by the technological and, hence, military vulnerability of such a society, and its inevitable counterpart, the omnipresence of the police.

In addition to the immediate changes in society and the slow biological and ecological damage inevitably initiated by a plutonium economy, there is also, according to a French engineer–politician, an inevitable major accident involved. In a written note to Ivo Rens, Professor of History in the Faculty of Law at University of Geneva, this French graduate of the elite Ecole Polytechnique, who wishes to remain anonymous, has stated:

The Super Phoenix will have a plutonium core containing 4,600 tons of molten sodium of which 3,200 tons, the primary circuit, will be highly radioactive and 1,200 tons, the secondary circuit, will be in contact with the pipes of the water circuit in the steam generator.

Sodium burns spontaneously when in contact with water, creating a fire that can be extinguished only with difficulty. To prevent such fires, water and sodium must be kept apart in the heat exchanges, and that separation depends on the integrity of the piping carrying the two circuits.

Yet such pipes are prone to cracking, especially when transporting corrosive sodium at high temperatures. Moreover cracks have already appeared in the water-steam heat exchangers of the Fessenheim Light Water Reactor. In fact, each of the four steam generators of the Super Phoenix reactor under construction contains some 3,000 welds. These welds must remain good throughout the operation of the reactor if leakages of hot sodium into water are to be avoided.

It is thus inevitable that sooner or later the secondary circuit will burn, leading to a fire in the primary circuit, when as a result of the heat generated by the combustion of 4,600 tons of sodium, all the safety devices protecting the reactor core will be destroyed. Even in the absence of a nuclear explosion, a cloud containing twelve to fifteen thousand tons of radioactive sodium carbonate and bicarbonate enriched by several hundred kilos of plutonium oxide dust, would contaminate tens of thousands of square kilometres, causing the death, over a shorter or longer period, of hundreds of thousands of people.[88]

The conscious or unconscious policy of determining national sacrifice areas is relentlessly extending to the international level. Will other nations, rather than parts of our own nations, be sacrificed? This, of course, is a new colonialism. The addiction to nuclear technology becomes more visible as nation-states become more irrational in their choices. The cost of this addiction is ever more apparent as local stories merge to form a global pattern of national insecurity and reckless endangerment of present and future generations.

On 19 January 1982 a French protest group launched five Soviet-made rockets, hitting the main building of the nuclear breeder reactor under construction in Creys-Malville. Officials say the hole in the concrete wall is 'minor'. No casualties were reported.

The Alternatives

Small is Beautiful, by E. F. Schumacher,[89] introduced the concept of appropriate human-scale technology especially in the developing world. It is still considered to be a classic on the subject of energy policy.

Another articulate critic of nuclear addiction is Amory Lovins. He introduced the term 'soft energy path' in 'Energy Strategy: The Road Not Taken?' in 1976,[90] and *Soft Energy Paths: Towards a Durable Peace*[91] in 1977. He and his talented wife, Hunter Lovins, published 'Energy/War: Breaking the Nuclear Link' in 1981.[92] Amory Lovins was able to demonstrate quite clearly at the Energy Conference in Paris on 16–18 September 1981, initiated by the Group de Bellerive, that France's nuclear policy would not solve her oil dependency problem. Oil is primarily (90 percent) used for transportation and space heating, two areas in which 'electrical power stations are certainly uncompetitive, if not altogether impractical'.[93] In spite of this, Electricité de France is attempting to persuade French industrial and residential customers to switch to electricity for heating in order to create a market for its excess nuclear-generating capacity. Lovins demonstrated that the human need for light and heat could be better met with conservation and soft energy technologies, more surely reducing oil consumption and more effectively increasing national and international security and employment.

Charles Komanoff of Komanoff Energy Associates, another speaker at the Paris Conference, supported Amory Lovins's scientific presentation, saying that the 'uranium for oil swap' ignores both the high relative price of electricity and the very limited scope of direct replacement.

Dr Karl Morgan, who also spoke at the Paris Conference, a long-time opponent of the liquid metal fast breeder reactor, reflected after the conference that, 'The French scientists did not really hear what was being said.'

The nuclear 'in-group' has been elitist for so long that outside ideas have become a foreign language. Their decisions are so surrounded by rationalisations that they have lost touch with reality. The consequences of this massive societal addiction are beyond the grasp of the masses of people whose lives and land and future are at stake.

The French Parliament approved on 6 October 1981 the energy plan of the Socialist Prime Minister Maurroy. It reinstated the reactors at Cattenom to the original 15,000 MWe capacity, called for construction of six new nuclear plants in 1982–83 and 'consultations' on the construction of nuclear plants at Chooz and Golfech.

It established a 'high level' scientific commission to study the expansion fo COGEMA's Cap de la Hague reprocessing plant. The commission was to have its report ready in six months. No indication of how the members of the commission would be chosen was given. Its function was to be purely advisory.

The promised public debates on nuclear power apparently will not be permitted. Energy Minister Hervé proposed that they be replaced by the creation of an information bank on energy and of advisory regional energy agencies. This means, of course, that information or propaganda flows outwards from government to governed with inflow restricted to advice from the minister's friends and former classmates designated as heads of regional agencies.

Prior to the decisions of the French Parliament, the people carried out two public demonstrations of their frustration at the collapse of the democratic process. At Golfech, more than 5,000 demonstrators broke down fences at the construction site of the nuclear plant. They set fire to the Electricité de France buildings and damaged heavy equipment. Damage estimates varied from 3.5

to 15 million francs. Police action and tear gas brought the demonstration to an end.

A peaceful demonstration of 4,000 people took place in Paris on 3 October 1981, sponsored by the national anti-nuclear co-ordination group, the political ecology movement, and some left-wing political organisations. The group called for a stop to the nuclear programme and a truly democratic debate on energy policy. Although the demonstration was planned to include a march to the Palais Bourbon, where Parliament would debate energy policy, it was disbanded by the organisers when they realised that several hundred 'autonomes' (infiltrators) had joined their ranks and begun clashes with police.

Most French activists concede that national political action has not changed the French government's nuclear policy. The system seems to have its own rationale and it operates on an inner dynamism of dependency which is not easily broken by pre-nuclear-age democratic mechanisms. The forces governing national choices are international.[94]

Media Coverage

Associated Press, Stockholm, Sweden
Swedish Vote Supports Use of Nuclear Power
Swedes voting in a referendum have backed completion of their nation's ambitious nuclear energy program.

Final returns from Sunday's vote showed 58 percent endorsed a 1975 government plan to put six more nuclear power plants into operation, while 38.6 percent voted to shut down within 10 years the six nuclear generators already in operation. An additional 3.3 percent cast blank ballots.

Voters in the referendum had no way to support unrestricted use of nuclear energy or an immediate shutdown of the nation's reactors. The only way to protest the alternatives was to cast a blank ballot.

A total of 74.3 percent of the nation's 6.3 million eligible voters cast ballots.

The future of nuclear energy became something of a national obsession following the nuclear accident at the Three Mile Island reactor in Harrisburg, Pa., a year ago this week. Center Party Prime Minister Thorbjorn Falldin won office last year on an anti-nuclear ticket and campaigned hard to scrap nuclear power through the referendum. However, after the vote he promised he would abide by the national consensus.

The Communist Party and environmental groups, which joined Mr Falldin's Centrists in a 'peoples' campaign' against nuclear power, said they would continue to fight it despite the vote.

Hinting he, too, may not have given up the fight altogether, Mr Falldin said he would continue to keep a close watch on nuclear plant safety and raised doubts about the activation of two of the four additional reactors that have been completed.

'They lack a waste processing contract,' he said.

Sweden has put six nuclear reactors into operation since it began construction five years ago. Four other plants have been completed and are awaiting start-up, while two more are under construction. The six working reactors now provide 22 percent of the nation's electricity.[95]

Sweden has never been involved in the development, testing or production of nuclear weapons. It has no known petroleum reserves, so little coal that it has not mined it in fifty years, and a 300,000-ton uranium deposit as yet unmined. Sweden began entering the nuclear age in 1972 and by 1977 had the highest per capita commercial nuclear energy use in the world. The Swedish voters startled the world in 1976 by unseating the Social Democratic Party for the first time in forty-four years, in protest against Premier Olaf Palme's aggressive nuclear power programme. It had been a strictly peaceful nuclear energy programme with no weapon overtones, and most of the world assumed that the programme enjoyed popular support. Ironically, Olaf Palme was re-elected to power six years later on an anti-nuclear-weapon platform.

The Swedish nuclear reactor plans of the early 1970s called for twenty-four reactors. Some industrialists, for example Hans Werthen, chairperson of Electrolux, were optimistically calling for fifty reactors. With the defeat of the Social Democratic Party, the number of planned reactors dropped to the twelve already constructed or in progress.

After the fall of the Social Democratic Party a coalition government with three constituent parties was formed. When this new government was faced with a decision on whether or not to allow operation of two newly completed nuclear reactors in 1978, the coalition, unable to agree, resigned. The debate became even more heated after the 1979 nuclear reactor accident at Three Mile Island, Pennsylvania in the USA. These were, briefly, the events which led to the national referendum in Sweden in April 1980.

The people of Sweden were given three choices in the referendum vote:

Line 1: continue operation of Sweden's six operating reactors; fuel four more reactors already constructed; and finish the two reactors now partially constructed.

Line 2: proposed the same actions as in Line 1, with municipal ownership of power plants in order to facilitate development of conservation and alternative energy sources. Proponents of this option hoped the development of alternative sources would allow for shutting down the nuclear reactors within 25 years.

Line 3: phasing out all Swedish nuclear power within ten years.

The vote broke down as follows: *Line 1*, backed by the Swedish nuclear industry and the Conservative party received 18.7 percent of the vote. *Line 2*, which called for a 25-year planned phase-out with municipal takeover of electricity companies received 39.3 percent of the vote. *Line 3*, which called for a 10-year phase-out, received a 38.6 percent vote. (3.3 percent of the voters cast blank ballots.)

The Associated Press report of this vote as a pro-nuclear 'victory' was grossly biased, yet such bias could not be recognised by readers without further information. The style of the AP report mimicked the straightforward reporting of facts which the reader expects.[96]

This 'moulding' of public opinion has become a highly developed art, propounding and endorsing misperceptions and frustrating the information-sharing necessary for informed judgment and democratic decision-making. It also prevents the feedback from reality on which the future of human life on earth depends.

Nuclear power, perceived by governments and industries as necessary for their military, economic and political strategies, is being defended and promoted regardless of the destruction of human health or truth-telling. Many nuclear proponents trustingly base their support for the industry on the partial and misleading propaganda which they have accepted as factual.

Paid advertisements for nuclear power are another means of deceiving the public with misleading and partial information. The advertisement reproduced on p. 267 was written by 'America's Electric Energy Companies', a rather nondescript conglomerate.

The second paragraph juxtaposes a statement that there 'are *safe* methods' for nuclear waste disposal with a list of countries with '*successful*' programmes. It mentions one proposed *safe* programme. No definitions of 'safe' or 'successful' are given. There is no explanation for the change of terminology or the implied equivalence.

Exactly what is the 'successful' twenty-year nuclear waste solidification programme in Canada, extolled by the advertisement? In a fact sheet prepared by Ontario Hydro, Canada's leader in nuclear energy, in November 1978 and updated in 1984, we find:

Q: If Atomic Energy of Canada Ltd does not come up with a waste management facility, will Ontario Hydro continue to use the swimming pools [to store spent fuel rods]?

A: There is no doubt that a geological repository for radioactive waste can be constructed since the concept for it has been adequately demonstrated in nature. Studies are being carried out by Hydro and AECL on the development of technology and facilities for the disposal of high level wastes.

   Since 1962, Hydro has safely stored irradiated fuel on-site in water-filled bays. Interim storage of irradiated fuel can be accommodated safely in existing and additional water-filled bays for several decades, if necessary.[97]

Storage of spent fuel in water-filled bays at the reactor site is the most temporary of available technological solutions to high level radioactive waste.

Discussion of nuclear waste has sparked an often heated dispute over nuclear power in the Canadian Medical Association. This began after the federal government issued a Green Paper[98] on the subject in 1978. Resolutions proposed and/or adopted by the physicians range from condemnation of nuclear development until a responsible waste disposal programme is in place, to acceptance of plans to use natural plutons (rock formations which are part of the Canadian shield) and glassification as a 'good long-term storage plan'.[99] Most subsequent Canadian Medical Association resolutions call for further study, especially in the disposal of reprocessed nuclear waste. There is clearly no well-tested disposal method in place in Canada, and no consensus on what a good plan ought to be.

There is no evidence that the nuclear waste situation is any better in Great Britain or France. In fact, it was reported in Ireland that the European Economic Community was seeking land between Kinsale and Macroom as a repository for European nuclear waste. Great Britain has long been discharging low and medium level

nuclear waste directly into the Irish Sea through a pipe-line 1½ miles long at the Windscale (Sellafield) plant in Cumbria.

The Swedish programme is still untried on a commercial scale.

The 'small amount' of commercial waste 'safely isolated' in the USA refers to the liquid waste resulting from reprocessing of spent fuel rods. In the USA, the amount from commercial reactor waste is sometimes called 'small' with respect to the amount from weapons reactors. The amount of reprocessing waste is measured in gallons, and therefore depends on the amount of dilution. In terms of curies of radioactivity, weapons waste and commercial reactor waste in the USA are about equal, with commercial waste growing at a faster rate. The military waste at Hanford is more diluted, hence there is a greater volume of liquid.[100] Under President Reagan's proposal to reprocess commercial waste for weapon use in the USA it will become useless to try to distinguish waste by source: all commercial waste will be military waste.

West Valley, New York

US commercial waste was reprocessed between 1966 and 1972 at West Valley, New York. The 600,000 gallon liquid high level waste residue is now stored in two large carbon steel tanks. The reprocessing plant was managed by Nuclear Fuel Service, a subsidiary of Paul Getty Oil Inc. It operated unsuccessfully between 1966 and 1972 when it closed supposedly for expansion. The average radiation exposure of workers at this plant increased steadily from its beginning and had reached 7.15 rem by the time it closed.[101] This average exposure for employees exceeds the maximum permissible nuclear worker exposure level of 5 rem per year.

After it became known that the nuclear processing plant had been built near an active earthquake fault, Nuclear Fuel Service declared bankruptcy and prepared to abandon the installation – leaving a residue of 600,000 gallons of high level radioactive nuclear reactor waste in two underground tanks. The carbon steel tanks are expected to last about thirty to thirty-five years, while the radioactive contents will remain dangerous for millions of years.

The table below gives the concentration of alpha-emitting radioactive particles in reactor fuel waste relative to natural soil. Plutonium and other actinides heavier than uranium, produced in fissioning, are alpha particle emitters.

Relative Concentration of Alpha Particles in
Various Types of Waste[102]

Natural earth	1
Uranium tailings, coal ash or fertiliser	50
Radium concentration in uranium mill tailings	500
Spent nuclear fuel rods	10,000,000,000

The radioactive waste at West Valley, a lovely fertile farm and
dairy area of New York State, contains as much radioactivity as
does the Hanford waste, but in a more concentrated form. If the
tank at West Valley springs a leak, the waste will be discharged into
Buttermilk Creek, Cataraugus Creek and Lake Erie – just upstream
from the water intake for the 1 million people in the Buffalo
metropolitan area. About 11 million people downstream, both in
the United States and Canada, will have seriously polluted drinking
water if the tanks are breached. The US federal government began
an experimental programme in 1982 to solidify this nuclear waste at
tax-payer expense.[103] Citizens are pressing to have the solidified
waste then removed from West Valley and placed in an as yet
unbuilt federal nuclear waste repository.

The 'America's Electrical Energy Companies' advertisement
goes on to discuss permanent waste storage. The text refers to 'most
of the left-over fuel' without noting that only a small part of the
nuclear industry waste can be called 'fuel'. It also mentions
recycling for 'future use' without referring to either weapons or
breeder reactors as recipients of this 'fuel' for 'future use'.

The stainless steel drums do not, of course, last for millions of
years, nor does the advertisement really claim that the geologically
stable sites coincide with the 'potentially acceptable sites' identified
by the US Geological Survey. The correspondence of the two is
implied by the juxtaposition of sentences.

The final section of the advertisement deals with the image of
'moving forward' rather than backward, and counterposes 'nations
moving forward with nuclear power' with national strategies to
reduce oil dependence. The reader is left to assume that nuclear
power reduces oil dependency, whereas the fact is that nuclear
power is used only to produce electricity, while oil goes primarily to
producing petrol and other petrochemical products, such as plastics,
fertilisers and medicines.

The most misleading part of the advertisement, however, is the
first paragraph which contrasts the electrical power generated for

Globe and Mail, 30 March 1984

AROUND THE WORLD
Reactors need dismantling

Vienna – more than 100 nuclear reactors around the world are close to retirement age and will soon have to be dismantled, raising the problem of how to dispose safely of the ensuing nuclear waste, according to a United Nations agency. The International Atomic Energy Agency said it was preparing a special safety guide to deal with this problem. About 80 nuclear research and test reactors have either been retired or are in the process of being dismantled in industrial countries. But of 260 test and research reactors in existence, 110 are now at the end of their 15- to 20-year lifespan. Twenty of these are in the developing world, where there is a lack of experience in the tricky dismantling processes.

750,000 people with high level radioactive waste amounting to 'only a four-foot cube'. It fails to mention the nearly 200,000 tons of uranium ore tailings waste resulting from production of one year's fuel. This waste contains thorium, radium, lead, bismuth and polonium in radioactive forms, and emits radon gas. It also fails to mention that the reactor waste could not actually be stored in a four-foot cube.

The waste has an extremely high temperature. It is still capable of fissioning and must be surrounded by sufficient water to absorb the neutrons and prevent a chain reaction from starting spontaneously. The space needed to store a four-foot cube of fission waste is about 10,000 times the actual size of the cube. This is roughly the size of a football field with a depth of 10 feet.[104]

Nuclear proponents usually speak of the 'size' of the waste, not the size of the storage area needed, and not the weight of the waste, which is 20 to 30 tons.[105] They also fail to mention the fact that this waste is produced at each plant for every year of operation.

Nuclear Workers' Strike

Daniel Cauchon and other workers from the Cap de la Hague nuclear reprocessing plant in France began a fast in the village church at Octeville to call attention to their strike for safe working conditions. They were interrupted by an order of the French government requiring a return to work in response to a 'state of emergency' at the plant. The management had appealed to the government after water tanks at la Hague became overheated and a gas explosion causing widespread contamination was feared.

The la Hague reprocessing plant had a history of broken pipes, dangerous build-ups of radioactivity, boilers buckling because of exposure to acids and intense heat, leaky valves, etc. Official records of worker contaminations show an increase from 280 during 1973 to 572 in 1975, the last year for which figures were released prior to the 1976 strike. In 1973 an explosion occurred at la Hague similar to the one which managers felt was imminent in 1976, hence government action to break the strike was credible.

The strike of September 1976 had been precipitated by two events, a serious accident in the summer of 1976, and rumours that the French government intended to turn over the management of the plant to private industry. These events triggered the first sit-in by workers at any nuclear installation in the world. The workers were demanding that the plant be closed, decontaminated, and fitted out with modern safety equipment.

The management obtained a court order to end the sit-in within twenty-four hours, but the strike lasted several weeks, providing an opportunity for the first national news coverage of the la Hague operation. Daniel Cauchon described the strike:

It was something of a big celebration. We had crawled out of the darkness into light. For the first time we met colleagues and saw that they were not robots, but comrades and friends.[106]

The union accepted the fact that safety reasons demanded that some workers remain on the job at la Hague, so the strike was never complete. They were not, however, prepared for the government declaration of a 'state of emergency', and have never known for certain whether or not this was really necessary. When the 'expert' threatens a serious radiological consequence, the 'non-expert' must conform even at the cost of loss of civil rights and loss of improvement in working conditions. Elite decision-makers in charge of a life-threatening technology command blind obedience.

As the workers feared, management of the la Hague plant was transferred to private industry, but they managed to force the French Nuclear Safety Board temporarily to suspend licensing of a proposed new reprocessing plant at la Hague. Hearings on the safety of workers and of the environment were begun.

Bernard Laponche, a physicist working with the French atomic authority, in late 1977 began to expose what he called the cover-up of serious safety problems which had plagued la Hague's whole history. Laponche's statements led the internationally known author Robert Jungk to interview workers at la Hague. He wrote:

> The workers at la Hague gave me a glimpse of the world's most frightening working conditions. They have sacrificed not only their health, but their rights to free speech and self-determination. They refer to themselves as 'radiation fodder' – the cannon fodder of the new technology. All fear that after a few years of working there they will be so much 'waste' dumped on the unemployment rolls, or, worse yet, the hospital. They have no faith in any sort of compensation for claims they might make years later for the delayed consequences of radiation overdoses. To date the record seems to confirm their pessimism.[107]

La Hague lies on the Peninsula of Cotentin, an isolated area of western France, too rocky to be an important farming area. When the la Hague nuclear fuel processing plant was first proposed, the local people were told that the plant would produce television sets. Soon rumours elaborated on this and the plant was described as also producing washing machines, refrigerators and other appliances. No one opposed the plant.

The 540-acre site was obtained and construction began. Two parts of the construction raised questions in the minds of the local people: the 100-metre (300-foot) chimney and the extraordinarily long pipe which ran far out to sea. This plant would obviously be releasing dangerous gases and liquids.

The mayor and town council repeatedly denied the fact that the plant would be hazardous or nuclear-related. Finally construction reached a point where scientists from Marcoule, France's first plutonium production installation, began to move in and the true purpose of the plant could no longer be denied.

'Official' information about la Hague stressed the safety of the plant which, it was claimed, posed 'absolutely no danger'. The high stacks would disperse dangerous toxic gases and the pipe would send liquid waste far from the French shore. Although it was not publicly discussed the local community realised that in case of a serious accident and possibly after the useful life of the plant, the peninsula of Cotentin could be isolated from the rest of France, 'sacrificed' to its polluted condition, and declared uninhabitable.

After an accident at the la Hague plant in the late 1960s, the French government had to buy up all the local milk because it was contaminated with radioactive iodine. The famous *beurre de la Hague* (la Hague butter) suffered a market decline in France and was renamed *beurre de Val de Saire*.

By 1975 French fishermen were reporting that the flesh of some fish had turned black. Government reports stated that radioactivity in fish caught at la Hague was five times greater than those caught at Cap Fréhel 62 miles away.[108]

Finally the people of the region formed a Committee Against the Atomic Pollution of la Hague. They requested assistance from Parisian scientists who dissented from France's nuclear commitment, and, with their help, documented radiation contamination of the land outside the plant boundaries to levels 10 to 20 times the legally permissible limit. These measurements were later confirmed in the presence of a notary.

The committee publicised the fact that between 185 and 203 per 1,000 deaths in the region around Cherbourg were due to cancer, while only 155 to 163 per 1,000 deaths were due to cancer in the region around Saint-Lô and Coutances.[109]

In 1976, the local people learned that six nuclear reactors were being planned for nearby Cap de Flamanville. High-tension lines would carry the power to Paris. The project was kept secret at first then submitted to a referendum. It is said that just before the voting an ore pit nearby was closed. The men who had been laid off were promised construction jobs and nuclear power plant jobs if the referendum was passed. The vote was 425 for the reactors, 248 against.

When the government illegally moved in machinery to test the soil, before they acquired the site, 200 peasants protested and occupied the site. They stayed almost a month until they were forced out by 250 armed troops accompanied by German shepherd dogs.

The farmers were required to obtain orange passes from Cherbourg in order to walk on their own fields. Barbed wire was erected around the proposed nuclear plant site and land was systematically purchased from farmers at very high prices. One small band of farmers formed a land pool of 161 acres, calling their organisation Groupement Foncier Agricole (Agricultural Lands Pool) to resist government buying. The French government immediately made plans to confiscate their land.

The waste from the la Hague reprocessing plant is placed in steel containers which are then deposited in grey concrete barrels. These are buried deep in the ground, near a site believed to be the Celtic necropolis. This historic site will be lost to history, too contaminated for archaeologists ever to explore, although no discussion of this loss seems to have ever been publicly permitted.

This waste disposal at la Hague is supposed to be 'temporary storage' for here as elsewhere the toxic material will far outlast the steel and concrete containers. Retrieval will be extremely dangerous and has never been tried.

The most dangerous nuclear waste at la Hague is dumped into steel containers 300 feet (100 metres) beneath the floor of the Fossé Nord Ouest, a low building 600 yards (550 metres) from the main plant. This building must be kept air-conditioned for centuries to come to prevent melting of the steel containers and release of the contents. The steel containers must be replaced every thirty to forty years.

A uranium enrichment plant is also planned for the French coast south of la Hague.

As these buildings one by one become too radioactive and deteriorated to use, they will be abandoned as unusable and dangerous monuments. It is questionable how long France and other nations will continue their vigilance over this waste. It is also questionable whether the people of Cotentin will be able to live much longer near the toxic nuclear centre at la Hague.

Meanwhile the private energy consortium COGEMA, which manages la Hague, is actively seeking new customers. In 1979, it obtained a contract to reprocess 1,600 tons of Japanese nuclear waste, at a cost of $1.62 billion (US). The West Germans are expected to pay COGEMA 2.5 billion Deutschmarks over the next few years to reprocess 1,750 tons of reactor waste. As more European countries and developing countries such as Brazil, Argentina and Iraq contract to send fuel to la Hague for reprocessing, the local French people and the nuclear workers have begun to realise that nuclear industrial development cannot be stopped by local or even national political action.[110] Decisions are being made in the boardrooms of transnational corporations and at top level meetings of governments and strategic planners.

Consensus within the Nuclear Establishment: The Australian Experience

Because historically nuclear technology is a recent phenomenon, the commonality of thinking within the industry can be partially explained by similar educational experiences, the restricted access to sensitive national security information and the like-mindedness which develops in in-group associations. Australia, though geographically remote from the US, Canadian and European nuclear centres and politically distant from the USSR, will serve as an illustration of the effective building-up of a pro-nuclear 'political will'.

The Australian Atomic Energy Commission (AAEC) was established in 1953 at a period of serious uranium exploration and mining in Rum Jungle in the Northern Territory. The uranium was ostensibly for the British weapon programme. The AAEC built two nuclear research reactors at Lucas Heights near Sydney, where they assembled a large number of research and technical staff. The Australian Atomic Energy Act and other restrictive legislation gave the Australian government power to inflict stiff fines and jail sentences for even speaking out against uranium mining. Although on the books, this legislation has not been used because of the strong anti-uranium stand of Australian trade unions.

Australia has its share of public advocates of nuclear weapons and technology, one of the most vocal being Sir Ernest Titterton, who took his academic degrees at Birmingham University in England. Between 1939 and 1943, Titterton was employed as a research officer with the British Admiralty, working on both radar and nuclear fission. He was part of the British atomic bomb team sent to Los Alamos, New Mexico in 1943–47, and he participated as senior member of the Timing Group for the first nuclear explosion at Alamagordo, New Mexico in July 1945.

Titterton was adviser on instrumentation for the Bikini weapon test in 1946, and later became head of the electronics division at Los Alamos, New Mexico. After four years at Los Alamos, Titterton returned to England to direct the research of the Atomic Energy Research establishment at Harwell. He went back to Australia in 1955, taking the chair of nuclear physics at the Australian National University in Canberra.

It had been a 'heady' twenty years of secrecy, high danger and excitement. He returned home a hero and an expert, privy to information and experiences not communicable to other people, bearer of information potentially a threat to national and international security in the nuclear age. It is hard to exaggerate the camaraderie and bonding among the men who participated in the birth of the nuclear age and who shared the sense of its apocalyptic horror. Their burden of secrecy and the popular aura of super-intelligence awarded to them guaranteed them special treatment and a unique self-image with respect to their peers in society.

Sir Ernest Titterton served on the Australian National Radiation Advisory Committee between 1957 and 1973, and the Atomic Weapons Test Safety Committee in the years 1955–73. He has received many academic honours and was knighted in 1970. He has written on a large range of subjects including energy sources and technologies and their hazards, and on the role of public opinion in nuclear decision-making. His 1956 book, *Facing the Atomic Future*,[111] and other popular writings, are frankly pro-nuclear propaganda. For example, he assumes that the global population will increase from 4 billion to 12 billion or 15 billion, with each inhabitant requiring as much electrical power as the average US citizen is projected to 'need' in the year 2000. This, of course, produces a staggering projection of global energy needs. He fails to mention that the links between gross national product (GNP), per capita energy consumption and standard of living are highly tenuous. They depend on the degree of planned obsolescence, on militarism, pollution/health balance, centralisation and complexity of societal structure and concomitant erosions of human and civil rights. Most population estimates for the year 2000 are a more modest 6 billion, and per capita energy need in the USA is modified considerably as conservation and efficiency reduce waste.

Sir Ernest Titterton unreservedly supports a plutonium economy and breeder reactors which he finds

safe; capable of generating electricity even more cheaply than the thermal nuclear stations; to have less impact on the environment than wind power, solar power or wave power.[112]

Breeder technology is of course untried on a commercial scale, predicted to be highly prone to serious fires and catastrophic accidents and hardly more benign than the low technology solar options. It could only be 'cost competitive' if supported by taxes rather than electrical rates.

Unlike most Western nuclear technologists, in 1965 Titterton spoke openly of the link between weapons and reactors. He favoured Australian development of the entire weapon cycle. He later modified his opinion and conformed to other nuclear proponents, saying that opposing nuclear reactor proliferation on the ground of possible diversion of the plutonium for weapons 'is a non-argument'.[113]

As Chairperson of the Australian Atomic Weapons Test Safety Commission, Titterton regularly assured the Australian people that they were in no danger from the British nuclear tests on their land, or the French tests in Polynesia.[114] As in other countries, the significance of the radioactive fallout was never communicated to the public.

Two British nuclear tests were carried out on the Monte Bello Islands off the west coast of Australia, and four tests, 'Operation Buffalo', near Maralinga in South Australia in 1956. Dr Hedley Marston, director of the division of biochemistry and nutrition of CSIRO, undertook to collect thyroid glands from grazing animals and analyse them to determine the extent of radioactive contamination. Marston was able to detect radioactive contamination in grazing animals within a 1,000 mile east-west band. He found heavily contaminated areas 1,500–2,000 miles away from the weapon test site on the north-eastern seaboard and in central western Queensland.

At Rockhampton, with a population of over 40,000, the levels of thyroid radioactivity after the second Monte Bello tests increased 100-fold above those observed after the first test – over all, a 3,000-fold increase over pre-test levels.[115]

The same areas received further fallout from the Maralinga tests in September and October of 1956. When Sir Ernest Titterton was asked to comment on Marston's findings he refused, reasserting that the tests were carried out safely.

Sir Ernest Titterton is not the only strongly pro-nuclear voice in Australia. He is supported by Sir Philip Baxter, a native of Wales, who also took his degrees at the University of Birmingham in England, receiving a PhD in mechanical engineering in 1928. Baxter became research director of the General Chemicals Division of Imperial Chemical Industries and a specialist in the chemistry of uranium. In 1941 he produced Britain's first uranium hexafluoride.

Baxter was a consultant to the British and US war-team which produced the atomic bomb, and he spent the year 1944–45 at Oak Ridge, Tennessee leading the British research team. He continued as a consultant to the British nuclear programme between 1945 and

1950, then emigrated to Australia where he became a professor of chemical engineering at New South Wales University of Technology (later the University of New South Wales), and consultant to the developing uranium mining industry.

The Australian Atomic Energy Commission was formed in 1953 and John Philip Baxter was named Deputy Chairman. He became Chairman in 1957, on a part-time basis, changing to full-time Chairman in 1969, a position he held until his retirement in 1972. He served with Titterton on the National Radiation Advisory Board until it was disbanded in 1973. Sir Philip was knighted in 1965.

Like Titterton, Baxter favoured the development of nuclear weapons and nuclear power in Australia. He claims that nuclear power does not cause pollution 'in the sense in which most of us use that word', with releases always within permitted limits.[116] Baxter suggests that all health hazards have been, are and will be taken care of by appropriate experts. He dismisses radioactive waste problems, stating:

> The required processes have been well established on a pilot scale and no problems need be expected.[117]

On the other hand, he continually proclaims against coal, claiming that its use for electricity production,

> apart from killing a very large number of people, would create environmental hazards which can only be described as horrific.[118]

It is disturbing that Baxter, a mechanical and chemical engineer, does not mention the large number of preventable coal-related deaths due to violations of health and safety laws for mining, and the ability of modern technology to significantly reduce noxious emissions from coal-fired plants. Also left unmentioned is the fact that deaths caused by the coal-fired plants occur during the lifespan of those receiving the benefits, while most nuclear-related deaths will be charged to future generations. The use of fossil fuel combustion must be reduced because of carbon dioxide problems, the same problems aggravated by forest depletion, but replacement with life-compatible renewable low technology alternatives is accepted by most scientists as the best long-term policy for energy production. It is basically its military connections which have made the nuclear path 'desirable' to national planners.

Unexpectedly, in 1978, Sir Philip reversed his stand against coal and became a vocal supporter of coal for synthetic liquid fuel.[119] Although he does not admit the fact, this rejection of coal for generating electricity and acceptance of it for liquid fuel suitable for transportation uses, reflects an understanding that nuclear power will not reduce oil dependency. The change in position on coal coincided with President Carter's announcement of the coal-based synthetic fuel programme in the USA.

Sir Philip Baxter favours a strong world government by nuclear nations and disfavours public debate. Ann Mozley Moyal describes him as follows:

He saw himself, and by extension the Commission (AAEC), as the central and sole source of policy proposals for nuclear development in Australia, and as the Government's Fountain-head of atomic knowledge and ideas.[120]

Sir Philip replied:

The concept that all decisions of government should be subject of public debate . . . is a dangerous heresy . . . The experts must in the end be trusted.[121]

While such a policy may be arguable when a technical problem demands a technical answer, it is not arguable when socio-political problems are systematically reduced to technical problems to be answered by technical experts.

As nuclear experts do in other countries, Sir Philip dismisses serious opponents to nuclear technology as ignorant, fearful, well-financed and communist-backed obstructors of progress. This is especially apparent in his response to Dr Diesendorf of the Society for Social Responsibility in Science. The designation 'expert' is limited to those who agree on the basic social questions which underlie technical solutions.

In Australia the nuclear establishment has clearly delineated who are the 'experts'. A pamphlet compiled by John Grover and widely disseminated to media and educational institutions called 'Uranium and Nuclear Energy Reliable Information Sources', contains as its introduction:

Attached is a list of informed persons in the Sydney area with a factual knowledge of uranium and nuclear issues. All of them are biased on the side of the facts but there is nothing wrong with this.

Persons listed by John Grover, January 1979

Sir Philip Baxter	– former Chairman, AAEC
Dr G. M. Watson	– retired, AAEC, Environmental and Pacific Health Division
Mr Leslie Kemeny	– Nuclear Engineer, University of New South Wales (Lucas Heights Reactor site)
Mr Les Nicholls	– Manager of Operations, Ranger Uranium Mines Ltd
Mr Tim Hooke	– Uranium Explorations Development Officer, Geopeko Ltd
Mr Geoff Sherrington	– Geochemist, Geopeko Ltd
Dr Paul Mailop	– Chief Metallurgist and Manager-elect of the Narbalek Uranium Mine
Mrs Charlotte A. Lawler	– Homemaker; wife of the Managing Director of the Queensland Mines Ltd (uranium mines) Mr J. W. Lawler
Mr John Grover	– Manager, Special Projects, Peko-Wallsend Ltd

All are financially dependent on uranium extraction and nuclear technology.

Whenever you need reliable research material without delay or people able to qualify the anti-uranium or anti-nuclear misinformation produced by the coterie of ex-scientists in the USA for the Australian movement, this list will be useful. [15 January 1979.]

I had first-hand experience of how this Australian 'reliable information source' works in February and March 1979. I was invited to Australia by six trade unions seeking information on worker health hazards in uranium mining, milling and transportation. They arranged my tour of Australia, where I spoke to workers, scientists, health professionals, church groups, and general audiences. The small nuclear research community at Lucas Heights expressed no interest in hosting a scientific discussion of the results of my ten years' research on radiation-related leukaemia and on the health effects of low level radiation which provided the basis of my public presentations. They never expressed scientific objections to my scientific journal publications, or objected to my concern about the health effects of radiation. They never even expressed a desire to have a dialogue about the Tri-State Leukaemia data base – 16 million people followed over a three-year period – which I had analysed.

While in Australia, I agreed to an interview for a radio talk show out of Sydney. During the interview, one of Australia's 'reliable experts', hidden in another room at the radio studio, listening to the interview, kept sending written messages to the interviewer. The fed-in questions were misleading for the audience, often off the topic, and prevented direct presentation of my major published scientific research findings. After I left the studio, the 'responsible expert' was allowed fifteen minutes air time to denounce what I had said without rebuttal. He attacked my competence and personal integrity. I was denied time to respond to the direct slander. The Australian audience was left with no way to distinguish fact from fiction.

When challenged to a public debate on Australian TV these 'reliable experts' refused. They are apparently only comfortable when their authority is unquestioned and the nuclear debate can be reduced to competition for public 'trust' rather than understanding. The general public has no way of knowing about their refusal to debate.

After I left Australia, the 'reliable experts' published letters denouncing me in most major journals and newspapers. Only the Editor of the *Australian Medical Association Journal* invited me to respond and he limited the number of words I might use. This is hardly scientific, honest or even mature behaviour on a question of major public concern.

As do many proponents of nuclear technology for weapons and power, Sir Philip Baxter dismisses 'anti-nuclear scientists', i.e. those whose research findings are unwanted, as part of an international communist conspiracy. This allows him to systematically fail to deal with such scientists, as 'not worth listening to'. It justifies not replying directly to their arguments. Capitalists talk about 'anti-nuclear communists' and communists of 'anti-nuclear imperialists'.

The following is a translation of a statement by Georges Marchais, the head of the French Communist Party, which appeared in *Der Spiegel* on 1 February 1981 (p. 16):

The French are the world-leaders in the atomic industry. That is why the Americans are doing everything to prevent their further development. I accuse the opponents of nuclear energy of being agents of American imperialism.

Sir Philip Baxter's writings between 1971 and 1975 reflect a preoccupation with impending global military disaster, recommending that Australia arm itself with nuclear weapons to protect itself from invaders and refugees after the holocaust. He sees over-population as one of 'the root causes of environmental deterioration',[122] concluding that the world would be better if it contained 1.5 billion rather than 3 billion people. His desire for zero population growth was, however, for the non-Australian world; Australia he saw as 'the last big continent which the white man has to develop and populate'.[123] His vision of the future seems to include self-sufficiency for Australia, and protection from refugees with 'shattered gene systems from radiation exposure' or who 'carry uncontrollable infectious disease' and from invaders fleeing nuclear destruction. National security measures for the protection of Australia should include, in his opinion, nuclear, chemical and biological weapons.[124]

Sir Philip's dreams of the nuclear military development of Australia have not as yet been realised. At least one major opposition force to such development has been the influential Australian trade unions. However, as the global military budget soars to $1 million (US) per *minute*, his predictions of nuclear war are increasingly credible.

United States–Australian nuclear co-operation has been growing since the Northwest Cape base agreement (1963) and the Nurrungar agreement (1969). The Northwest Cape Agreement includes a Naval Communications Station in western Australia for beaming nuclear 'go' signals to submarines in the Indian and Western Pacific oceans. The Pine Gap base, Northern Territory, is a joint research amenity which monitors and controls both the American geostationary 949–647 satellite system which provides inter-continental ballistic missile launch warnings to US Strategic Air Command and also the lower orbit satellite system, 'Big Bird', which is used for military surveillance of the Soviet Union, China, India and Indochina. The Nurrungar communication centre in South Australia transmits early warning and reconnaissance data via military satellite to US bases.[125] Covert US Central Intelligence Agency operations from the bases in Australia have been documented[126, 127] and the US presence has increased since the Soviet invasion of Afghanistan. In April 1980, the day after a meeting with the US Joint Military Chief of Staff,

Mr Killen . . . told Parliament that the Cockburn Sound offer had

been made to ensure the US is not lacking in support of its
Australian ally in the heavy burden which it bears in deterring
war and nuclear attack.

Cockburn Sound in Western Australia is being offered to the USA
as a new nuclear base. Apparently it will be especially designed to
service the Trident nuclear submarines.

Far from being a haven from nuclear holocaust, Australia has
become a prime nuclear target, 'co-belligerent with the US in its
confrontation with Russia'.[128]

This survey of the nuclear intellectual climate in Australia is
important for understanding the fanaticism of the pro-nuclear
movement. There is obviously no need to invoke a 'conspiracy
theory' or to assign a profit motivation to explain this global
phenomenon. A small number of crusading proponents shared an
incredibly moving experience – the development, testing and use of
the first thermonuclear devices. Their sense of personal survival
demanded 'living with' and 'controlling' this technology. The logic
was simple. One could not unlearn the nuclear science. An enemy
armed with this technology was unbeatable. There was nowhere to
go to escape the nuclear age. Therefore, like it or not, other people
must be cajoled into co-operation. Some co-operated willingly
because of their limited knowledge of human health. They
essentially trusted that the biological problems had been solved by
the experts. Those who opposed nuclear development were
discredited, economically penalised and kept busy responding to
lengthy, complicated government defences of the health and safety
aspects of the industry.

Elaborate national and international review committees were set
up to politically re-screen scientific literature which had already
been screened by scientific review. These political reviews were
made the new arbiters between 'dissenting scientific opinions'.
Dissenters from the arbitration became 'ex-scientists'. As the basic
scientific claims of nuclear proponents were more seriously challenged, the political review committees, like the United Nations
Scientific Committee on the Effects of Atomic Radiation
(UNSCEAR) and the US National Academy of Science Committee
on the Biological Effects of Ionising Radiation (BEIR), became
more prestigious. Scientists in other disciplines tended to accept
these national and international arbiters as unbiased judges,
especially because their publications were used as the basic
educational sources in universities.

It seems clear that this situation evolved gradually from decisions made by a group of 'special persons' who shared the peak experience of a thermonuclear explosion, an experience not easily shared with outsiders. The experts were primarily physicists, chemists and engineers; they were born between 1890 and 1920 and they had no scruples about using science to promote war efforts. This biased subset of the larger scientific community shared both the awesome secrets and the heady honours and responsibilities of the nuclear age.

For the most part no one broke ranks, with the notable exception of Dr Robert Oppenheimer who was eventually ostracised because of his opposition to the hydrogen bomb, and Leo Szilard, the Hungarian-born physicist who had suggested the nuclear chain reaction experiment in the first place.

The majority of Manhattan Project scientists, while fearing the awesome power of nuclear science, also had an overwhelming conviction that it must be pursued at all costs. These latter have dominated national science advisory apparatus, trained and

Nuclear Weapon Tests Prior to 5 August 1963

Year	USA				USSR	UK	France
	Pacific	Nevada	Other	Total			
1945	3			3			
1946	2			2			
1947							
1948	3			3			
1949					1		
1950							
1951	4	11		15	2		
1952	2	8		10		1	
1953		11		11		2	
1954	6			6	1		
1955	1	17		18	4		
1956	17	1		18	7	6	
1957		31	1	32	13	7	
1958	35	39	3	77	25	5	
1959							
1960							3
1961		9	1	10	50		1
1962	36	62		98	39		
1963		25	3	28			
Total	109	214	8	331*	142	21*	4

* Two tests were joint USA–UK tests.

selected their successors, and attracted supporters by their strong personal conviction and clarity of vision.

The Manhattan Project may well be classified as one of the major mind-altering events of recorded history. It contained the seeds of true peace as well as the seeds of catastrophic war. It was an extraordinary experience of group science, producing a collective result far surpassing the sum of the component parts. It was an international accomplishment: Albert Einstein, a German Jew who defined matter as 'frozen energy' in 1917; Niels Bohr, a Dane who described atomic structure; Ernest Rutherford, a British physicist who suggested an atom could be caused to disintegrate; Otto Hahn and Franz Strassman, German physicists who discovered that the uranium atom could be caused to split when subjected to a slow stream of neutrons; Leo Szilard, a Hungarian physicist who suggested the chain reaction to sustain fissioning; and Enrico Fermi, an Italian physicist who achieved the first sustained nuclear chain reaction below the football stadium of the University of Chicago in June 1942. The total process took about twenty-five years. Scientists would never again work in physical and intellectual isolation. The possibility of intellectual collaboration for mega-scientific projects far outshadowed in importance any previous mega-construction projects undertaken by humans, such as castle-building, fortification of cities, pyramids, etc.

For those who experienced the thrill of accomplishment when large amounts of money and human labour were pooled to achieve the nuclear bomb, this was perceived as the beginning of a new 'style' of group science. The science 'establishment' was spawned. A well-funded elite could engage in scientific research on a full-time basis.

I call this experience a seed of peace because, unlike co-operation for war-making which had fuelled the movement towards fortified city-states and nations, co-operation for science *could* be directed towards the alleviation of human suffering and deliberate peace building. A hint of this can be seen in such clichés as War on Cancer or War on Poverty. Unfortunately, the scientific co-operation to date has remained primarily under the control of the military segment of society and is not yet free enough to be a major force for peace. Even the medical sector concentrates on 'curing' the effects of military science, namely through cancer research and recombitant DNA. Preventive health action would require phase-out of all war.

Another notable difference between construction mega-projects and science mega-projects is that the latter require free co-operation, while the former can be forced.

Opinion Moulders

The Bilderbergers continue to meet and informally build a consensus among first world national leaders, in spite of their more visible offspring, the Trilateral Commission. In May 1983 Canada hosted a meeting of 100 people at Montebello in Quebec. Prime Minister Trudeau, former US Secretary of State Kissinger, banker David Rockefeller, former West German president Walter Scheel, *Time* Editor-in-Chief Henry Anatole Grunwald and the Editor-in-Chief of the Montreal LeDevoir, Lise Bissonnette, were in attendance. Others in attendance at this heavily guarded meeting were Donald McDonald, former Canadian Cabinet Minister and Head of a Royal Commission studying the Canadian economy; Allan MacEachen, Canadian Minister of External Affairs; Conrad Black, Chairperson of Argus Corporation Ltd; Paul Volcker, Chairperson of the US Federal Reserve Board; former French Premier Raymond Barre; European banker Evelyn de Rothschild and Lord Carrington, former British foreign minister.[129]

Agenda items for the 1983 meeting included East–West relations, Containment, Detente or Confrontation, Protection and Employment, and Risks in Banking and Finance. One wonders at the human costs of the armed truce which has held the world in terror since 1945. These costs are being managed quietly by the Bilderberger elite.

Spaceship Earth

While sophisticated think tanks spent long hours trying to grasp the interdependence of the global economic system, one dramatic picture served instantly to awaken millions of people to reality. In 1972, a spaceship photograph of the beautiful blue marble, earth, floating uniquely and serenely against a black hostile universe, dramatically spoke for the fragility and finiteness of the human habitat.

Almost simultaneously a report *The Limits to Growth*,[130] by Donella and Dennis Meadows, Jorgen Randers and William Behrens III, based on a computer projection of global resource availability, appeared in paperback form. The publication went through nine printings and two editions between October 1972 and September 1975. *The Limits to Growth* grew out of an informal international association, sometimes referred to as 'invisible college' and known as the Club of Rome, which began to meet in April 1968.

> None of its members holds public office nor does the group seek to express any single ideological, political or national point of view. All are united, however, by their overriding conviction that the major problems facing mankind [humankind] are of such complexity and are so interrelated that traditional institutions and policies are no longer able to cope with them, nor even come to grips with their full content.[131]

The Club of Rome approached the global agenda quite differently from Bilderbergers and Trilateralists. They attempted a non-political study of the main problems and their interacting aspects, producing a computer simulation of the global system on which they could test economic and social changes.

Basic problems common to all nations were:

> poverty in the midst of plenty, degradation of the environment, loss of faith in institutions, uncontrolled urban spread, insecurity of employment, alienation of youth, rejection of traditional values, inflation and other monetary and economic disruption.[132]

These were basically the human fallout caused by Bilderberger and Trilateral planning and the unprecedented international arms race. While politicians sought to control these unwanted side effects by information control and external force against the victims, the Club of Rome attempted through systems analysis to design a society which would alleviate the basic global system malfunctioning.

So the Club of Rome backed Potomac Associates to produce a computerised global systems model and test out various future scenarios as a preparation for the book. The Club members identified two positive feedback loops, i.e. areas growing without restraint, namely population and industrial capital. Both were experiencing exponential growth. The negative feedback loops which would eventually stop the growth were identified as pollution of the environment, depletion of non-renewable resources and famine. They identified some activities as short-term helps to relieve the pressure of the runaway growth, but could find no reason to believe that the global system, unaltered, would not collapse. Given this world view, it is not difficult to identify some aspects of modern life, such as improved medical care, as 'negative', since they favour population growth. Even war and natural disaster could be seen as 'positive' since they reduce population. The report called for the stabilisation of population size so that the number of births would equal the number of deaths. They also wanted a limitation on industrial capital investment such that it equalled depreciation, and the diversion of excess industrial capacity into the production of consumer goods.

These drastic measures could not alone prevent the collapse of the world system. They would have to be matched by curbs on the negative feedback loops. This would entail a drastic reduction in per capita pollution and in the depletion of irreplaceable resources. Only such a steady state, dynamic equilibrium could be expected to support a global village.

> It seems possible, however, that a society released from struggling with the many problems caused by growth may have more energy and ingenuity available for solving other problems.[133]

In such a steady state the arts could flourish along with other non-polluting human endeavours such as religion, basic scientific research, education, athletics and social interactions. These are essentially non-consuming and non-polluting activities which could flourish as leisure time increased.

While there have been many criticisms of *Limits to Growth*, it is a sobering book which raises many new questions. Even those persons most optimistic about new technological 'solutions' to global problems were forced to agree that they would only delay the collapse problem. The dynamic nature of the growth–collapse system was perhaps the strongest motivating force for change set forth in this book. With exponential growth within a finite system we cannot put off change to the next generation.

The Club of Rome had put its finger on the process of global economic and ecological collapse and yet it hardly grasped the subtle biological collapse through damage of the gene pool which had already begun. A slightly damaged gene pool means offspring physically unable to cope with pollution as well as their parents have coped. This means that for a 'steady state' of quality of life, pollution level must be lower per capita than it was for the previous generation.

For many people the elaborate computer simulation undertaken in preparation for *The Limits to Growth* was unnecessary. The picture of spaceship earth, a closed ecological garden to be shared with others, told the whole story.

In a prophetic message in 1854, Chief Seattle, when he surrendered Indian land in the territory of Washington to the US government stated:

> The shining water that moves in the streams and rivers is not just water, but the blood of our ancestors. If we sell you our land, you must remember that it is sacred and you must teach your children that it is sacred . . .
>
> The rivers are our brothers, they quench our thirst . . .
>
> The air is precious to the red man, for all things share the same breath – the beast, the tree, the man, they all share the same breath . . .
>
> This we know. The earth does not belong to man: man belongs to the earth. This we know. All things are connected like the blood which unites one family. All things are connected.
>
> Whatever befalls the earth befalls the sons of the earth. Man did not weave the web of life; he is merely a strand in it. Whatever he does to the web, he does to himself . . .
>
> The white man too shall pass; perhaps sooner than all other tribes. Continue to contaminate your beds, and you will one night suffocate in your own waste.[134]

This lesson: 'All things are connected' appears to be as difficult for humans to comprehend as was the discovery that the earth was not the centre of the solar system. The problem of radioactive waste is related to those connections.

The 1970s was a period of growing awareness of the implications of the nuclear age. With deepening understanding comes initial confusion because of the contradiction between desired outcomes like peace and economic growth, and actual outcomes like the nuclear arms race and environmental disasters. As we begin to contrast causes and effects as they were perceived prior to the Second World War with causes and effects as they are perceived in today's technological world, we can hope for new directions and ways of coping to emerge. The old historical wisdom has outlived its usefulness. The hard work of unpacking the myths and bursting bubbles of false hope must precede the changing of ill-founded belief systems, outdated paradigms and destructive behaviour patterns into new ways of living and enjoying life in a post-war and post-industrial global village. Failure to measure up to this change may mean outright obliteration of the species or, at the least, delayed maturation.

In a world come into adulthood, exotic mega-projects or wars which endanger the shared biosphere are suicidal and omnicidal. The time of constant growth is over. It is now time to bloom.

PART FOUR

A Time to Bloom:

Towards Change

I will open up rivers on the bare heights,
 and fountains in the broad valleys;
I will turn the desert into a marshland,
 and the dry ground into springs of water.
I will plant in the desert the cedar, acacia,
 myrtle, and olive;
I will set in the wasteland the cypress,
 together with the plane tree and the pine,
That all may see and know, observe and
 understand,
That the hand of the Lord has done this,
 the Holy One of Israel has created it.

Isaiah 41:18–20

One of the several remarkable happenings of 1972 was that magnificent snap-shot sent back by astronauts of a beautiful blue marble against a backdrop of hostile black space. The concept of 'spaceship earth' was both attractive and frightening, highlighting the fragility and uniqueness of earth as well as its comforting habitability. This idea of smallness, preciousness and solidarity became etched on our consciousness, helping to shape the 1980s.

In the same year, the Club of Rome issued its report *The Limits to Growth* on the predicament of humankind, and the United Nations convened the World Conference on the Environment in Stockholm. The growth of world population and the limited nature of world resources became a focus of consciousness. The problems of the accumulation of highly toxic wastes in our planet home began to disturb thoughtful and responsible citizens.

Without much political power, the United Nations had been developing quietly over the first three decades of its existence. It continuously expanded to include new nations, as one by one they followed the lead of Gandhi's India and achieved political inde-

World Conferences Sponsored by the United Nations

1972	Stockholm	World Conference on the Environment
1974	Rome	World Food Conference
	Bucharest	World Conference on Population
	Caracas	World Conference on the Sea
1975	Mexico City	World Conference on Women
1976	Geneva	World Conference on Employment and Basic Needs
	Vancouver	World Conference on Human Settlements – HABITAT
1977	Mar del Plata	World Water Conference
	Nairobi	World Conference on Desertification
1978	Buenos Aires	World Conference on Technical Co-operation among Developing Countries
	Geneva	Law of the Seas Conference
	New York	First United Nations Special Session on Disarmament
	Alma-Ata	International Conference on Primary Health Care
1979	Rome	World Conference on Land Reform
	Vienna	World Conference on Science and Technology
		International Year of the Child

pendence. As the developing nations gained more seats in the UN General Assembly, first world nations withdrew to drawing-room diplomacy, playing down UN accomplishments and blocking its potential for world leadership in the usual political sense. The UN Security Council, with the 'big five' nations' veto power, became both obstructive and anachronistic.

The United Nations developed its own style of calling for international consciousness through single issue world conferences, international years dedicated to oppressed or forgotten groups of people, and special sessions on disarmament. Unlike the Bilderbergers, the Trilateral Commission or the Club of Rome, the UN does not have the backing of big money, the media or personally powerful members. However, many concerned people have formed on a volunteer basis a supportive network of Non-governmental Organisations (NGOs) affiliated with the UN and providing it with a research and communication network. Though often not aware of

1972	Santiago	⎫	World Conference on Trade and
1976	Nairobi	⎬	Development – UNCTAD
1979	Manila	⎭	
1980	New York		Special Session on Economic and Social Issues
	Copenhagen		Mid-Decade Conference on Women
1981	New York		World Conference on the Law of the Sea – UNCLOS
	Nairobi		World Conference on New and Renewable Sources of Energy
			International Year of the Disabled
1982	New York		Second Special Session on Disarmament
			World Conference on the Elderly
1983	Vienna		Meetings on International Communication
1984			International Youth Year
	Mexico City		International Conference on Population
1985	Nairobi		World Conference ending the UN Decade for Women
	Geneva		Third Review of the Parties to the Treaty on Non-proliferation of Nuclear Weapons
1986			World Peace Year

the catalytic role of the UN, people generally *did* know of the conferences and have become aware of the major global survival questions.

Much of the conceptual framework of the UN Conference on the Environment in Stockholm was prepared by Barbara Ward and René Dubois, and they suggest a few basic truths on which to build:

'We can murder the biosphere.'
'Ultimately our planet is a finite and limited system.'
'We cannot run a functioning planetary society on the totally irresponsible sovereignty of a hundred and twenty different countries.'[1]

Barbara Ward gave one of the major addresses of the Conference, calling for a planetary system which is 'complex, human and loveable'. Her vision included respect for local diversity and autonomy, with essential global unities needed to achieve the best possible co-ordination and distribution of limited resources. She identified quite clearly the obstacles to the normal growth of the earth system into a peace-based equitable human habitat. The first obstacle was the resistance of people in developed countries to the sacrifices needed to stabilise the world. Sometimes these sacrifices are merely a second or third family car, or a third television set. Material goods had become the standard of 'success' in developed countries and in the elite class in developing countries, and the basis of an individual's self-worth in the eyes of peers.

The second obstacle to a hopeful outcome was the illusion that because world leaders talk about a problem, they are doing something to correct the situation. This rhetoric produces a mesmerisation which blots out of mind starving children, festering cities, dying oceans and a wasted earth. Settled habits, existing institutions and established interests prevent action on behalf of survival. People seem generally incapable of even thinking about the ultimate calamity caused by uncontrolled population growth, pollution of air, food and water, resource exhaustion, social alienation and nuclear holocaust.

World Issues Ranked in Order of Seriousness by the Global Futurist Network

1.43	Nuclear weapons proliferation
1.70	World poverty
2.00	Starvation
2.00	Nuclear war
2.02	Minority rights
2.10	Pollution
2.11	Potential nuclear accident
2.19	Overconsumption by rich nations
2.42	Depletion of water resources
2.45	Depletion of arable lands
2.47	Overpopulation
2.51	Lack of long-range planning
2.92	Inflation
3.17	Depletion of mineral resources
3.19	Unemployment
3.32	Energy shortage

Rank: 1 Very serious
    4 Moderately serious
    7 Not very serious

Nations were preparing for wars of utter and total destruction in which they would gain no land, no resources, no culture or science, and in which they stood to lose what little resources and good they now had. 'In fact if we are greedy about this delicate planet, we shall simply have no planet.'[2]

During the 1970s, more and more women were coming to see the connections between the oppression of women, the nuclear arms race, violence, concentration of money and power, and greed. The international tensions were not an inevitable result of 'human nature', they were, rather, a blatant failure to control ignoble human drives, especially the drive to oppress the physically 'weak' within the nation-state and between nations, the age-old 'rule of the fist'. The situation was basically reversible since the change required was one of attitude. Simply protecting others from our unruly aggression, rather than being overly preoccupied with protecting ourselves and our possessions from others would be sufficient.

We had in the past formed towns and cities and nations, without destroying family ties and traditions. Were we not capable of forming a global village? Could we not build a society based on peaceful co-operation? Could not the global village be multi-cultural and multi-lingual, preserving the best of national pride?

Food

Is a world confronted with the impossibility of 'constant growth', and with the obvious global failure of the trickle-down theory – by which the prosperity of a few provides a higher standard of living for all – capable of change? The theory had appeared to work in the USA and Europe, but the impoverishment and resource depletion of colonies was the sad residue of the first world's 'economic growth'. The costs of growth had been externalised to the colonies. How could the costs of planetary economic growth be externalised? The options seemed to be to charge the cost to future generations (pollution, destroyed gene pool) or to find an alternative to the growth ethic itself. Perhaps growth could cease, and blooming begin.

A time to bloom is one in which the fundamental human needs for food, shelter, clothing, medical care, education, work and social dignity are met, so that the tremendous human potential for creativity, art, science and culture is freed. The ability to 'bloom' is an untried human potential. In the past it has been starved and unexpressed because of the struggle for mere survival and elementary social functioning. It is also a potential which in itself makes modest demands on the material resources and money of the world community. In the 1980s the 'hardware' global infrastructure and technological requirements for communication, equitable distribution of goods and services, medical and social assistance are largely in place. The human infrastructure of solidarity is not as completely developed, but a beginning has been made. The stage is set for a global celebration – but the people have no food with which to celebrate, little hope to give them joy. The money for food and celebration of life is being heedlessly squandered on weapons and unneeded technological mega-projects.

The United Nations Food and Agriculture Organisation (FAO) reported in January 1981 that for the second year in a row the world community consumed more grain than it produced, and grain reserves are now at about 14 percent of one year's global consumption. This reserve is far below the level set by FAO as a minimum for global food security.

In the face of this fragile and extremely vulnerable world food situation, developed countries are paying farmers to reduce crop production, and developing countries are raising cash crops for export to provide money for armaments. The rising costs of fertilisers, transportation and shipping are expected to cause a doubling of 1970 food prices in the market places of the world by the year 2000. This prediction has already come true in some countries of the world.

Governments are using food as an instrument of force both to subdue citizens within the nation and as a bargaining tool between nations. Inflated food prices together with high unemployment help keep workers docile. Those malnourished in childhood tend to be of lower mental ability and so are more easily exploited in adulthood. Food availability and prices differ widely between the various ghetto-like nation-states, with poorer states frequently paying more.

Good farmland is being used for tobacco or other cash cropping to the detriment of the health and survival of the population. Tobacco, for example, so depletes the soil of nutrients that the land cannot be used for other crops for two or three years after the tobacco has been harvested. Further, cigarette manufacturers prefer 'flue-cured', i.e. smoke dried, tobacco. In Kenya about three acres of trees are felled for flue-curing for every acre of tobacco planted. In the Tabora district of Tanzania over 500,000 acres of forest have been destroyed over a twenty-year period to produce 120,000 tons of tobacco. The forests, of course, take much longer than three years to regenerate. The resulting bareness of the treeless soil causes it to be eroded still more by wind and rain. The forests also provide about half the atmospheric oxygen needed to sustain life. It goes without saying that the global community has no need for tobacco – a lung and heart disease-causing drug – such that this gross destruction of forest and reduction of food and quality of soil in developing countries could be justified.

First world analyses of deforestation in third world countries tend frequently to blame the victim, i.e. they assume that firewood gathered for cooking is the cause of the problem. When the problem is thus defined population growth becomes an easy target of development policy. Apart from failing to deal with the underlying economic inequity and ecological–health perspective, such an approach also fails to see large families functioning in poorer countries as social security, unemployment insurance, health care and old age assistance. Without a stabilised global economic and social system there can be no voluntary control of procreation. Force only serves to destabilise further the global system and to intensify human suffering.

According to a United Nations Food and Agricultural Organisation, all suitable land globally will have been developed for farming by 1985. The only hope for humankind to have an adequate food supply will be better management of present farmland. This in turn leads to rethinking our addictive militarism. Poisoning of the land through war, through the dumping of defoliants and of chemical, radioactive chemical and biological warfare materials, inevitably precludes this future possibility for the earth.

Instead of constantly reacting with stop-gap measures to mitigate the harmful effects of militarism and poor land management, positive encouragement to develop new symbiotic relationships between humans and their environment is required. A new social integration would grow out of the diversity thus generated. Achieving the global village phase of social evolution depends enormously on the global sharing of the humanisation process by each local component of the earth family, i.e. on co-operation as opposed to competition or confrontation. We must condemn the rule of the fist as suicidal for the human family.

An Integrated Farm System (IFS) makes use of nature's own processes of photosynthesis and bio-degradation to provide increased food production together with energy and fertiliser self-sufficiency in a confined space. Such farms have been established in Fiji, Papua-New Guinea, Thailand, India and Sri Lanka.

M. Amaratunga has described a 2½-acre IFS in Sri Lanka. The available land was allocated as follows:[3]

2	percent shelter for humans
0.5	percent shelter for animals
1	percent other buildings
0.5	percent bio-gas plant and setting pond
2	percent algae ponds
5	percent fish ponds
9	percent pathways and roadways
80	percent agriculture

The farm has five cows, three calves and about 100 ducks.

Anaerobic* digestion of cow dung more than doubles the available nitrogen hence it is a better fertiliser than compost. The bio-gas plant on this small farm digests human and animal dung and other organic waste such as the non-edible parts of plants. The gas produced is methane, suitable for use in cooking, lighting, motor fuel and a small (1 kilowatt) electrical generator.

The bio-gas plant also provides improved sanitation and weed control. When the sludge is retained for thirty days or more, all weed germination stops and human pathogens are killed. Plots fertilised with the sludge are virtually weed free. Rural epidemics of bowel disease cease.

The protein content of dried algae is between 40 and 80 percent and it can be a valuable source of animal feed. On this small farm, the algae pond yields about 500 kg a year.

The annual yield of fish protein from the fish pond is about 150 kg. The pond is also used to sustain about 100 ducks.

It is obvious that the Integrated Farming System brings together experts in many fields: agriculturists, botanists, mechanical and chemical engineers and inventors. It provides an alternative energy to wood, thus reducing deforestation; it improves the soil and the crop yield, reduces rural health problems and above all, provides people with a healthy spirit of self-reliance. Diversity of skills and

---

* Anaerobic bacteria have the ability to use oxygen in compounds for metabolism rather than free oxygen available in air. They do not require air in order to live.

Solar-powered Village

Bou Saada, one of 1,000 Socialist Villages being constructed by the government of Algeria to provide better housing for the rural poor, is using solar energy to meet the needs of its 1,500 inhabitants. Bou Saada, on a high plateau, about 125 miles south of Algiers, is being built entirely of locally available materials. Water and solid waste is being recycled through a biogas plant, and solar collectors provide for electrical energy needs. Solar technology is being harnessed for new farming techniques and new high-yield crops are being developed.

This project is being supervised by the Algerian National Scientific Research Organisation (ONRS). Also participating in the planning is the United Nations University.

Algeria is a country of over 10 million people, having an annual per capita income of $207 (US). It has an illiteracy level of 81 percent and a lifespan expectancy of 35 years.

---

knowledge is employed in the service of a unified, humanised and reasonably designed lifestyle, which is sustainable into the foreseeable future.

For all its benefits, the IFS concept is not welcomed in all rural areas. The farming village with farm plots in the surrounding countryside frequently does not have a suitable arrangement of buildings for utilising the bio-gas producer. It is difficult to modify what already exists as this requires financial investment. A farmer may not wish to lose valuable land for a bio-gas plant for a farm co-operative, unless some additional agricultural land is allotted to him. Bio-gas plants located in villages can pose problems of space and odour. The plant also requires water mixed with cow dung in a ratio of 5:4 by volume. In rural areas with a limited water supply, this presents serious problems.

Co-operative ventures such as an IFS group pose other social problems. Distribution of gas produced in the bio-gas plant may in practice be inequitable, with families farthest from the plant receiving the least gas. The system is also female oriented, providing the greatest perceived benefit for the woman who receives gas for cooking. She need no longer spend most of the day gathering firewood. Since in such communities women are not usually the decision-makers, this liberation tends to be a low priority in village planning. But none of these problems would be insurmountable given a strong enough motivation to change.

Use of bio-mass can meet a still wider range of human needs: charcoal or densified wood pellets can serve as a substitute for fuel oil; sunflower oil or latex from euphorbia plants can serve as diesel oil; ethanol from sugar cane, methanol from eucalyptus or methane from water hyacinth can serve as petrol and feedstocks. The water hyacinth is a fast-growing weed which thrives even in polluted water. Brazil and Sudan have estimated that they can meet all their energy needs with bio-mass, suitably managed and developed. India has concluded that organic waste alone could contribute 40 percent of its energy needs. These countries need assistance to accomplish the conversion of bio-mass technology. Such conversion presupposes global security and co-operation. It also raises questions about the priority given to loans going for third world purchase of first world nuclear technology.

Without improvement in rural living the phenomenon of rural migration to cities, estimated to be 45,000 persons per day in 1970, will continue. The food crisis will worsen, as will the problems of urban slums and violent crimes.

The Oceans

Human blood contains sodium, potassium and calcium in almost the same proportion as is found today in the oceans. The seas are sometimes called nature's cradle, since all living earth creatures have been nourished by its life-supporting nutrients, both in the evolutionary sense and in the pre-birth sense. More than 70 percent of our planet is water.

The oceans determine climate, produce water vapour, humidity and air pressure and control wind and rain. The plant-plankton and algae of the oceans are eaten by the animal-plankton, producing abundant oxygen to serve the needs of all higher animals of sea and land, including humans. About half of the atmospheric oxygen comes from the oceans and half from the forests. Although people can survive famine by hoarding food, it is less easy to hoard oxygen. If the plankton dies, half of the atmospheric oxygen will disappear, and the ocean fish, a major source of protein, will die. Given the present major world problem of deforestation, the death of the oceans would undoubtedly lead to the death of most higher life forms on earth, including humans.

Even nuclear weapons may seem not too catastrophic compared

In United Nations Environment Programme (UNEP) Regional Seas Report and Studies No. 16 of 1982, 'The Health of the Oceans', it is stated with respect to the North Sea pollution:

> There has been *no detectable effect* on marine fish populations *due to pollution*, even in those areas of the North Seas which are subject to the highest inputs of contaminants. However, shellfish stocks in some estuaries have probably been affected. Additionally, seals and porpoises living in some parts of Waddensea have high concentrations of PCBs and mercury in their tissues, and populations of these mammals have been declining continuously in recent years. [Emphasis R.B.]

Of course 'detectable' depends on the quality of information-gathering and analytical techniques, and 'due to pollution' is at best a matter of judgment. Pollution of the North Sea derives from highly industrialised coastal regions of the United Kingdom and Scandinavia, off-shore and on-shore oil operations and refineries, sewer sludge, atmospheric input of metals due to rain-out, and radionuclides carried by tides from Windscale, Dounreay and Cap de la Hague.

to an active volcano, a tidal wave or an earthquake. After these natural events, the rain cleans the air, the land and the rocks, washing refuse and debris into the rivers which carry it to the sea. Bacteria busily break down complex chemicals into simple components, making them again available for food. The silt feeds the plankton, keeping the ocean clean. Currents and upwellings keep biological nutrients circulating in the seas.

However, as humans are beginning to realise, the natural cycles and ocean-filtering system work only on 'natural' materials, now usually called 'biodegradable'. Much modern human waste, such as plastics or fission products, is non-biodegradable or toxic to living systems. Humans mimic nature rather crudely by this standard.

Although there are many intakes for the oceans, the only outlets are rain and plankton. Whatever waste humans produce on earth, if not naturally and harmlessly recycled, will remain on earth lessening its ability to support life. Usually waste travels to the ocean 'sinks', is taken up gradually by plankton which are eaten by fish, and eventually it arrives back on the human dinner table. The burial of waste in the earth merely delays the movement of the waste into the earth's waterways for transport to the sea.[4]

All marine life depends on the plant-plankton for its basic food supply, and the plant-plankton depend on the sun (photosynthesis). An estimated 90 percent of all marine life is found on the continental shelves, where sunshine, plant-plankton, and trace minerals which have been washed out to sea by the rivers, are readily available. The important breeding grounds for marine life constitute only 8 percent of the ocean surface, less than 1 percent of the ocean volume. The continental shelves are also the regions of greatest human pollution.

In recent years, the oceans have been subjected to both deliberate and accidental pollution with extremely toxic materials. This includes pesticides, herbicides, and fission products. Much of the pollution is still working its way through the ocean recycling system, and so has not yet had its impact on the human diet. The pollution of oceans also includes air-borne fumes and particulates which are carried by the wind and deposited in the sea with rain.[5] Much of the deliberate polluting of the oceans has been shrouded in secrecy and ignorance.

According to a UPI report of 2 January 1981, George Earle IV, a lieutenant-commander in the US Navy during the Second World War, broke thirty-three years of silence and called a press conference in his Vermont home. Earle, stationed at Mustin Field outside Philadelphia after the war, recalled three secret missions on 16, 20 and 22 October 1947, when he was asked to fly half a dozen large metal canisters of radioactive material to a site 100 miles from Atlantic City in New Jersey and drop them into the ocean. The canisters weighed 2 to 3 tons, and were checked with Geiger counters before take-off. Earle was told to fly low so that the canisters would not break open. The trips were never entered in the flight log.

After leaking canisters of radioactive waste were found off the coast of San Francisco, Earle contacted the navy and the US Nuclear Regulatory Commission suggesting that they investigate the Atlantic dumping. His letters were never answered.

Eventually a story unfolded of twenty-five years of barge and plane disposal of radioactive trash in fifty ocean dumps off the US coast. The largest radioactive waste dumps lie within a few hours' boat ride of New York, Newark, Boston, Los Angeles and San Francisco, in the prime coastal fishing areas. A retired navy officer, totally unaware of environmental recycling, dismissed the problem curtly: 'The sailor pissing over the side of the boat is a more important source of pollution.'

Many of the canisters failed to sink, so sailors punctured them with bullets. Sometimes the sea was rough, so they dumped the canisters in the San Francisco Bay and returned quickly to port.

As in other matters related to national security, the Atomic Energy Commission repeatedly lied about the hazards. Jackson Davis, biology professor at the University of California's Santa Cruz campus, turned up a study done by Pneumo-Dynamics in 1961. The firm's researchers determined that 36 percent of the nuclear waste drums were damaged and four of nine concrete blocks were completely demolished. The official Atomic Energy Commission press release stated that more than 94 percent of the 162 canisters tested had remained intact.[6] The lie, of course, rests on the definition of 'intact'.

Air pollution and ocean dumping have been based on a philosophy of dilution and dispersion. It was thought, wishfully, that the winds and water would spread the toxic radioactive chemicals so that no individual person could receive a 'harmful' dose, i.e. a dose above the damage threshold. It is now known that there is no threshold and the same number of deaths will be caused whether the toxic material is diluted or not. The difference is that with dilution the health problems are spread over a larger population and it is more difficult for the victim to trace the health damage back to the polluter. One hundred cancers in a population of 3,000 are easier to 'see' than 100 cancers in a population of 3 million people.

Nearly 50,000 canisters of radioactive waste were dumped near the Farallon Islands, 23 miles from San Francisco's Golden Gate. In a report released by the US Environmental Protection Agency in 1980, the bottom sediment of the ocean was reported to be 2,000 times higher than 'background radiation', i.e. radiation due to global nuclear fallout. Levels near New Jersey were found to be 260,000 times higher than background, and plutonium was found in the edible parts of fish.

About 120 miles off the Delaware–Maryland shore the US Navy deliberately sank a nuclear submarine, *Sea Wolf*, containing its entire nuclear reactor with about 33,000 curies of radioactive material. The reason given for the sinking of this submarine was fear of fire in the sodium cooling solution. It may have been an aborted attempt to build a breeder reactor-powered submarine. The breeder reactor uses liquid sodium as a coolant.

Because of secrecy and a deliberate non-keeping of records, it is difficult to estimate the total amount of nuclear garbage and radioactive debris already dumped in the oceans. European radioactive garbage is dumped in the Irish Sea and North Sea at a rate of about 5,000 tons a year. It is thought that 65,000 tons were dumped between 1967 and 1979. Past US military dumping into the oceans is estimated to be about 11,000 tons. The US Navy is now proposing sea burial for about 100 of its outmoded nuclear submarines.

The Japanese planned to dump 100,000 tons of radioactive waste into the Pacific Ocean every year beginning in 1982, but were prevented (or delayed) by growing citizen opposition in Micronesia and Japan. In a report issued on 15 February 1981, the Japanese Union of Scientists Against Nuclear Power Plants stated:

> According to our research, the full scale nuclear dumping will have a danger of having as many as 6,000 people die of cancer every year in Japan and the Pacific Islands.[7]

The report states correctly that the ocean is a 'complicated moving substance', with strong currents and upwellings. In fact, as has long been known, the ocean seems designed to recycle the waste material deposited on the ocean floor so that it re-enters the food chain, the web of life. The ocean cannot be expected to treat radioactive waste differently. However, dumping waste into the ocean delays its impact on human health. The health costs are manifested only in the next generation.

Non-deliberate polluting of the oceans such as almost occurred with the sinking of the French freighter *Mont Louis* off the Belgian port of Ostend, 25 August 1984, with a cargo of 240 tonnes of uranium hexafluoride, is more difficult to prevent than deliberate military carelessness. International shipping of uranium is essentially uncontrolled and unregulated, and even passenger ferries apparently transport it across the English Channel (London *Times*, 27 September 1984).

Another 1972 treaty, ratified by about fifty countries, was drawn up in London at the International Convention on the Prevention of Marine Pollution by Dumping of Wastes and Other Matter. Recently two small Pacific countries, Naura and Kiribati, proposed an amendment to the dumping treaty to prevent all dumping of radioactive material into the oceans. This amendment was presented for vote at the seventh consultative meeting of the treaty countries in London on 14 February 1983. It was by far the largest and best-attended meeting in the history of the treaty organisation. Spain, Portugal, Sweden, Denmark, Norway, Finland and Iceland supported an amendment for a two-year moratorium on all nuclear waste dumping at sea. This was voted in by 75 percent of the member nations. Opposition to the resolution came from the United States, Great Britain, Japan, the Netherlands, Switzerland and South Africa, all of whom had vested interests in the dumping of radioactive waste. The US delegation, headed by John Hernandez, Deputy Director of the US Environmental Protection Agency, took a position in opposition to the moratorium in spite of recent US legislation establishing a similar domestic moratorium.[8]

As mentioned already, the USA was proposing to rid itself of older nuclear submarines by sinking them in either or both Atlantic and Pacific waters, and was also working on a plan to jettison high level radioactive waste into the floor of the ocean. The LDC appointed a committee to decide whether or not this 'emplacement' would be considered 'dumping'.

Due to growing pressure from the public and from international environmental groups, Holland appears to be withdrawing from ocean dumping. Previously Dutch ships dumped nuclear waste from Switzerland, Belgium and Holland. However, it appears that Belgium will assume the dumping role if Holland pulls out. Belgium has never signed the London Dumping Convention Agreement, hence action by those who signed the agreement may be futile for stopping sea dumping.

Countries attending the 1983 LDC but abstaining from voting on the nuclear dumping resolution were: the USSR, France, the Federal Republic of Germany, Brazil and Greece. The motion by Nauru and Kiribati has been tabled until the two-year study to be conducted during the dumping moratorium is completed.

To date most of the ocean dumping of radioactive materials has been classified as low or medium level nuclear waste. This means nuclear fuel rods and reprocessing waste, with its high concentration of alpha particles, have ordinarily not been dumped. However, some ten countries are now developing and proposing experiments in ocean floor burial for high level radioactive waste. Future massive leaking of this poison into the food chain appears to be inevitable unless there is a halt to all production of nuclear waste or development of secure land burial.

In a meeting held in Guam on 17–19 November 1982, participants of the Nuclear Free Pacific/Micronesian Education and Solidarity Conference drew up several resolutions. The tenor of their concern can be gleaned by the following excerpts:

> We view the ocean as our farm and source of livelihood and categorically reject all forms of nuclear-related activity which endanger the natural balance of the ocean in even the slightest degree.

> The experience of the people of Hiroshima, Nagasaki, the Marshalls and Muroroa is an indictment of all nations who subscribe to the continued proliferation of nuclear weapons.

> All nations responsible for the destruction done as a result of their nuclear testing in the various countries and islands must be made to pay the very people they have inflicted damage upon.

Even as the Nuclear Free Pacific meeting was being held, part of the island of Muroroa in French Polynesia, its coral pedestal blown apart by a French nuclear detonation, was slowly sinking into the Pacific waters. Plutonium and other long-lasting poisons from some ninety-six nuclear detonations on the island, washed into the oceans by cyclones, are mixing freely with the currents and entering the spawning grounds for the coastal fisheries of South America and the Pacific Islands. The French military testing programme has quietly prepared to move its grim operations to another Polynesian island, Fangataufa.[9]

While this illegal dumping appears to undermine and render useless the moratorium of the London Dumping Convention, some important gains have been made. The terms of reference given for the two-year study of ocean dumping, to be conducted during the LDC moratorium, require proof of the safety of dumping. In the past, dumping was assumed to be safe unless proven harmful. The LDC 1983 session will be remembered for shifting the burden of proof to the polluter, and for failing to rubber-stamp the wishes of the United States and Great Britain. Discouragement with reckless nuclear pollution in the past has not been allowed to serve as a licence for future international policy. One avenue of externalisation of costs to future generations is under serious international scrutiny.

Meeting in Funafuti, Tuvalu, on 27 August 1984, the leaders of fourteen South Pacific nations, including Australia and New Zealand, endorsed plans to make the South Pacific a nuclear-free zone. The approved draft principles ban the acquisition, storage, manufacture and testing of nuclear weapons and the dumping of nuclear waste in the ocean. Some members, for example New Zealand and Vanuatu (formerly New Hebrides), ban nuclear ships from their ports.

Women's Consciousness

The systematic and educative work of United Nations conferences, and the growing awareness amongst community groups of possible local ecological and military-related disasters, has been leavened by the political awakening of women. Traditionally charged with birthing and assisting the dying, women have been generally more attuned than men to the signs of sickness and death in the earth's biosphere. A strong global rallying cry was heard in *Silent Spring* by Rachel Carson.[10] The theme has been elaborated and particularised by such internationally known leaders as Helen Caldicott of Australia (and later of the United States), Petra Kelly of the Federal Republic of Germany, Solange Fernex of France, Marie Thérèse Danielsson of French Polynesia and many others. Women, less hampered by society's economic and social censures because they have less in the first place to lose in these areas, are freer to speak and mourn for the dying earth system.

Elizabeth Kübler-Ross, in *Death and Dying*,[11] has clearly outlined the stages of comprehension and coping that occur in an individual experiencing death. Her identification of stages in this process has analogies in social consciousness of the species death threat as experienced by people in the nuclear age. Never before has humankind been required to deal with the reality of the death of our species as seriously as it must in the 1980s. By roughly gauging the proportions of people in a given population who have reached each of the stages described below, in dealing with this threat of species death, we can ascertain the overall coping stage of a region, a nation or even of our whole world. The stage in which the largest proportion of the population is found will determine the political, economic and social profile of that nation, region or world.

The basic stages, as extended from the stages of response to death and dying on an individual level identified by Elizabeth Kübler-Ross to the concept of species death and dying, can be characterised as follows:

1. *Denial stage*: The individual focusses on positive aspects of society such as modern travel and communication, breakthroughs in medicine and superfluity of material things, as signs of human

vitality and well-being. He or she denies that this may be a carefully constructed and fragile façade, easily sabotaged by nuclear war, ecological disaster or terrorism. Individual responsibility for the earth system's health is not taken; rather, the stance of the individual is to enjoy today and to hope calamity does not strike within his or her lifetime. There is general numbness, helplessness and a feeling of paralysis when nuclear war is discussed.

2. *Anger and rage*: Once the nuclear reality overcomes the human defence mechanisms which protect and insulate the psyche, rebellion against the harsh truth is a natural reaction. It may show itself in unwanted and unexplained tears which well up when seeing cherry blossoms or looking into the trusting eyes of a happy child. It may erupt in social protest, violence towards oneself or others, refusal to bear children, or a hedonistic attitude toward life. The façade of 'well-being' crumbles, personal efforts appear futile and the individual is close to giving up the struggle for life and health. It is a stance which is more in touch with reality than is the denial stance.

3. *Barter*: When rage has played out its energy, the individual usually finds some external action which relieves the inner urgency to 'do something'. This action may be token or ineffectual in real terms but it relieves the inner drive and somewhat dissipates the feelings of anger and rage. Because the action is in tune with some, but not the whole, reality, it may well be ineffective and/or inappropriate. For example, the individual may join a peace study group or help to distribute pamphlets once a month. A person might decide to read about nuclear war and become politically more active on the issues. This is a step closer to reality, a more personally satisfying stance, but still one step short of the total commitment needed to either halt the species death process or enable the earth to 'die with dignity' – whatever this may mean.

4. *Acceptance*: An individual death process, such as cancer, does not always end in a swift death. Sometimes there is a partial recovery and some or many years of fruitful living. This usually reflects a whole-hearted 'conversion' on the part of the patient: positive thinking; wholesome daily habits of rest, play and work; consciousness of the nourishing qualities of food, air and water; carefulness in avoiding physical and mental stress; development of wholesome interpersonal relationships within the family and with peers; professional assistance in minimising the strength of the cancer and maximising the healing capability of the body's own defence mechanisms. It is a process which demands the total

attention and co-operation of the individual and his or her primary community. So, too, honest coping and the healing of a dying earth must eventually demand total human attention, loving gentleness and care, drastic changes in lifestyles and priorities and a permanent change in human attitudes and values.

Women are conspicuously involved globally in assisting people through these stages of comprehension and psychic growth. For people reaching the fourth stage of coping there appears again a unique human potential to 'blossom' and to find new solutions to perennial interpersonal problems. Physical limitations, as any mature person has experienced, cause us to acknowledge and cultivate human interdependencies. They also encourage develop-ment of our personal inner resources and of a broader view of our relationships with other spacial–temporal realities. They 'mellow' us, making us more compassionate and tender, more conscious of our own and others' vulnerability. Hence we may become better planetary citizens, able to listen to the 'powerless', able to imagine a world not organised around the rule of the fist.[12]

The global village will probably evolve according to its own inner dynamic. Parents do not plan the characteristics of their children, they provide living germ plasm and the environment within which a viable fertilised ovum can develop. The problem of allowing the human race and its earth habitat to move out of a constant growth, expansionist mode of being and acting, into a mode of fruitfulness signalled by blossoming is the essential problem of the 1980s. Since I believe that such a movement requires the 'death' of old modes of thinking and acting, with a simultaneous birthing of the new mode, I will approach the subject in two ways. First I will address the

---

Small-scale Technology Amenable to Local Control

In China's south-western province of Szechwan there are more than 400,000 gas-generating sealed stone fermentation pits. The pits are connected by plastic pipes to cooking stoves and gas lights. The blue flame is without smoke and the lights burn with the brightness of a 100-watt electric bulb.

The pits are 'fed' crop residues, weeds, tree leaves, manure and water. The fermented compost is regularly removed to be used as high-quality fertiliser. Some pits are being used to drive small internal combustion engines, water pumps, rice threshers, flour mills and electric generators.

movement towards death and dying, and those organisations which assist in this phase. Then I will address the birthing, the signs of new life now visible. No one can predict when or how the crossover from forced growth to blooming will take place. It is only possible to watch the life process with awe and wonder, trying both to understand and to assist in the miracle of life's unfolding towards a global village.

The First Stage: Overcoming Denial of Species Death

Coming to terms with the reality of species death has been one of the commonalities linking numerous anti-nuclear and peace groups globally. A complex human interaction is taking place. Not only is each person at a different stage of perception of the problem, but, from the vantage point of that perception, many are trying to influence the whole society.

Let us first focus on movements within NATO and the Warsaw Pact nations, the most powerful and influential minority of nations in the global community. 'Better dead than red' and 'better active today than radioactive tomorrow' represent two basic fears: the death of human freedom and the death of human life. The problem, as it is perceived, is that by blindly grasping either one, we stand to lose both. The arms race causes human rights violations; human rights violations cause the arms race. To oversimplify for the sake of clarity: NATO justifies its militarism because of the Warsaw Pact nations' internal repression of civil liberties; the Warsaw Pact nations judge internal control to be necessary because of NATO capitalistic aggression. The East fears Western military strength; the West fears Eastern denial of civil liberty. This causes Eastern military escalation and Western suppression of dissent.

These two 'deaths' have been graphically presented in film and art and by vocal organisations such as Amnesty International and International Physicians for the Prevention of War (called Physicians for Social Responsibility in some countries). The global village dies from internal bleeding or external violence.

Yet few speak of the arms race as forcing human rights violations and loss of civil liberties in the arms-addicted countries of East and West alike. Few speak of the violence exercised by governments against their own citizens, and the standing army they require to keep 'order' within and to force co-operation with funding military programmes at the expense of human need. The confrontational stance towards neighbouring countries which fail to provide all the resources and co-operation a nation perceives as 'needed', gives it an excuse to develop further its fist-power and oppress its own citizens even more. Punishment, whether it is imprisonment, torture, silencing, boycott or war, is a manipulative instrument of force used both within and between countries. It narrowly defines the areas of 'acceptable blooming' for the individual citizens of the punished or punishing nation. It escalates violence and death.

Violation of human rights within a country also fuels the arms race. Put on the individual level, if one's neighbour is heavily armed in order to control his household, and believes he has life and death power over his wife, children and servants, then one will feel obliged to arm oneself in fear of the violence extending outward from that household to oneself at some future time. No amount of assurance that the internal violence of this neighbouring household has no relationship to external peaceful relations with other households will be believed.

At the East German Leipzig Film Festival in 1982 two anti-war films from the West received awards. One called 'If You Love This Planet', made by Terri Nash for the National Film Board of Canada, graphically presented the horrible insanity of nuclear war and the intense human devastation it would cause. A Swedish commentary on the women's peace camp at Greenham Common in England, called 'Let Us Fight for the Sake of Preservation of Life', also received an award. These women have been enduring severe hardships, camping in the open near the site of deployment for 96 NATO Cruise missiles, in order to draw attention to the problem and force public debate of the issue in Britain.[13]

Film and TV producers from fifty-six countries attending the festival formulated a peace call to artists throughout the world:

Let us close ranks! Let us fight together in the worldwide peace movement: Let us say No to new NATO missiles in Western Europe! Stop the arms race! Arms limitation! Disarmament![14]

The Western peace movement clearly provides hope for Warsaw Pact countries, especially those countries being used to buffer the Soviet Union. The East fears Western military strength.

While we might have wished to see both Warsaw Pact and NATO weapons specifically mentioned in the. peace call, most military analysts agree that the new NATO missiles, the Cruise and Pershing II, represent both an escalation of the nuclear arms race and a change in NATO military philosophy from deterrence to war-winning. The much-discussed Russian SS20s are generally seen by military strategists to be modernisations of the older SS4 and SS5 missiles rather than an escalation of the arms race.[15] From an arms control point of view, the statement flowed from a correct perception of the reality.

Soviet peace initiatives have always been regarded with suspicion in the West for two reasons. First, it is obvious that in a world in which two differently organised social systems exist, the 'weaker' of the two systems would stand to gain the most from a time of peace. Peace would provide time for growth, consolidation and perhaps expansion. Obviously the Soviet Union wishes to be freed from the fear of any external force exerted on it by other nations hostile to its social plans. Second, since the Soviet social experiment required far-reaching changes in human behaviour, social and political organisation, standards of conduct of human affairs and patterns of thought, it was imposed on its citizens with force. How could the Western observer be sure that this policy of force used within the country would not become an external policy of force against the 'free world' if the 'Russian bear' was allowed to grow unrestrictedly? Hence the Soviet Union's many peace initiatives are rejected out of hand, and the Western nations become locked into a suicidal arms race to 'stay ahead' with the Soviet Union always a few steps behind on the same insane road.[16]

While Warsaw Pact nations see the problem as the arms race and the aggressor as NATO countries, their own people struggle to gain some political control of the Soviet social experiment. This is most visible in Poland's Solidarity trade union movement, in Afghanistan and in the political defection of mainland China. Film producers and writers within Warsaw Pact nations could usefully direct their attention to both the threat to peace posed by totalitarian regimes which fuel Western fears of loss of freedom and the fear of nuclear destruction of life by global war or environmental collapse. If the East focusses only on the Western grass-roots peace movement as a sign of hope, it will fail to uncover the death-dealing actions against human freedom and peace of the leaders of its own nations. The Western peace movement must be used in the East to reduce fear and also to relax the level of internal state control.

As Eastern nations draw hope from the Western anti-nuclear movements, so Western nations draw a certain amount of hope from citizen action within the Warsaw Pact nations. A Polish Pope has given some visibility to the integrity and inner strength of the

Polish people often seen by the West as oppressed by a cruel totalitarian regime. In particular, the vigour of the Catholic Church in Poland causes Western believers to modify their fears for the survival of faith within professedly atheistic national states. Film producers and writers in the West would do well to focus on positive grass-roots initiatives within Warsaw Pact countries, for these are the actions which give hope in the West. Unmasking Western national governments' use of secrecy, manipulation and punishment of citizens to ensure support for their programmes of escalation of the arms race, would also be helpful in the West. As has often been said, truth is the first casualty of war, and the world is now in the first stages of the Third World War. It is important to reduce the level of the West's fear of loss of freedom due to Eastern aggression. This fear fuels militarism. The West needs to be awakened to its own suppression of civil and human rights in the name of protecting them.

It is also important to see how Western militarism leads more and more into totalitarianism and Eastern totalitarianism leads more and more into military competition and destruction of its own social goals. World disarmament has been one of the stated goals of Soviet foreign policy as framed by Lenin himself[17] after the 1917 Socialist Revolution. In 1950 the Soviet Union publicly called for the first World Disarmament Conference and it has made numerous peace initiatives since then. Disarmament is not the root problem.

The global sickness has a name. It is called violence, whether manifested within a nation or between nations. Its presence puts a premium on strength and makes 'blooming' an expendable luxury. It rears its head as the devaluation and distortion of women, either through virtual slavery, exaggerated equality with males (military service, heavy manual labour) or rape. It leads to rape of the land, violence towards the poor, oppression of the weak. It thrives on feats of extraordinary power, mega-projects, technological ego-trips and requires the passive co-operation of the weak and ignorant. It is unable to survive in the face of truth, human solidarity, compassion and non-violent action.

In the past, violence in both its internal and external forms was displaced from smaller social groupings to larger social groupings, from family to city, city to nation, nation to national alliances. The development of national loyalty has been an achievement with extraordinary potential, one which gives hope that humans have a capacity, as yet dormant, to form a global village. Family ties may be considered 'natural' and tribal society basic to human social development. Yet national loyalty, built on bonds of common language and culture, has achieved a strength sufficient to motivate sacrifice of lives to preserve it. A look at the transition from tribal to non-tribal society may serve to clarify this death to life evolution.

In the historical transition from tribal to city organisation, the patriarch of the tribe lost his life/death power over members of the tribe. He lost the right to mobilise an army to attack another tribe or family within the city-state. Obviously, an army only commissioned to keep 'order' within the family is still an army. It would be a constant source of threat to others within a city-state.

As 'keeping of the peace' became a civil function, mediated by a neutral police force, social arbitrators and courts of law, the physical prowess of the head of the tribe decreased in social importance. The older way of holding a family or tribe responsible for crimes committed by any of its members gave way to a more focussed direct dealing with unacceptable social behaviour in individuals. Inter-tribal feuding, revengeful attacks on the family, children, servants or animals of one's offenders came to be regarded as barbarian behaviour within the city-state. It was in the common interest of the tribes to accept this diminution of tribal power and vest it in a stronger co-operative city-state, capable of even greater violence against other city-states. Humans began to move from inherited leadership to elected or democratic leadership; from martial law to just legal systems and voluntary co-operation. Such social growth is obviously not yet fully accomplished. In fact, social justice is still quite primitive and uneven globally. Its development is fraught with problems. Desire to maximise one person's freedom and creativity must be tempered by the rights of other people to the same freedom to 'bloom'. Privacy must be respected in some areas, but it reduces social co-operation if carried to an extreme in all areas of life. Agents for social change cease to bring benefit when change becomes anarchy. Ownership of property which can give freedom and dignity can also result in gross oppression. All the internal problems of managing a large number of people with various types of needs, gifts and expectations, have been aggravated by the constant threat of external aggression from other city-states.

From the point of view of government, a war can become a strong rallying call for internal social unity and co-operation. It can engage the energy of the young and unruly, provide work for many, inspire willingness to endure hardship and sacrifices, fuel patriotism and heroics, and, if one's city-state wins the conflict, add material resources, culture and slaves to the society's wealth. The deaths, while difficult for family and friends, mean more wealth to be shared with fewer people. The 'losers' become second-class citizens, non-sharers in the wealth, and sources of violence within the society as they rebel against their fate. This pattern of using violence to enhance national control is visible today as, for example, when both Argentina and Great Britain sought to 'cure' domestic problems by engaging in the Falklands war.[18] Even this encounter had a nuclear dimension (exposed in the *New Statesman* in August 1984), in the shape of a British Polaris nuclear missile submarine sent to the South Atlantic to be ready for action 'if needed'. This was in direct violation of the Treaty of Tlatelolco prohibiting nuclear weapons in Latin America, of which Britain is a party.

While some violence was eventually renounced, as, for example, slavery and a father's right to kill his child, other forms of violence were seen as 'socially useful', as, for example, torture and imprisonment, or killing children by sending them to war. As women sought and achieved some liberation from their position of social inferiority under rule-of-the-fist law, they were again subdued by exaggerated treatment as the physical equals of males in heavy work or war, or by rape, or treated as sex objects.

Violence cannot be used to stop violence. Violence within a society will breed a violent society; and a violent society will be corrupted by violence within its ranks. The violence directed at others will be self-destructive. The way of violence is death.

The unmasking of the human species' terminal illness must involve dealing with violence: personal, family, city, national and global. This multilayered approach is seen in the methodology of Gandhi: truth-telling, removing the social stigma from untouchables and non-violent action against unjust social structures. He understood violence.[19]

Disarmament talks between the United States and the USSR cannot bring about an acceptable 'peace'. They merely buy time for an armed truce. The so-called nuclear 'peace' since 1945 has consisted of some 150 wars and numerous nuclear near catastrophes. The chances are that this violent 'peace' will crumble with disastrous consequences within the next few years.

While the physicians' movement to prevent the planet's terminal illness is impressive, it must also look to the violence which breeds and condones such extraordinary social brutality. Rape crisis centres, on the other hand, need to understand their place in the broader peace movement and movement against the rape of the earth.

The death process is far advanced. Unmasking violence in every cell of the global body is like pursuing a metastatic cancer which is in an advanced stage. There is no single drug or treatment which will remove the malignancy. The global 'immune system' needs to be mobilised as it never has been mobilised before.

Relieving the stressful situation in the centre of the global cancer – nuclear Europe, and its remote 'primary tumours' in the United States and the USSR – will go a long way towards removing the stresses in developing countries which are struggling for national survival as their resources feed the first world arms race. As they become caught in borrowing money for 'development', their poor either succumb to starvation and dehumanisation, or become 'guerillas' fighting for a place in the sun. Internal 'unrest' becomes an excuse for national internal repression, generating more violence and human slaughter. Martial law becomes the rule, rather than the exception, and first world 'assistance' of repressive regimes compounds the problems.

Seeing the terminal illness of the planet as violence raises the possibility of social 'surgery' to save the planet. Transnational corporations have straddled the violence of human societal organisation, frequently selling weapons to both sides in a conflict. In the past, they could expect to gain irrespective of which protagonist won in a conflict situation. They have the most to lose if all organised violence is relegated to history books. Their diversion of scientific and human resources, together with their provocateur role, will have to be curbed if the global village is to mature and survive. Unmasking this dimension of the problem will also call for co-operation between important layers of the social strata. Disengaging socially harmful behaviour from socially helpful activities requires extremely delicate non-violent transformation. It is like surgery to remove a tumour while at the same time saving the organs. The fragile human components in the picture need food, shelter, jobs, and health care throughout the delicate operation. Otherwise the operation may be successful, but the patient will die. The United Nations Educational, Scientific and Cultural Organisation (UNESCO) has been in the process of developing a code of behaviour for transnational corporations. This is apparently one of the reasons for US withdrawal from UNESCO and refusal to contribute financially to its work, effective from the end of 1984.

Those organisations which are systematically helping the human community to 'see' that it is heading toward species death include: the Society of Friends – which deals directly with fostering non-violence in society; Amnesty International – which deals with social behaviour within nations; and anti-nuclear movements – which deal with violence between nations and against the bio-sphere.

Auxiliary to these mainstream activities are solidarity groups, actions for corporate responsibility of multinationals, human rights groups and various media efforts to communicate the problems. All are vital to this phase of revealing reality and counteracting community attempts to deny the species death problem. Some groups specialise in single issues within these various areas, such as opposition to the MX missiles or aid to Nicaragua. Co-operation between single-issue and multi-issue groups is helpful.

Unless consciousness-raising is followed by equal efforts to deal with fear and rage, however, all this global awareness will be useless for global health. This brings us to the next stage of coping with species death and dying, namely the stage of frustration, rage and anger. Who is trying to deal with the awakened human conscious-ness?

The Second Stage: Dealing With Rage

What is the difference between throwing people into the flames and throwing flames at the people?

So spoke Dorothy Day during the Second World War when most Christians were bravely, and thoughtlessly, giving their lives to preserve democracy and thereby granting their governments the 'presumption of justice' in their conduct of the war. It was a just cause, was it not? We had to defend democracy and freedom in return for sharing in its benefits. Did not St Augustine and St Thomas Aquinas, recognised Doctors of the Christian Church, preach the just war theory? Besides, the United States and its European allies had been attacked; they had to defend themselves. Self-defence was morally right. In her book *The Long Loneliness*,[20] Dorothy Day explains her pacifisim, which began in 1917 during the First World War. Although she did not know that the US State Department's charge that German submarines had attacked and sunk a peaceful passenger ship (the Cunard liner called *Lusitania*) was manipulative and misleading, Dorothy Day had grasped the imperialistic nature of the war. She knew it was about power, not life or peace.

318   *A Time to Bloom*

The British-owned *Lusitania* was advertised in American newspapers as a normal passenger ship. In reality it was carrying guns, munitions and other military supplies from New York City to Liverpool in England. German intelligence services in the United States and spies aboard the *Lusitania* reported this to the German ambassador in Washington, and he in turn reported it to the US Department of State. Robert M. Lansing, Counsellor for the State Department, chose to ignore the information.

The German Embassy then formally asked the US government to dissuade American citizens from sailing on the British ship. The United States was a neutral country at that time, and since the British ship was carrying arms, it was liable to attack from German submarines. Germany and Great Britain were already at war with one another. The Germans went so far as to purchase advertisements in major US newspapers, warning civilians of the danger. Although Lansing knew better, he and even President Woodrow Wilson over whom he had influence, denounced the German appeal as an attempt to intimidate the United States. They urged Americans to ignore the warning.

On 7 May 1915, the *Lusitania* was attacked by a German submarine off Ireland, torpedoed and sunk. Civilian casualties were 1,198, including 128 Americans. This 'wanton attack on an innocent vessel' became the rallying cry under which loyal Americans marched off to fight a just war – the war to end all wars. Robert Lansing succeeded in keeping the truth about the *Lusitania* suppressed even during an official inquiry. Relatives of victims were cheated out of any compensation for their loss because the subterfuge was judged necessary to preserve national morale in time of war.[21] Dorothy Day had understood that it was not the nation which was being preserved; it was, rather, the life/death power over the people of the nation vested in its leaders which was being so desperately and ruthlessly preserved.

It was an older and wiser Dorothy Day who recognised the pre-war brutalisation period in the early 1930s for what it was, as major Western nations, having caught their breath after the First World War, began to prepare for the Second World War. On 1 May 1933 she founded *The Catholic Worker* newspaper, and began to use it for speaking out strongly against totalitarianism and anti-semitism. Hitler began not with the gas chambers and mass extermination of the Jews, but by limiting the hours when they were allowed to shop for groceries. Jews could only shop between 4 and 5 p.m., after the better fruits and vegetables had been taken. Most non-Jews did not protest. Why should they? Many liked the preferential treatment. Many even laughed when the Jews were forced to clean the streets of Vienna with toothbrushes on the Sabbath, and kept silence when they were forced to wear identifying stars. Many who dissented were too afraid to protest. A small minority of non-Jews prepared to risk their lives to save as many Jews as possible, and by their courage they kept alive hope for human survival on the planet earth.

The year 1940 found Dorothy Day in Washington speaking before Congress on behalf of the rights of conscience of young men who wished to be conscientious objectors in the military draft. When she was censured for this by a Catholic priest (who had been lobbying to protect seminarians and religious brothers from military draft), Dorothy replied: 'I am speaking for lay people. They are the ones who fight the wars.'

In the Second World War Americans again went off to fight, believing themselves the innocent victims of aggression, not knowing about the Pearl Harbor manipulation intended to 'strengthen their will to fight'. For Dorothy Day war not only interrupted her traditional works of mercy – feeding the hungry, clothing the naked, teaching the ignorant – it reversed them.

In spring 1957 the City of New York carried out an air raid drill. The compulsory civil defence exercise was supposed to prepare the population for an attack which included the dropping of a hydrogen bomb. Little was said about the fact that a hydrogen bomb is about 1,000 times more powerful than the bombs which levelled Hiroshima and Nagasaki. There would be no levelling of cities in the Third World War. Rather, these powerful new bombs would create holes 300 to 500 feet deep and half a mile across. In New York, the ocean and river waters would quickly rush in to fill the giant hole. The fire storms which followed the blast would suck the oxygen out of any air raid shelters far enough away from the epicentre to survive blast and flood. Humans and animals would be blinded by the flash; hair and clothing would spontaneously catch fire; bodies and automobiles would be hurled through the air as missiles in the gale force whirlwinds; huge buildings would crumble; death would be by evaporation or burning or starvation and grief.[22]

Dorothy Day, with twenty-eight other people, defied the compulsory civil defence drill orders and sat in New York's Central Park. They could not participate in the pretence, the charade which said nuclear war was survivable. It was a crime to prepare for nuclear war, so Dorothy Day and the others quietly awaited arrest and went to jail.

The non-violent force of truth penetrated into the minds and hearts of many other Americans as news accounts of her protest were carried across the country. As authorities tried to reinforce their 'control' by continuing the periodic civil defence drills, the numbers of protesters who strolled through Central Park in defiance multiplied. Finally, Governor Nelson Rockefeller called off the confrontation, and the civil defence drills ceased.

Four years later, in 1961, Thomas Merton published his first moral statement in *The Catholic Worker*, a poem describing the Jewish holocaust; the poem ends:

Do not think yourself better because you burn up friends and enemies with long-range missiles without ever seeing what you have done.

Merton, Jim Douglass, Dan Berrigan and countless others of that generation learned from Dorothy Day to direct anger and frustration against evil and injustice, not against the role-playing public.

Alexander Solzhenitsyn clearly echoed this insight in *The Gulag Archipelago*[23] when he realised that had the roles been reversed and had he been the judge, jailer or torturer rather than the prisoner, he would have played his part as enthusiastically as they had. Solzhenitsyn was expelled from the USSR in 1974 and has never ceased to speak out against the violent treatment of dissidents there. The struggle against the mindless mistreatment of citizens within the nation-state is the protest call which awakens human consciousness in the East.

Protest demonstrations become the liturgies through which angers and frustrations are fruitfully channelled, friends are made, life is joyfully celebrated and the seasoned peacemaker shares experiences with the young and restless. Protest demonstrations force local attention on the violence which surrounds us, whether it be the Gulap Archipelago, a local shipyard building Trident submarines, or a nuclear power plant which serves as an excuse to perpetuate the 'peaceful atom' myth.

Demonstrations engage the heart and mind, provide a milieu in which new non-violent tactics can be developed and explored, and often ground personal commitment to the point of being able to sustain martyrs. Demonstrations are also vulnerable to outside manipulation by provocateurs, are likely to be feared and misunderstood by outsiders and are threatening to civil authorities. The action must have a clear and reasonable goal, the behaviour of the demonstrators must be acceptable to the ordinary citizen and non-threatening physically to the opponents. The 'victory' of such action lies in its compelling moral rectitude which wins over fair-minded observers.

The Shibokusa women of Japan are the descendants of farmers who were allotted a piece of land on the northern slope of Mount Fuji in the seventeenth century. The land was stony and the soil poor, but with years of cultivation their families had established radish and bean farms, and even a silkworm industry. In 1936, when Japan began in earnest to prepare for war, the army took over the land of the peasants, declaring Shibokusa land their drill and manoeuvre ground. After the Second World War, the US Army took over the land and established a base there.

On 20 June 1955 seventy farmers staged a protest demanding to be allowed to farm their land. All their efforts to use legal and political avenues of recourse had failed. By accident, as protesters were being carted off to jail, the jeep in which the chief of police was riding crashed and he was killed. The chance happening drew national media attention to the protest, and Tokyo officials, wishing to avoid or quiet the questions being raised nationally, promised the farmers an additional 50 hectares of land to make up for the land usurped by the military. The people planted pine trees on the new land, but when the trees were grown, the government again declared that the state owned the land.

The women had long been accustomed to hardship. As they tried to eke out what little food and income they could from their remaining land holdings, their husbands were forced to seek jobs in the cities or as itinerant workers to support their families. The women decided to act together for the preservation of life. They built small cottages on or around the military installations on their land and began to force the soldiers to confront their own behaviour and its effect on the lives of the farming people they had displaced.

In 1970 a squad of 1,000 riot police turned out to evict a small group of women from one of the cottages. The women had dressed in white, 'an appropriate way to face death', and were waiting for the police. They had some live hand grenades which they had picked up on forays into the camp, trying to discover the nature of this military invader entrenched on their farmland. As they talked together about living and dying, they decided to surrender and stay alive so that they could build more cottages and continue their resistance.

Police riot gear is, of course, useless when women agree to surrender, to absorb the violence and consistently give back life. The women now have about fifteen cottages. They regularly disrupt military exercises, popping up and walking around the mock battlefield; planting scarecrows to decoy the troops. 'The police have realised that we are physically easier to arrest than men, but more trouble afterwards. We never give our name, age or anything. We just say we are so old we can't remember when we were born or who we are . . .'[24]

What do the women accomplish? Is their suffering in vain? Who can know what happens in the mind of a young man when a woman old enough to be his grandmother appears on the mock battlefield and asks him why he prepares to kill his brothers and sisters with whom he shares the earth? 'This land used to be green, why do you destroy it? Why must we die to let you learn to kill?'

Mount Fuji is a peace symbol in Japan, and on her northern slope men prepare for war. Dr Nakagawa, a professor at the Kobe University, said to me: 'We were taught that there were only two realities, the mountains and the sea. Neither one should ever be soiled with human blood or radioactive pollution.' The uneducated Shibokusa women have learned this lesson from the earth.

Both demonstrations and sustained protests allow newcomers to engage their energies during the period of rage which follows closely upon awakening. The Quakers and other traditional Peace Churches have been consistently helpful in developing non-violent styles of protest demonstration, and in engaging the energies of those newly awakened to nuclear peril.

In the spring of 1978, I travelled to New Hampshire in the USA to a small district courthouse, for a trial of one of the protesters arrested at the Seabrook Nuclear Power Reactor Demonstration. The action had been organised by the Clamshell Alliance, and more than 1,100 protesters had been arrested. The civil authorities decided to try the protesters one at a time with a six-person jury for each case. Everyone knew this was impossible in the small New England town. The first defendant was a Seabrook resident, a clean-shaven man with a wife and young child. His defence was that the harm posed to his family, his land and his fishing rights by this large nuclear reactor, was greater than the harm posed to society by his trespassing. The power plant was located on the short New Hampshire sea coast, in a rural area. The electricity would be transmitted to Boston, in Massachusetts, to the south.

As I listened to the witnesses, I found that the demonstrators had gone through a rigorous training, conducted by the American Friends Service in Cambridge, Massachusetts, prior to the action. In groups of twelve to fifteen, they learned to know one another well. They practised through role playing, shared with their group probable behaviour under stress, talked about their fears, their needs, their homes. These 'affinity groups', as they were called, made decisions by consensus, not voting. After the members had become comfortable with one another, they began planning their protest action, each one freely deciding whether or not to risk being arrested, whether or not to co-operate with the arresting process, what needs or fears they would have if they were sent to prison. As it turned out, most of those who were imprisoned were kept in the large open floor area of national armouries, without chairs or beds or any 'normal' prison furnishing, for about fourteen days. Authorities had set up a differential bail system which appeared to be discriminatory and punitive, so most of the protesters refused to avail themselves of the bail until one amount was set for everyone. But this gets ahead of the story.

Each affinity group had members who had decided *not* to be arrested. They were prepared to take care of family needs, medicines, lawyer contacts, and other personal wishes of those arrested. Their jobs were to be supportive.

Protesters learned how to build latrines which were ecologically acceptable near the protest site and discussed other human problems which might cause them to act in an irresponsible way because of their large numbers and the primitive on-site conditions. Police officers testified at the trial to the fact that protesters left the reactor site grounds in better condition than they found them. No papers or trash were scattered around, no damage was done to persons or property.

The protesters also developed a manual of behaviour which they pledged to follow, and which they shared with police and other civil authorities beforehand. There were no quick movements, no surprises which might provoke violence.

As I listened to the witnesses for the defence, I learned that the ages of the protesters ranged from twenty to seventy years, and that they came from all walks of life, with various levels of technical and professional background. The local people supported their action. In New England, most of the important decisions are made at the town meeting, and Seabrook as well as nine other surrounding towns had voted against the nuclear reactor. The Governor of New Hampshire declared the town meeting votes 'merely advisory' and proceeded with construction. This was the first remembered incident of a New England government overriding the vote of a town meeting. Even a large Boston newspaper which supported the nuclear project, both because Boston would receive the power and because Boston financiers received a fine interest rate guaranteed by government for their investment, gave front-page coverage to the violation of civil rights. The American Civil Liberty Union took an interest and lawyers came from Boston to try to assist in the cases.

I was the first expert witness for the defence. After I was sworn in, the attorney established my credentials for the court: PhD in mathematics with application in medicine; cancer researcher specialising in the health effects of radiation; had testified before the US Congress and in licensing hearings for nuclear power plants before the US Nuclear Regulatory Commission. He then laid legal grounds for my testimony: the defendant had been influenced in his decision to protest against the nuclear plant by hearing me speak and reading my publications. The judge accepted my credentials as an expert witness. All during the morning, the attorney for the state had been jumping up and objecting to almost everything said in the courtroom. Frequently the judge accepted the objection before the attorney spoke. This time the judge asked him if he was going to object. He obliged, and the judge ruled that anything I might say was irrelevant to the case, and I was dismissed, without having said a word. Helen Caldicott was the next witness. She was sworn in and her credentials presented to the court: medical doctor specialising in cystic fibrosis, a genetic disease; expert in medical effects of radiation; had testified before the US Congress. The judge accepted her credentials as an expert witness, the attorney for the state objected to her testimony as irrelevant and she also was dismissed without having said a word.

During the lunch hour one of the reporters, a woman in her sixties, told me that she had been doing controversial stories for major newspapers all her adult life, but had never seen anything like the paranoia which surrounds the question of radiation and health.

It appeared to her that as official information becomes less credible, government spokespersons are increasingly called 'eminent' and their committees called 'prestigious'. Scientists and physicians who try to report on health effects of radiation are silenced and 'discredited'. Their research funds are cut off and 'critiques' of their work circulated behind their backs. Media people

are warned that these scientists are unreliable sources of information, that their work is not 'acceptable to their peers'. All this was
disturbing to the woman because she thought, as I did, that the peer
review which scientific works receive prior to their publication in
professional journals is the relevant peer review. Acceptance for
publication and presentation at scientific meetings at which there is
free discussion and opportunity for criticism means that the science
has passed the usual peer review process. Many of the so-called
'controversial' scientists also informally seek review of their work
among colleagues working in the same, or in closely related fields.
What then happens is that the work of these scientists is rejected by
government-appointed committees, national or international,
which perform a second, more political, review. The reporter was
obviously both disturbed by the court's silencing of Helen
Caldicott[25] and myself, and also afraid to incur the backlash she had
experienced on other occasions if she interviewed us and tried to
report the radiation health problems in the press.

As we waited for the afternoon session of the trial to begin, the
young defendant spoke to the observers, explaining that he was
Christian and would appreciate people joining with him in praying
the 'Our Father'. He apologised if this was offensive to anyone, and
asked that those who were not Christian silently pray in their own
way. It was very moving to feel the pathos of that prayer quietly
spoken in the alien courtroom environment.

More details of the protest unfolded during the afternoon.
Organisers had held an educational-style demonstration in a nearby
park, concurrently with the protest at the plant site. This was a
'legal' protest, with police permit, and funds donated for site
clean-up, comfort and first aid stations and other amenities. Those
who had not gone through the training session had attended this
demonstration. Only people 'recognised' as part of an affinity group
had been allowed on the reactor site. This had proved to be an
effective remedy against police or security guard infiltration, and
against possible outside provocateurs inciting to violence.

During the time when protesters were detained in the armouries,
affinity groups took turns entertaining the whole assembly of
prisoners. They performed mime, wrote songs, searched out
creative new strategies and made friends with the guards. Local
authorities tried to increase the repressiveness of the detention by
forbidding the prisoners fresh fruit because of a possible 'health
hazard'. After the word got out, the rubbish bin taken out daily to
be emptied would return filled with fresh fruit.

The judge found the young man guilty of criminal trespassing, because he had stood on the parking lot of the reactor site, and gave him the maximum penalty, two months in jail. Later that week a Catholic nun, a high school science teacher, and four others were found guilty and given the same harsh sentence. Eventually the state dropped charges against the rest of the protesters, having made 'examples' of a few, and the anti-nuclear movement realised the role of courts and jails in the exercising of power over others.

In autumn 1977, about a year before the Seabrook demonstration, the US Government Accounting Office stated that the official radiation monitoring system in the USA did not measure radiation exposure for 40 percent of the population and provided only an 'educated guess' of the level of exposure for the other 60 percent. The report warned: 'levels of radiation are increasing and affect not only the health of current populations, but of future generations because of genetic damage.' This report, blaming the problem on lack of resources, staff and knowhow in government agencies, received little publicity except among the aware persons who seek out such information. Seabrook was generally viewed by the uninformed public as a protest of naive college students or irresponsible 'hippies'.

By 1984 the Seabrook nuclear power plant had cost $2.6 billion (US), was drowning in unpaid bills and threatening to declare bankruptcy. Several other nuclear plants are on the 'critical list' for financial reasons, and the large Washington (State) public power supply system defaulted on its nuclear projects in 1983.[26]

The spring following the Seabrook action, 1979, saw the release of another US Government Accounting Office (GAO) study: 'Areas Around Nuclear Facilities Should Be Better Prepared for Radiological Emergencies.' The GAO report, which had been in preparation for months, was released on 30 March 1979, two days after the Three Mile Island (TMI) accident. After the accident, the government generally ignored its own reports and attempted to restore the image of nuclear power. They prepared a 'travelling show', a group of scientists who would hold closed symposia, i.e. no other scientists allowed to speak, on the TMI accident throughout the United States, Europe and Japan. These government-paid scientists insisted that radiation releases at TMI were much too low to have caused the health problems reported by people living near TMI and by area farmers about their animals. All such problems were blamed on the media and irresponsible scientists who caused panic.

After Three Mile Island, citizen protests in the United States, Europe and Japan grew in size. Many reached and even exceeded the 500,000 level. Outstanding among the many demonstrations which I have participated in, however, was the peace rally in New York City, on 12 July 1982, at the opening of the Second United Nations Special Session on Disarmament. This convocation of 1.3 million people – representing a large number of languages and cultures – resembled a global family picnic. A large number of children, elderly people, and those in wheelchairs or with physical disabilities participated. The police were relaxed, some even wearing Japanese paper cranes over their uniform, and the first-aid station reported a quiet day with only a few minor complaints treated. I sat on the lawn in Central Park and watched a man move from group to group asking for a 'joint'. People just smiled and offered him a share of their sandwiches or nuts.

I reflected on how far the art of public protest had come since the early Seabrook days. People had discovered that their fellow humans are naturally gentle, orderly and non-violent. It is the exception, the disturbed individual who causes difficulty. Even worse, it is the deliberate provocateur. Why should we all live in fear? Perhaps there was hope if we all co-operated.

Hearts were strengthened that day. Everyone who came was changed, confirmed in a new mode of behaviour. Each person went away with a picture of what might be, and a realisation that many other people cared and were angry and wanted to live in this fragile world. The 1.3 million able to be present in New York probably represented another 130 million or more unable to travel. The anti-nuclear movement had achieved a veritable majority in many areas of the world, but did not yet control power.

Protest demonstrations can be fruitful outlets for social rage, but there are other, more gentle, ways to begin to release the grief, rage and anger which follow nuclear awakening. In 1977 a psychiatrist, Carol Wolman, and a peace activist, Natalie Shiras, began to use the model developed by the feminist movement to explore people's understanding of nuclear issues. How do you feel about people who work to produce nuclear weapons? How do you feel about people who oppose war? Do you expect war in your lifetime? How long do you think life as we now know it will survive?

When analogous questions have been used in women's groups to release feelings, break down isolation and despair, women have spontaneously surfaced reasons for hope and helped one another to channel anger safely. The dialogue generally ends with increased courage and conviction. Far from being invigorated, small groups of people using this technique on the nuclear issue became more withdrawn, changed the subject, felt guilty for their emotions, and displayed a variety of defence mechanisms, some more pathological than others. It was obvious that the social prohibition against discussing or feeling fearful about nuclear realities was even greater than that which surrounds questions of women's liberation. Some psychiatrists believe that this is the reason why there has been so little discussion of nuclear issues in the USA until very recently. It may also account for some of the widespread hedonism and nihilism in that country. Admitting our inability to cope with the thought of species annihilation is perhaps as upsetting as thinking about the nuclear holocaust itself.[27]

In response to the blocks to dialogue, many small groups are starting at a less threatening point and beginning to share their first awareness of the nuclear bomb, their feelings at the time, their subsequent coping or non-coping with the reality. These small sensitive questionings begin ever so gently. People now avoid such questions as: What do you think about nuclear war? A better question is: When do you first recall learning about nuclear weapons? How did you feel?

The atom-splitting event has deeply affected human security in ways quite different from military affairs. The earth on which we rest is now known to be fragile. Children, who have not developed coping and defence mechanisms, are most seriously affected. They describe weird nightmares and fears that the earth itself will disappear and there will be nothingness.[28]

I have nightmares not about a nuclear war but about nothing being there afterwards. Nothing is in the dream, really, like a fog. Try to picture nothing. This world is not the way it is meant to be. (Stacey, aged fourteen years.)[28]

Dr Robert Jay Lifton has described the reaction to nuclear realities as 'psychic numbing', the wilful diminishing of one's capacity to feel.[29] It was first described by the survivors of Hiroshima as a way of being able to walk past the dying, leave behind those who would be burned alive, refuse water to those consumed with thirst, cremate one's own children, and fulfil the various other 'tasks' necessary for survival. The American Psychiatric Association Task Force on the Psychosocial Impact of Nuclear Developments, chaired by Dr Rita Rogers, undertook a study of

---

Soviet Youth More Afraid of Nuclear War

Recently a group of Harvard University psychiatrists, sponsored by
Harvard and by International Physicians for the Prevention of
Nuclear War, went to the Soviet Union and conducted interviews with
350 school children. Some results showed Soviet children seemed to
have deeper fears of a nuclear first strike and concern for their own
survival than previously assumed . . .

The average age of the Soviet students was 12.7 years. A
comparable study in the USA questioned youths of average age
sixteen. Ninety-nine percent of the Soviet youths reported they were
very worried about war, compared with 58 percent of US youths.
Only 6 percent of the Soviet youths said the two nations would
survive a war; 22 percent of the American youths thought survival
was possible.

Toronto Star, 6 December 1983

---

these effects on children. Dr William Beardslee, who is on the
psychiatric staff of Boston Children's Hospital, and Dr John Mack,
well known for research on adolescent suicides, co-authored the
report.[30] They interviewed more than 1,000 Boston grammar and
high school students. The threat of annihilation had made it almost
impossible for children to comprehend natural death. They tend to
equate their own death with annihilation from an external source.
Existence becomes unpredictable, the earth unreliable. The world
is perceived as insane and life at the mercy of some whim, the
'button pusher', or even a malfunctioning computer. Beardslee and
Mack expressed concern about the impact of nuclear realities on
children's 'ego ideal', i.e. the image of their best self or what they
would wish to be, the integrating image which lies at the centre of
the personality.

Schools frequently fail to assist children and thus add further to
their dilemma. Wars are still taught as the lynch pins of history;
bloody warriors are presented as glorious heroes; violence and
conflict as inevitable results of 'human nature'. At the same time
national patriotism is stressed and the United Nations is described
as weak and ineffective. Robert K. Musil, who directed the
television documentary 'Shadows of the Nuclear Age: American
Culture and the Bomb', stated:

it was with that awful knowledge [of a nuclear blast] – we were not safe at all – that I experienced duck-and-cover drills and developed an early disillusion with, even disdain for, authority.

Teachers can do much to stop the unhealthy shielding of children from the reality of the nuclear crisis. As Helen Caldicott has said so well, 'to know is terrifying, but not to know may be fatal.' Children already know a great deal about this reality.[31] Family and classroom sharing, sensibly geared to the age of the children, is preferable to an alienating silence. Families and friends can discuss the meaning of real security in contrast to living always in an armed camp; the rule of the fist in contrast to living in friendship; competition in contrast to co-operation; the enemy images we create; war as 'natural' or even helpful for solving problems; other ways to resolve conflict.[32]

There are many organisations which help to direct anger and rage into positive directions, including the American Friends Service, the Fellowship of Reconciliation, the International Mobilisation for Survival, the Nuclear Freeze movement, the Campaign for Nuclear Disarmament, and Stop the Cruise. The variety, scope and geographical distribution of these organisations and movements, at least in the developed world, is sufficient to meet the needs, sensibilities and styles of just about everyone. We can imagine that solidarity is the life-line for people in countries with oppressive totalitarian regimes, so global outreach must be on the agenda of all peace groups.

The Third Stage: Barter

The scene is the Meistersingerhalle in Nürnberg, 18 February 1983. The occasion is the Tribunal against Nuclear Weapons of Mass Destruction in East and West, convoked by Die Grünen, the Greens, a political party in the Federal Republic of West Germany. Many earlier events made this historic occasion possible.

In New York City on 11 June 1982, eve of the massive peace rally, one of the jurists exclaimed:

Nowadays we have to enforce international law before the event if we want to survive. There will be no Nuremburg tribunal to judge the crimes against humanity after World War III because after a nuclear war there will be no winners.

In a war it is winners, not losers, who hold tribunals. There was a tribunal in Japan in the 1960s which formally condemned the atomic bombing of Hiroshima and Nagasaki as a crime against humanity. Needless to say, this received little publicity and had no effect on the nuclear nations.

Petra Kelly, young vocal spokesperson for Die Grünen since its political debut in 1978, was born in Nürnberg. The rise of Hitler, the war, and the subsequent trials against war criminals had been etched in her mind during her crucial growing-up years. It was primarily through her vision and years of hard work that the tribunal of 1983 was realised. The date was carefully chosen. In the Berlin Sports Hall, forty years prior, on 18 February 1943, Joseph Goebbels, propaganda secretary for Hitler, had given his historic speech:

Do you want total war? Do you want it, if necessary, more total and more radical than we can imagine today. . .

Even the Nazis, themselves on the brink of developing the atomic bomb, presented their aggression as a way to 'settle European problems' and prevent a nuclear holocaust. It was for them a war to end all wars, because they saw their victory as giving them command not only of all of Europe, but of the scattered European colonies in Asia, Africa, North and South America. Germany's power would then far exceed that of either the United States or the Soviet Union. Germany would attain 'peace'. Its power would be unchallenged in Europe and elsewhere.

Between 1 September 1939, the day when the open hostilities of the Second World War began, and 10 May 1940, the day Winston Churchill became Prime Minister of Great Britain, warring parties were threatened with US intervention unless they adhered to the international law which forbade bombing of civilian populations and unarmed cities. The Germans, having illegally bombed Poland, ceased bombing expeditions but were restive. Churchill was also known to favour bombing. Apparently by accident, on 10 May 1940 the German Air Force bombed a German city, Freiburg. Embarrassed, Goebbels, the propaganda minister, publicly blamed the attack on the British Royal Air Force. Both countries immediately launched into air attacks, claiming it was permissible to disregard international law because the other side had done so. The barbarity ended in the horrible fire bombings of seventy German cities, including Dresden, and seventy Japanese cities, including Tokyo.

The Nürnberg Principles, 1946

1. *Principles of International Law Recognised in the Charter of the Nürnberg Tribunal and in the Judgment of the Tribunal as Formulated by the International Law Commission, June–July 1950.*

Principle I
Any person who commits an act which constitutes a crime under international law is responsible therefore and liable to punishment.

Principle II
The fact that internal law does not impose a penalty for an act which constitutes a crime under international law does not relieve the person who committed the act from responsibility under international law.

Principle III
The fact that a person who committed an act which constitutes a crime under international law acted as Head of State or responsible government official does not relieve him from responsibility under international law.

Principle IV
The fact that a person acted pursuant to order of his government or of a superior does not relieve him from responsibility under international law, provided a moral choice was in fact possible to him.

Principle V
Any person charged with a crime under international law has a right to a fair trial on the facts and law.

The atomic bombing of Hiroshima and Nagasaki ushered in a period when civilian populations would be held permanent hostage under threat of annihilation. Civilian targets were 'preferred' since missile silos would be assumed empty in a retaliatory strike.

The Nürnberg Tribunal of 1983 was an attempt to restore international law as superior to any national law, and to avert an even more devastating holocaust than was unleashed on the 'developed' world in 1939.

There is no judge at a People's Tribunal, only a jury. The people themselves must judge; must free themselves of propaganda and develop a sense of solidarity with the people of all nations. The Nürnberg jury consisted of six persons:

Principle VI
The crimes hereinafter set out are punishable as crimes under
international law:

(a) Crimes against peace:
    (i)   Planning, preparation, initiation or waging of a war of
        aggression or a war in violation of international treaties,
        agreements or assurances;
    (ii) Participation in a common plan or conspiracy for the
        accomplishment of any of the acts mentioned under (i).

(b) War crimes:
Violations of the laws or customs of war which include, but are not
limited to, murder, ill-treatment or deportation to slave-labour or for
any other purpose of civilian population of or in occupied territory,
murder or ill-treatment of prisoners of war or persons on the seas,
killings of hostages, plunder of public or private property, wanton
destruction of cities, towns, or villages, or devastation not justified by
military necessity.

(c) Crimes against humanity:
Murder, extermination, enslavement, deportation and other inhuman
acts done against any civilian population, or persecutions on
political, racial or religious grounds, when such acts are done or such
persecutions are carried on in execution of or in connection with any
crime against peace or any war crime.

Principle VII
Complicity in the commission of a crime against peace, a war crime,
or a crime against humanity as set forth in Principle VI is a crime
under international law.

Professor Dr Ossip K. Flechtheim, specialist in international
law and politics. Dr Flechtheim served as section leader for the
US prosecutor for war crimes in Nürnberg, 1946–47, and now
resides in Berlin.

Herman Veerbeck, Catholic priest from Holland. Father
Veerbeck is the chairperson of the Dutch Radical Democratic
Party which promotes original christian principles and an
independent position for Holland in Europe.

Reverend Dr John F. Steinbruck, from Washington, DC,
USA. Reverend Steinbruck manages a refuge for homeless
victims of the US military budget and financial priorities.

Vladimir Lomeiko, publisher of the Soviet literary magazine
Literaturnaja Gazeta.

Professor Dr Seiei Shinohara, philosopher and author. Professor Shinohara has been active in the Japanese peace movement since 1952.

Freda Meissner-Blau, a journalist, ecologist and feminist from Vienna, Austria.

Professor Dr George Wald, nobel laureate in biology, was later substituted for Luise Rinser who was unable to attend.

It did not really matter which former General or Colonel one listened to, on the evening of 18 February 1983. The stories were similar. Nuclear weapons make no sense, even from a military point of view. Gert Bastian, a German ex-General, was the most eloquent and had a humble honesty about him which clearly told the audience he was neither naive nor angry – those stages were long past. The main question was how to live with the present nuclear reality; how to stop a holocaust which made no sense and brought only the destruction of everything we hold important. Niño Pasti, an ex-General from Italy, explained the uselessness of talks about arms control. Several times he murmured, 'There must be a whole other way to do it.' A former Soviet Colonel, Professor Daniil Proektor, said much the same thing, but he described attempts to negotiate with the United States as coming up against an 'iron wall'. I could not help but think of the 'iron curtain' imagery used in the United States to describe the Soviet Union. The jury questions to the eight retired military leaders were respectful but forthright. The Soviet delegates wished to avoid discussion of human rights, but it kept coming up. They defended their roles in the nuclear arms race as self-defence. Although it seemed obvious that the roughly 1,000 observers, participants and press agreed that the Soviet Union had not initiated the arms race, or even been responsible for most of its escalation since 1945, they still felt it must be held responsible for its intent to defend against or revenge attack on its land by threats of massive retaliation against civilians. By eliminating the human rights issues, i.e. violence within the nation, the Soviet Union was at least contributing to the fear and violence between nations. The United States government and the Soviet government had both refused official participation in the Nürnberg Tribunal. There was general audience appreciation for members of both superpower nations who were willing to speak on behalf of or in apology for their absent governments. Empty chairs, saved for absent official delegations, stood in mute testimony throughout the three days to government's inability to enter into a change process.

The clearest example of those in the barter stage with respect to species death are military officers and high government officials. There can be no doubt that they know the basic facts about the weapons and have contemplated every strategic detail of an attack and counterattack. Such information, and the fear initially experienced and then overcome, can partly explain their patronising view of 'ordinary citizens'. They see them as either blissfully ignorant or as venting their emotions in useless protests. These enlightened officials have passed through those stages and taken a more realistic stance.

There seem to be two identifying signs that indicate that an individual has been trapped in the barter stage, rather than moving through it to the stage of holistic comprehension of the nuclear reality, a comprehension which frees us for total acceptance of reality and fruitful action. This less than whole-hearted acceptance stage, called barter, is marked by a holding on to the old lifestyle, behaviour and belief framework, within which the dilemma has no solution, together with habitual lying to those most closely affected by one's decisions. It was retirement which had set the military men free to share their knowledge and concern.

There is no way within the accepted format of arms control discussions of dealing with international security in the true sense. Military officers cannot question whether or not war should be relegated to history books, along with castlebuilding, colonialism, cannibalism and slavery. Heads of nations cannot negotiate away their national sovereignty. The world 'system' is stuck in a counterproductive death spiral. It cannot be 'saved' by military, political or technological action within the present behavioural framework. Each of these is locked into a role which has become both irrelevant and destructive.

In the barter stage of understanding species death one accepts some, but not all, of the reality. One looks for 'magical cures', or in the nuclear case a technological fix, which will preserve the status quo and make the problem go away. In the face of catastrophe we buy a little time.

I remember a cancer patient who had just undergone surgery to remove his voice box and most of his jaw. He was holding a cigarette to the tiny tracheotomy opening in his throat through which he breathed. He obviously knew intimately both cancer and pain, but from a physical point of view he was acting inappropriately. He no doubt wanted the surgery to make him well, but he could not go one step further than that and stop smoking. The surgery bought a little time to continue his destructive habit. It was not a cure.

Politicians on the national level in nuclear countries, as well as their civilian and military leaders, know well the nuclear peril. However, from the point of view of survival of the species, their coping with this knowledge is inappropriate and even counterproductive. Writing them long letters describing the horror of nuclear war is almost certainly useless. They know the horror, but their response is inappropriate.

One can at least hope that the present international failure to deal effectively with the escalating threat of nuclear holocaust will awaken and alarm the general public, forcing systemic changes.[33] The many ways in which this nuclear threat warps public policy and the decision-making processes are largely due to the inability of the competitive nation-state system to contain and foster conflict resolution behaviour. If this is a correct analysis of the global situation, namely it is a societal structure problem, rather than an innate human flaw which leads to species suicide, then one can begin to sort out appropriate and inappropriate responses to the knowledge and anger which results from discovery of the nuclear peril. The possible behaviours of the global system of organisation may be characterised as leading to breakdown, cultural disintegration, chaos and eventual death, or leading to breakthrough to new behaviour and relationships; what I have called 'a time for blooming'.

Let us first look at the breakdown process, the inappropriate behaviour, the following of myths or beliefs which eventually dissolve before the harsh face of reality. If we begin with the assumption that human nature is basically violent, that it always has been and always will be violent, we have placed this image in the centre of collective consciousness as an 'ego ideal'. Our 'best society' is believed to be violent, selfish and aggressive. From this premise we move to a second assumption, namely, that wars are inevitable. Internal dissent is subversive or traitorous and internal control must be strengthened. Nations make war, nations are too 'big' to be controlled, nations will always make war in order to expand, to ensure access to resources, to obtain what is needed by their citizens. In this world view, arms control agreements can only delay the conflicts. Some day one 'power' will rule the world. There may even be a desire to have the war, determine the 'winner' and enjoy the 'spoils'. The wartime accumulation of land and culture, resources and citizens has in the past produced the great nations of the developed world. Building through wars is also very much a part of the developing world pattern – more than 150 wars have been fought since 1945.

It is at this point that reality intervenes. The difficulty posed by *nuclear* war is that the winners are really losers; the land, the resources and the people are destroyed, not accumulated; and the survivors will face the task of trying to rebuild a severely crippled world with their own crippled bodies and minds. Humans can rebuild cultures, but not air, water and food.

This leads to the concepts of either limiting the battleground, as for example, by a first strike, or displacing the battleground, as in the 'star wars' scenario.[34] The first-strike scenario, which has been part of US strategic planning since at least the time of the Berlin Wall crisis in the early 1960s, and which was made formal and official under President Jimmy Carter in 1979,[35,36] has a fatal flaw. Obviously, in a first strike by the United States against the Soviet Union, it would be imperative to destroy the military command centre, the so-called C^3I plan: command, control, communication and intelligence. This would make it impossible for the Soviet Union to surrender. Each surviving Soviet military base, plane and submarine becomes an isolated 'enemy', likely to retaliate in any way and at any time. The 'gains' or 'accumulations' of war come from surrender, not utter destruction. Star wars might afford a colossal 'last hurrah' for the participants, but they will hardly affect the conduct of affairs on earth. How can we imagine a lone space pilot returning to say he now 'owned the earth?' On the other hand, in a star war any nation would be best advised to keep its expensive 'toys' on the earth where they are safe. Can a government, relying on surveillance satellites, expect respect and co-operation globally? As one American remarked, not even the local police force could control the South Bronx area of New York City.

How do normally intelligent, responsible people become caught up in these fantasies? Ira Chernus, author of *Mysticism in Rabbinic Judaism*, has suggested two basic fantasies which underlie the inappropriate responses to nuclear peril: the fantasy of the hero survivor and the fantasy of complete annihilation.[37]

These fantasies surface in the language of images of our different cultures and religions, but it is not difficult to see common patterns. Science fiction stories revel in the fantasy of the heroic survivor. War is seen as a cleansing away of past 'sins', a conversion experience, bringing humans to their senses and ushering in a period of rebirth and revitalisation. It is seen as a chance to start human life all over again, without the mistakes of the past. This fantasy of 'a clean slate' appeals to many people.

Those who develop and test nuclear weapons are more likely to experience an 'embrace of nuclearism', described by Dr Robert J. Lifton as a conversion experience characterised by total immersion in death anxiety followed by rebirth into a new world view.[38] This describes the behaviour of those who shared the Alamagordo experience, the testing of the first nuclear bomb in July 1945, and the passionate crusading type of pro-nuclear advocate always ready to defend the technology and persuade others. Those who embrace nuclearism experience a relief from death anxiety which they find energising and an experience they would like to share with others.

It seems similar to the exhilaration experienced by the stock-car racer, the mountain climber or the tight-rope walker who has 'conquered' the fear of death. Yet that fear of death serves a good normal purpose in life. Both high-ranking political and military leaders have been forced by their structural role in society to pass through the reality of nuclear death and overwhelming catastrophe and to create a psychic world within which they can plan and act. They become 'daredevils' or are said to practise 'brinkmanship'. Their attitude towards nuclear war is probably dual: either they will be 'heroic survivors' or they will experience 'total annihilation'. Probably Jimmy Carter is a good example of a national leader who was unable to assume a bravado stance. He appeared 'weak' or 'vacillating'.

The opposite of the heroic survival fantasy is the fantasy of immediate total annihilation of the world. This is commonly expressed as: 'the world may be blown up', 'everything on earth will be destroyed', 'no living thing will remain'. Some even clothe the fantasy in piety: 'God will destroy the world because of its sinfulness', 'This may be God's plan for the end of the world', or 'Jesus described the last days, it is our fate'. The main message of this fantasy is that the individual has neither responsibility nor power over the catastrophe. There is no motivation to act to avert the impending annihilation. Reality is not so simple. For most people the nuclear dying will not be fast. It will take days, and months and years. There will be a lack of food and shelter, no medical care, no communication or churches or town meetings or businesses. There will be few children, and these will lack basic good health.

Serious religions have never taught that God created life for ultimate frustration, or that humans were basically flawed, moving inevitably towards mass suicide and omnicide. Religions generally issue a call to choose Life, not Death.

We can only conclude that in the enlightened person who has passed through the anger stage of awareness of species death, the fantasies of heroic survivors who rebuild the earth or of quick and painless annihilation of the earth, are essentially avoidance mechanisms which free the individual from nuclear responsibility. They serve to keep the societal status quo, and allow the slow backstep and forwardstep dance of death with the fascinating awe-filled nuclear holocaust. The individual is kept functional by his or her belief in a personal future. He or she will be either a heroic survivor or will experience total annihilation. Most likely neither is true.

The rectifying homeostatic control mechanism for assisting societies to become freed from the barter stage of nuclear coping will not come directly from consciousness-raising efforts or protest demonstrations. These actions are in fact frequently viewed as subversive to governments, who have great fears of 'arousing' the public. Demonstrations are viewed with alarm, seen as easily becoming anarchic. In fact, frequently those who have passed through these first two stages of confronting nuclear death will try to shield others from them (through secrecy and manipulation of information) and will severely punish demonstrations or emotional outbursts. After all, they *know* the facts and have managed to 'deal with these facts rationally'. People can be hired to shore up the rationalisations, spinning out wordy reports and learned briefs to divert the opposition and give 'respectability' to official plans. They believe their actions are justified and therefore any means required to support them are justified.

If, as a society, we are able to break out of this barter phase with its half acceptance of reality, half escape into illusion, it will be due to the careful patient building of a consensus in various social and political groups, which make an impact on the national power structures from within and from without, simultaneously. In this category of action I would place the Green Party, the International Physicians for the Prevention of War (and Physicians for Social Responsibility), the major Churches and the World Council of Churches, the trade unions and international labour organisations, the global network of feminists, international organisations of indigenous people and the international solidarity movements. Unlike the protest groups, which have arisen to fill a need in time of crisis, these groups are part of the infrastructure of a well-functioning society. As they become increasingly international in their thinking and acting, they are developing the infrastructure for the global village. They are forcing the present national and international power-brokers to understand that the world belongs to everyone, and if it can be healed that healing will have to be the work of all.

In November 1982 Petra Kelly collapsed and lost consciousness in a Munich taxi cab and had to be rushed to hospital. This frail young woman had been exhausting herself in anti-nuclear activities for many years, motivated through watching her younger sister die of leukaemia. Cancer was a reality for Petra Kelly, not a statistic. At age thirty-five, Petra became a leader and one of the chief spokespeople of the Green Party in Germany. She has a degree in political science from the American University in Washington, DC, and a Master's Degree from the University of Amsterdam. At the age of twenty-four she was hired as administrative adviser at the European Community headquarters in Brussels. Her political campaigning for the parliamentary elections in Germany brought her to Munich in November 1982 for public appearances, and her collapse resulted from non-stop speaking engagements, press interviews and public lectures. Petra recovered in time to preside over the Nürnberg Tribunal in February 1983 and won a seat in the Bundestag, the German Parliament, in March 1983. The Greens won twenty-seven parliamentary seats, roughly 5.6 percent of the vote, in their first major win after five years' uphill struggle for recognition. There are Green Parties in France, Holland, Belgium and British Columbia (Canada) and they are developing a new approach to exercising political power. In the 1984 German election the Greens secured more political gains in regional governments.

The Greens tried to avoid making any member a superstar. They hoped to rotate the twenty-seven seats in the Bundestag after two years. There were 'stand-ins' or executive assistants for each person serving prepared to take up the political office when asked to do so. The party platform stressed peace, disarmament and responsible use of the biosphere. Members of the Green Party in the Bundestag kept only about one-sixth of their monthly salary. The rest of the money went into a special fund used for the development of new ways to live responsibly and compatibly on the interpersonal and global levels. They planned to continue using blockades, peace marches, hunger strikes and other non-violent tactics to bring about social change.[39]

In 1984 Petra Kelly left the Green Party caucus, although she remains an Independent Member of the Bundestag.

As an outside observer, and a friend of Petra since 1978, I could not help but notice her instinct for healing our social system. First she deliberately reduced the power, money and secrecy now at the disposal of political leaders. Simultaneously, she began to help develop healthier relationships between people and the living earth system, assuming responsibility for societal action and mobilising the energy and the common sense of people. On one occasion, after receiving an invitation from the Soviet Embassy to speak in Moscow, Petra answered that she would come only if she were allowed to organise a peace demonstration in Red Square. She never heard from them again. In her insightful way, she had touched the centre of the global problem. The symbolic problem with the Soviet Union is force – curtailment of political freedom of its own citizens. The West will arm for fear of losing its freedom even if it loses its own freedom in the process. There is no disarmament without universal recognition of human rights and human dignity. There is no peace without sharing the trappings of power.

At Nürnberg, after three days of hearing evidence, the jury reached a verdict.

The jury had listened to old women as well as old men. The Germans called themselves simply German, rejecting the 'East German' and 'West German' identifications. Music, poetry, drama and flowers relieved the heaviness of the dialogue. The judges and those who must act on the verdict were the people of the world.

At the end of March, President Reagan announced his space-war scenario, and the US–Soviet disarmament talks in Geneva seemed to be deadlocked. On 28 March 1983 the United Nations released a report on violence by governments against their own citizens. 'All classes of people, rich and poor, peasants, urban workers, professional classes, religious groups and ethnic minorities and majorities, have been affected.' The executions are carried out with little or no legality, and their number is conservatively estimated at 2 million since 1968. The report, submitted to the UN Human Rights Commission in Geneva, stated that the executions were most common in areas experiencing wars and disturbances between countries – violence within and violence without mutually feeding on one another. Amnesty International, the credible and humane voice for national sanity, confirmed the UN finding and said that political killings have reached 'nightmare' levels globally.

The Physicians' groups are important not only for their educational role with respect to the public, but also for their awareness of the fragility of life and the need to protect future generations. Their

Nürnberg Declaration, 1983

Faced by the imminent destruction of our planet, the jury of the Nürnberg Tribunal against First Strike and Mass Destruction Weapons in East and West has passed the following declaration:

1. Any use and the threat of using atomic, biological and chemical weapons is contrary to international law and criminal.
2. The very planning and preparation of their use, whether limited or unlimited, is contrary to international law and criminal.
3. In particular, any strategy and any kind of atomic build-up which render a first strike possible are contrary to international law and criminal. The deployment of Pershing II and Cruise Missiles in Western Europe is therefore contrary to international law and criminal.
4. The deployment of weapon systems of any kind in outer space is contrary to international law and criminal.
5. Any attack on nuclear power stations is contrary to international law and criminal.
6. Further proliferation of the nuclear industry throughout the whole world presents an intolerable risk for nations and promotes a nuclear build-up.
7. Proliferation of any technology leading to nuclear war is contrary to international law and criminal.
8. Squandering human resources, raw materials and sums of

actions in opposing senseless 'civil defence' fantasies provide a healthy dose of reality for public officials. Their growing personal awareness will no doubt find many other insightful ways to focus human attention on the increasingly pathological behaviour of national leaders.

Church authorities also command respect within society, and their increasing involvement in the formation of a just, equitable and participatory society, as a prerequisite to peace, is most helpful. The US Catholic Bishops' Peace Pastoral has been a welcome sign.[40] One can only hope that the Physicians and Church leaders do not themselves become entrenched in a 'barter stage', able to denounce violence and injustice in governments but unable to come to terms with their own oppressive behaviour. Just as countries which violate the human rights of their own citizens will never be trusted, so too institutions and professions will need to examine their own behaviour both internal and in relationship to others, if a 'time for blooming' is to be achieved.

money running into billions on arms at the expense of the most pressing social needs of our time and also at the cost of all destitute people, especially in the Third World, is contrary to international law and criminal.

Faced by arms development and the increasing risk that Europe may become a nuclear Auschwitz, we call upon mankind to rid itself of all means of mass destruction. The governments must be compelled to ensure safety not by deterrence based on ever-increasing potentials to destroy, but by arms renunciation and by consistent disarmament.

Not violence and weapons, but reason and trust must determine the fate of our planet, as the arms race dehumanises politics, leads to militarisation of thought and action and prevents peoples from living together in an environment free of violence.

Freda Meissner-Blau, Ecologist (Austria)
Prof Dr Ossip Flechtheim, Political scientist (FRG)
Vladimir Lomeiko, Publisher (USSR)
Prof Dr Seiei Shinohara, Philosopher (Japan)
Dr John F. Steinbruck, Reverend (USA)
Hermann Verbeck, Priest (The Netherlands)
Prof Dr George Wald, Biochemist and Nobel Prize Winner (USA)

Nürnberg, 20 February 1983

International trade unions are also potentially constructive global organisations. Similarly, they need to look to their own internal structures of authority and priorities if they wish to remain viable. Hopefully in the future worker rights will be guaranteed by law rather than by yearly bargaining and the worker will be free to participate in the complex community and environmental concerns of society.

Feminist groups, especially the Women's Peace Camps, are developing decentralised styles of organisation which maximise small-group and personal initiatives. They are learning techniques for consensus decision-making, preserving unity in diversity, eliminating punishment as a manipulative device for retaining societal control and developing conflict-resolution skills. They reject war as a way to peace, choosing 'peaceful' acts instead. Their non-co-operation with official government and military stances poses one of the most fundamental challenges to outmoded social structures. As the old structures crumble, violence against women will undoubtedly escalate, frightening many feminists back into conformity.[41]

Movements of indigenous people, those who do not choose to die with the white man and his violent society, are also of great importance at this time. Likewise, the solidarity movements such as those which unmask external intervention and violence in El Salvador, Honduras and Guatemala, and the violence used against Poland, Chile, Nicaragua, Afghanistan and Cuba, serve to awaken global concern and dispel the fantasy of power and prosperity in the exploitative and violent superpowers. The role of these movements is extremely important if the leaders of nations are to move on from the paralysing fantasies of the bargaining stage of awareness of species death into a wholesome respect for the entire reality and total commitment to attempting to halt the death process, even if this commitment requires a reduction in a nation's power over the lives of its citizens and the loss of the 'right' to declare war.

The Fourth Stage: Acceptance of Reality

The 'acceptance' of species death as a fact is anything but a passive phenomenon. It is rather a profound contact with the entire reality, the fearful catastrophic power of annihilation together with the wilful violence which produces it. While the military and government leaders, believing they have 'accepted' the reality of the nuclear age, bargain for time, those who understand the sickness induced by violence and fear move beyond the bargaining and attempt to discover new human strengths that can overcome the violence within and restore health.

Not all members of the peace and anti-nuclear movement are in the second, 'protest' stage; although all see the demonstrations as important for developing human visibility and solidarity. Many who have reached a fourth stage of dealing with species death find peace demonstrations too shallow a response. Many military and government leaders have also moved beyond 'bargaining', but they usually wait until retirement to speak out because it is politically unacceptable for them to break ranks while in military or government office. One of the strangest human paradoxes is that we fear human censure and loss of job more than we fear nuclear extinction. Four-star retired Admiral Noel Gayler, veteran of three wars, had the courage to state flatly: 'Nuclear weapons have no military usefulness.' And a poignant chapter in human history is being written by Veterans for Multilateral Disarmament. During the summer of 1984 Soviet veterans welcomed the Canadian Chairman of this relatively new organisation to discuss the prevention of nuclear war. In October 1984 the Soviet Committee of War Veterans met in Belgrade with other veteran organisations from all over the world to discuss international security and disarmament. Thus while the young men train for war, the old men are laying a groundwork for a peace not built on mutual threats and global terror.

While the building of a non-violent global community will take centuries of human learning and activity, the turning away from the nuclear precipice must be accomplished quickly. 'How' to do this has millions of answers – one for each person who wakes up to the total reality. Solutions call for a changed lifestyle, non-violent parenting, a care for the earth system and for its produce. Lawyers, doctors, scientists, teachers, business executives, church leaders, blue-collar workers, trade unionists, all play a role. The general mode of action is non-co-operation with death, co-operation with life. Dan and Phil Berrigan lead the way in the USA to non-violent action against weapon production, women in England and elsewhere have set up peace camps at military bases, physicians have pointed out human inability to recover from nuclear war, politicians have organised people's tribunals, artists have spoken to the hearts and emotions of a confused and fearful generation, and bishops have issued pastoral letters. The hallmark of the fourth-stage person is an obsession with the problem which makes 'life as usual' pretences impossible. Such people cannot be silent. Prison sentences, scathing criticism, disbelief of family members, all these and more are overcome. The fourth-stage person discovers that his or her abilities to speak and act far exceed anything ever dreamed possible. These awakened people are precious for the health of the global society, they come from all nations and all classes; they speak strongly for non-violence, for solidarity, for concern. They understand that international violence is not *an* issue, but *the* issue.

It will not be possible even to outline the global village in this book. The imperative of turning away from violence, rather, must be our focus. It makes little difference whether the violence is within the nation or between nations. This turning away requires an examination of all human systems and decision-making processes in society. It means examining the feedback mechanisms whereby we test the health of our decisions. With these two understandings, decision-making and feedback, we can begin with new courage to

The Declaration of Universal Human Rights

Article 1
All human beings are born free and equal in dignity and rights. They
are endowed with reason and conscience and should act toward one
another in a spirit of brotherhood.

Article 2
Everyone is entitled to all the rights and freedoms set forth in this
Declaration, without distinction of any kind, such as race, colour, sex,
language, religion, political or other opinion, national or social origin,
property, birth, or other status. Furthermore, no distinction shall be
made on the basis of the political, jurisdictional or international status
of the country or territory to which a person belongs, whether it be
independent, trust, non-selfgoverning, or under any other limitation
of sovereignty.

Article 3
Everyone has the right to life, liberty and security of person.

Article 4
No one shall be held in slavery or servitude; slavery and the slave
trade shall be prohibited in all their forms.

Article 5
No one shall be subjected to torture or to cruel, inhuman or
degrading treatment or punishment.

Article 6
Everyone has the right to recognition everywhere as a person before
the law.

Article 7
All are equal before the law and are entitled, without any
discrimination, to equal protection of the law. All are entitled to equal
protection against any discrimination in violation of this Declaration
and against any incitement to such discrimination.

Article 8
Everyone has the right to an effective remedy by the competent
national tribunals for acts violating the fundamental rights granted
him by the constitution or by law.

Article 9
No one shall be subjected to arbitrary arrest, detention or exile.

Article 10
Everyone is entitled in full equality to a fair and public hearing by an independent and impartial tribunal, in the determination of his rights and obligations and of any criminal charge against him.

Article 11
Everyone charged with a penal offence has the right to be presumed innocent until proved guilty according to law in a public trial at which he has had all the guarantees necessary for his defence.

No one shall be held guilty of any penal offence on account of any act or omission that did not constitute a penal offence, under national or international law, at the time when it was committed. Nor shall a heavier penalty be imposed than the one that was applicable at the time the penal offence was committed.

Article 12
No one shall be subjected to arbitrary interference with his privacy, family, home or correspondence, nor to attacks upon his honour and reputation. Everyone has the right to the protection of the law against such interference or attacks.

Article 13
Everyone has the right to freedom of movement and residence within the borders of each state.

Everyone has the right to leave any country, including his own, and to return to his country.

Article 14
Everyone has the right to seek and to enjoy in other countries asylum from persecution.

This right may not be invoked in the case of prosecutions genuinely arising from non-political crimes or from acts contrary to the purposes and principles of the United Nations.

Article 15
Everyone has the right to a nationality. No one shall be arbitrarily deprived of his nationality nor denied the right to change his nationality.

Article 16
Men and women of full age, without any limitation due to race,

nationality or religion, have the right to marry and to found a family. They are entitled to equal rights as to marriage, during marriage and at its dissolution.

Marriage shall be entered into only with the free and full consent of the intending spouses. The family is the natural and fundamental group unit of society and is entitled to protection by society and the state.

Article 17
Everyone has the right to own property alone as well as in association with others.

No one shall be arbitrarily deprived of his property.

Article 18
Everyone has the right to freedom of thought, conscience and religion; this right includes freedom to change his religion or belief, and freedom, either alone or in community with others and in public or private, to manifest his religion or belief in teaching, practice, worship and observance.

Article 19
Everyone has a right to freedom of opinion and expression; this right includes freedom to hold opinions without interference and to seek, receive and impart information and ideas through any media and regardless of frontiers.

Article 20
Everyone has the right to freedom of peaceful assembly and association.

No one may be compelled to belong to an association.

Article 21
Everyone has the right to take part in the government of his country, directly or through freely chosen representatives.

Everyone has the right of equal access to public service in his country.

The will of the people shall be the basis of the authority of government; this will shall be expressed in periodic and genuine elections, which shall be by universal and equal suffrage and shall be held by secret vote or by equivalent free voting procedures.

Article 22
Everyone, as a member of society, has the right to social security and is entitled to the realisation, through national effort and

international co-operation and in accordance with the organisation and resources of each state, of the economic, social and cultural rights indispensable for his dignity and the free development of his personality.

Article 23
Everyone has the right to work, to a free choice of employment, to just and favourable conditions of work, and to protection against unemployment.

Everyone, without any discrimination, has the right to equal pay for equal work.

Everyone who works has the right to just and favourable remuneration, insuring for himself and his family an existence worthy of human dignity, and supplemented, if necessary, by other means of social protection.

Everyone has the right to form and to join trade unions for the protection of his interests.

Article 24
Everyone has the right to rest and leisure, including a reasonable limitation of working hours, and periodic holidays with pay.

Article 25
Everyone has the right to a standard of living adequate for the health and well-being of himself and his family, including food, clothing, housing, medical care and necessary social services, and the right to security in the event of unemployment, sickness, disability, widowhood, old age or other lack of livelihood in circumstances beyond his control.

Motherhood and childhood are entitled to special care and assistance. All children, whether born in or out of wedlock, shall enjoy the same special protection.

Article 26
Everyone has the right to education. Education shall be free, at least in the elementary and fundamental stages. Elementary education shall be compulsory. Technical and professional education shall be made generally available and higher education shall be equally accessible to all on the basis of merit.

Education shall be directed to the full development of the human personality and to the strengthening of respect for human rights and fundamental freedoms. It shall promote understanding, tolerance and friendship among all nations, racial or religious groups, and shall further the activities of the United Nations for the maintenance of peace.

Parents have a prior right to choose the kind of education that shall be given to their children.

Article 27
Everyone has the right freely to participate in the cultural life of the community, to enjoy the arts and to share in scientific advancement and its benefits.

Everyone has the right to the protection of the moral and material interests resulting from any scientific, literary or artistic production of which he is the author.

Article 28
Everyone is entitled to a special and international order in which the rights and freedom set forth in this Declaration can be fully realised.

Article 29
Everyone has duties to the community in which alone the free and full development of his personality is possible.

In the exercise of his rights and freedoms, everyone shall be subject only to such limitations as are determined by law solely for the purpose of securing due recognition and respect for the rights and freedoms of others and of meeting the just requirements of morality, public order and the general welfare in a democratic society.

These rights and freedoms may in no case be exercised contrary to the purposes and principles of the United Nations.

Article 30
Nothing in this Declaration may be interpreted as implying for any state, group or person any right to engage in an activity or to perform an act aimed at the destruction of any of the rights and freedoms set forth herein.

walk an unknown path only infrequently trodden by our predecessors.

Such junctures for change are not unknown in human history. Cannibalism, slavery, the building of armed cities and castles, and colonialism have all become socially rejected behaviours. In our day, war itself must be rejected and human rights must be universally recognised.

The basic stance between members of a family is love and trust. The basic stance within a nation is exchange, governed by laws of 'fairness'. The basic stance between nations is now threat, although some alliances – NATO, Warsaw Pact or non-aligned – combine exchange between nations and threat to nations not in the alliance. It will be necessary to extend the love–trust relationship globally, probably within global societies, churches or networks of like-minded people. Imagine various networks of trust based on church, profession, race, sex or other differentiating characteristics. The global anti-nuclear movement provides a good example of international trust relationships. The 'exchange' mode of relationships must be extended to all other people and nations, and the use of threat between people and nations outlawed. Threat and extortion are called 'organised crime' within nations. They need to be so labelled and resisted on the international level also.

Just as families are expected to be upright and law-abiding so should nations be held accountable. Just as family privacy is inviolable unless violence erupts within it, so too national integrity is integral to global integrity. However, when a nation practises violence against its citizens, or when it threatens or attacks other nations, its leaders must be 'restrained' and made to understand that such behaviour threatens all life on the planet.

On Making Decisions

First we must examine our human system models, since it is the human resolve to use nuclear weapons which constitutes the immediate threat to life on earth. The decision to use implies necessarily the decision to design, build and test these weapons, and to develop ever more terrible instruments of mass destruction. The global death toll since the dawning of the nuclear age, including slow genetic deaths, stands at 13 million people, at least. One must look at the ideas input, the decision-makers and the workers who implement these decisions. It is the ideas people who dream dreams and provide the material on which decisions are made. It is the failure of feedback which makes 13 million casualties 'invisible' within the global community.

Model 1 on p. 352 is a schematic representation of an abnormal human system, showing a disproportionate emphasis on the decision-maker. Such a model is frequently used in totalitarian states, paternalistic societies and churches. Such a system is bound to fail eventually for two reasons. First, the deficient input of ideas leaves the decision-maker unaware of all available actions, hence it limits his or her choice of actions. It also fails because the workers are considered slaves, and their feedback to the ideas people on the

Model 1: Authority-centred Human System

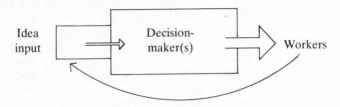

problems connected with the choices is not valued. In such a model, habit usually governs decisions and decision-makers become more and more defensive, harmful and removed from reality. Eventually the workers rebel because they are physically unable to carry out the decisions, or because they refuse to co-operate any longer with decisions so removed from their lives and value systems. It is directly opposite to the Biblical call for the leader to be the servant of servants.

An example of authority-centred decision-making which resulted from national preoccupation with maintaining maximum access, and if possible, exclusive access, to resources deemed 'strategic' was one which related to germ plasm, the fundamental building block of all food plants. In the 1970s international co-operation among agricultural researchers and seed storage centres was encouraged and a new agency, the Consultative Group on International Agricultural Research (CGIAR), was formed. CGIAR is housed at and funded by the World Bank. As the Green Revolution displaced more traditional food plants, threatening them with extinction, the CGIAR moved more strongly toward establishing seed banks, rather than trying to preserve crops in their natural environment. Besides the more obvious problems with preservation of grains, vegetables, tubers and tree crops in artificial environments – such as loss due to electrical failure or disease, genetic drift as the crops try to adapt to the new setting and loss of the benefits of natural selection in their natural habitat – seed banks have an even more serious flaw. Invaluable information gathered by farmers about their crops over the centuries, such as how the crops fit into the surrounding ecosystem, the control of pests, how and when they can be eaten, is lost.

Between 75 and 90 percent of the stored seed is held in the first world countries, although the greatest reservoirs of biological diversity are in the third world. Maize originated in Central

America and the Andes; barley, sorghum and millet in Asia Minor and Ethiopia; rice in south-east Asia and potatoes in the Andes. The seeds are separated from their natural environment and from the people who have preserved them into the twentieth century.

As large oil, chemical and pharmaceutical transnational corporations purchase these genetic building blocks and develop genetic engineering research tools, they try to 'protect their investment' by private property rights. They achieved 'patent protection' rights in the United States as early as 1930. In 1961 several European countries formed the International Union for the Protection of New Plant Varieties (UPOV). The United States, Israel and South Africa joined UPOV in 1978. The truth is that 90 percent of the value of these patented foods comes from their basic genetic endowment, another 9.9 percent is due to improvement over the centuries, and the corporations add about a 0.1 percent change and then declare that they 'own' the food and all rights to its seed.[42]

Idea input into this global seed bank system, therefore, is extremely limited. UPOV is a private, self-regulated agency designed to protect corporate vested interests. The farmers who carry real knowledge of the food – its growth, needs, and use – are not valued. The ecosystem within which the plant developed and thrived is not preserved. The decision-maker role is exaggerated, input into decisions impoverished and workers treated like robots. Sooner or later a food disaster must occur, causing widespread suffering and death.

Model 2: Think-tank Human Systems

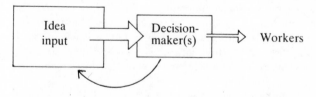

This model of decision-making over-emphasises the ideas people, the so-called 'think tanks'. These ideas are surrounded with both secrecy and prestige. The 'leave it to the experts' philosophy results in decision-makers allowing the input segment actually to make the decisions. This results in so-called 'rubber stamp' legislation.

The military and strategic sectors of the United States, USSR and other major militaristic nations are best described by this model.

Their 'ideas people', selected yearly from the most intelligent university graduates, are busy creating both the need and the new weapons to meet the need. Given their strategists' creative persuasiveness, government officials frequently rubber-stamp their funding requests. In this way the arms race becomes self-perpetuating and governments lose control. There is a corresponding 'brain drain' from the civilian sector which causes even more social distortion. Few people seem to understand the superiority of West German and Japanese consumer products in relation to their post-Second World War cessation of military development. The scarce supply of creative people cannot be used for military purposes without a corresponding deficit in the civilian sector and vice versa. Military priority means that health research, environmental studies, consumer technology and the arts can receive only 'spin-offs' from the military programmes. We should note that it was not the 'strategic experts' who suggested the nuclear freeze. It cannot be expected that their ideas will include making their own jobs obsolete. Strategic peace planners must become a part of official decision-making channels.

The Model 2 human system weakness accounts also for many of our global energy problems. As long as energy systems are viewed by strategic planners as spin-offs from military technology, the energy 'crisis' will never be sensibly dealt with. This logical error applies even to non-militaristic countries with completely peaceful intentions. Such countries are not in a position to re-invent the technology and re-think its impact on workers and the public at large. Not understanding the military imperative which underlies its development and use, they do not understand the human health effects estimates which are minimised or ignored in the research and development process. Ignorance of the effects will, unfortunately, not prevent their occurrence. Good intentions will not, unfortunately, insulate us against unwittingly co-operating with the oppressive violent system moving the human race towards annihilation.

There is another striking example of a think-tank human system. With little or no public discussion of the consequences, approximately 11,000 objects, detectable on radar, have been put into space. At the Thirteenth Conference on the United Nations of the Next Decade, 'Co-operation or Confrontation in Outer Space',[43] July 9–15 1978, experts predicted increased numbers of space collisions. 'It seems probable that a debris belt will be created, perhaps becoming significant before the end of this century.' This debris belt may significantly affect global weather and crop production.

Although the Outer Space Treaty of 1967 specifically banned the stationing of nuclear and other weapons of mass destruction in space, this treaty is now being construed by the United States and the USSR as allowing anti-satellite weapons. Most of the functioning satellites are being widely used by the military for information-gathering, verification and communication. In the future, they may be used for star wars.

Ironically, the USA military knows more about Soviet military hazards placed near civilian residences than do the people living near them. Similarly, the Soviet Union monitors US weapon depots and transportation, although this information is withheld from the American people at risk. Information which used to be withheld as a 'secret from the enemy', now becomes a way to protect government and military decision-makers from accountability to the people they represent. The idea input is totally biased by being limited to strategic planners who believe war is inevitable, and by those whose vested interests lie in creating needs for their products and services.

Model 3: Ineffective Human Systems

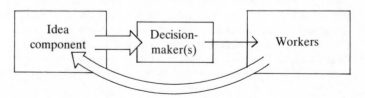

The Model 3 type of decision-making is typical of the United Nations, with its rich inflow of ideas and information, but its inability to affect the conduct of sovereign nations. Non-governmental organisations provide the UN with direct feedback from people of all nations, but outside these vital communication networks, few people even know of UN decisions or actions. The major global decisions are, unfortunately, made in the backrooms of military and/or industrial think tanks.

The decision-making weakness of the United Nations flows directly from the unwillingness of nations to relinquish some of their sovereignty. This unwillingness flows not only from human greed and desire for power over the lives of others, but also from the real existence of weapons of terror which could be used to coerce, and from fear of totalitarian states which abuse the rights of their citizens. UN decision-making is only possible if the levels of violence, terrorising and fear decline. This means that arms reduction and respect for human rights must be linked in the first and second worlds, and the interconnections of these injustices with third world development problems must be made clear.

United Nations' activities with respect to outer space provide a good example of ineffective decision-making. The space-related activities at the United Nations are co-ordinated by the Committee on the Peaceful Uses of Outer Space within the General Assembly, the Outer Space Affairs Division and the Centre for Natural Resources, Energy and Transport within the Secretariat. Specialised agencies such as the UN Educational Scientific and Cultural Organisation (UNESCO), the World Meteorological Organisation (WMO), the Food and Agricultural Organisation (FAO), the Centre for Natural Resources, Energy and Transport (CNRET), the International Civil Aviation Organisation (ICAO), the International Atomic Energy Agency (IAEA), the World Bank and others are involved with idea input and with implementation feedback into UN thinking and planning. Besides the Outer Space Treaty of 1967, there are other UN-sponsored documents codifying facets of space law. The Agreement Regarding Rescue and Return of Astronauts and Objects, which entered into force in 1978; the Convention Regarding International Liability for Damage Caused by Space Objects, which entered into force in 1972; and the Convention on Registration of Objects Launched into Outer Space, which entered into force in 1976, are hardly known by the public or taught about in schools.

In the last analysis, it will be people who keep agreements, and people who must demand that their nations enter into and respect international law. Unfortunately, people who understand and attempt to keep international law are usually labelled 'anarchists', unpatriotic, naive, or even criminals within their own countries. In the United States, the Soviet Union and other nations, citizens are sent to prison for obeying international laws which are seen to conflict with national interest, even when national leaders publicly claim to adhere to the international treaties.

In spite of United Nations' efforts, space, the last area properly called the common heritage of humankind, is rapidly becoming militarised. It is well known that both the United States and the USSR are conducting research on the use of directed energy weapons such as lasers and particle beams, yet the UN disarmament agency cannot deal with these developments as long as the programmes are called 'energy systems' by the nations concerned. The US Department of Defence has already spent more than $1.3 billion[16] to demonstrate a laser's ability to bring down missiles in flight. Particle-beam weapons rapidly heat a target, melting or cracking it with thermal stress. They would have a range of several thousand miles at the speed of light, and could destroy enemy vehicles without having to manoeuvre to intercept them. The Pentagon budget for 1984–88 includes $7 billion for building space weapons.[44]

The Soviet particle-beam weapon programme has been under way since about 1960, just as has the US programme. Since 1969 the Soviet Union has averaged about eighty satellite launchings per year, the majority being directed to military operations. Most of the satellites are engaged in photographic reconnaissance, electronic ferreting, radar calibration, military navigation and communication, ocean and land surveillance. Soviet strategists view the US space programme, probably correctly, as primarily military. In *Military Strategy* by V. D. Sokolovskii, quoted in the *Soviet Aerospace Handbook*, we read: 'The US imperialists have subordinated space exploration to military aims and . . . they intend to use space to accomplish their aggressive projects . . .'

This accusation is difficult to refute. In December 1978 a report entitled *US Civilian Space Programs: An Overview* was prepared by the Science Policy Research Division of the Congressional Research Service of the Library of Congress for the US House of Representatives Committee on Science and Technology. The report strongly suggests that the civilian National Aeronautics and Space Administration (NASA) be phased out and replaced by the US Department of Defence. The Pentagon expects to shift the space shuttle programme to exclusively military launchings 'some time after 1985'.[45] The Pentagon plan for improving just military communications calls for an $18 million expenditure. A new space shuttle launching pad is now under construction at the Vandenberg Air Force Base. The US military space programme is now funded at about the same level as the civilian programme, but the military manages to spend its own budget and also to usurp about 30–40 percent of the civilian programme.

Because of the weak decision-making component of the United Nations system, all important national decisions are made outside its aegis. The people of both superpower nations are also kept in ignorance of their national programmes and their consequences. While consumers are warned by environmentalists that spray-can deodorants can damage the ozone, military planners conceive a thirty-year Solar Powered Satellite (SPS) programme which will dump millions of tons of rocket exhaust including the ozone-destroying nitrogen oxides into the upper atmosphere. The USA-based SPS programme calls for 11,000 new rocket launches, each burning 20 million pounds of liquid hydrogen and liquid oxygen, and creating millions of pounds of water vapour, carbon dioxide, carbon monoxide and nitrogen oxides. The upper atmosphere cannot be considered immune to this massive pollution. Its effects on life are generally unknown. Harm to the atmosphere would decrease earth's protection from the sun's ultraviolet rays with corresponding negative effects on human health and on food. Flight through the microwave beams from the SPS to earth would have to be prohibited for aircraft and would be fatal for birds. Humans exposed to even 'stray' microwaves could suffer from cataracts, disruptions of biorythms and central nervous system damage.

Although all orbiting satellites will eventually plunge back into the earth's atmosphere to be vaporised, to pile up as garbage in our already polluted oceans, or to shower the earth with their secret military cargo, no international agency really governs their proliferation. The following breakdown of satellites on 12 January 1983 was identified by the North American Air Defence Command:

		%
Spain	1	0.02
Australia	1	0.02
Italy	1	0.02
France/Germany	2	0.04
Indonesia	2	0.04
People's Republic of China	4	0.08
West Germany	4	0.08
India	5	0.10
NATO	5	0.10
United Kingdom	10	0.21
Canada	11	0.23
European Space Association	13	0.27
France	25	0.52
International Telecommunications Space Organisation	28	0.59
Japan	48	1.00

		%
USSR	1,966	41.20
USA	2,650	55.50
Total	4,776	100.000

Other orbiting debris is rubbish from satellites which have outlived their usefulness. About 55 percent of the usable satellites belong to the USA, and about 58 percent belong to North America, Europe or Japan. The two superpowers own and control 96.7 percent of the satellites. When the public asks for solar energy, they are told that these satellites will supply unlimited solar energy for the future!

In a survival scenario no part of the decision-making can be secret since human survival in the limited biosphere is at stake. All three components are important and all contribute to the overall health of the system. There are at least two feedback loops: one for immediate reactions of physical or psychological stress, and one for long-term effects on the biosphere or human gene pool.

The best example of a functioning balanced system is the human body. Creativity, imagination and emotions supply the idea input. This is modified and made operative by the judgement tempered by personal and communal experience. The feedback from the nervous system is usually immediate, but there is also long-term feedback mediated by organic and hormonal disturbances, sometimes requiring interpretation by the professional physician or psychiatrist. We can easily elaborate on this systematic biological model which operates for each of us so smoothly for some seventy or more years, and even reproduces a vital copy or copies of itself to perpetuate the experiment. The disorder of a person who decides to attempt what he or she cannot do, or of the person who lacks imagination to live more fruitfully, provides an example of failure to use this system properly.

The simple balanced human system model can be applied to several levels or hierarchies of decision-making. Each of these must be brought into scrutiny if human control over the human experiment on earth is to be restored. The decision of Bell Telephone and of the Canadian government to ease a satellite into orbit with the assistance of the US shuttle programme will be called an operational decision'. These operational decisions hinge on the US national decision to launch the shuttle space programme itself,

Model 4: Balanced Human System

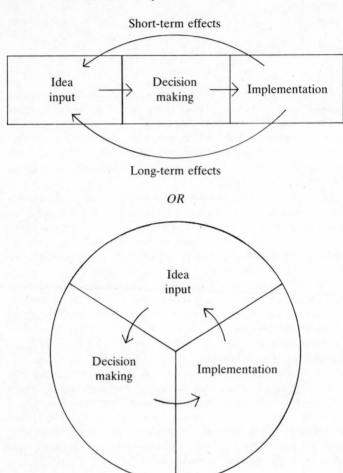

OR

which in turn hinges on the military decision to develop space for 'national security'. The operational decision, 'what will be done', is dependent upon an overall plan, 'what can be done', in this case liaising with the US programme. The planning level, 'what can be done', depends in turn on the policy level, 'what ought to be done'. Schematically the activity flows as follows:

	Idea input	Decision making	Implementation
Policy level 'what ought to be done'			
Planning level 'what can be done'			
Operational level 'what will be done'			

It is not hard to identify examples of distortion of this system, limits of control for each level, and places where the system could be improved with feedback.

One general observation can be made: what if Bell Telephone, Canada and other organised groups recognised the need for a communications satellite, and through UN agencies were able to develop a satellite programme designed to meet this need? This decision could then be researched and developed on the policy level of the United Nations, and implemented in such a way that nations and corporations shared in the planning, costs and benefits. It should not be necessary to graft civilian programmes on to military ventures and adapt them to fit military needs. This grafting not only reduces the importance of civilian priorities but also perpetuates, helps finance and even tends to legitimise the military prerogative to dominate the policy level of decision-making.

To use an analogy: in a home where the breadwinner is addicted to drugs or alcohol, the family's financial priority will undoubtedly be the addictive needs. Health care, education, shelter and even food will be secondary. The family response to this distorted and destructive habit is frequently passive support, i.e. avoidance of direct confrontation. An effort is made to win small concessions from the addict, some small financial contribution or sign of caring. The family nourishes an everlasting hope of reform fed by moments of caring and love shown by the addict. The dream of what the loved one could be overrides the ugly reality. The addiction becomes a problem for the whole family, not just for the alcoholic or drug addict.

Canadian co-operation with US military addiction may well be labelled as an example of this type of passive co-operation. Canadians have a basic kinship and sympathy towards Americans and they periodically exploit the US military addiction for their own

United Nations '2000' Fund

ARTICLE 112

All Members, determined to fulfill their pledge expressed in Article 56, unconditionally recognize the urgent need to reduce their military expenditures and to divert their financial and industrial means toward achievement of the purposes set forth in Article 55. To accomplish this goal, all Members shall comply with the following program:

1. Each Member shall disclose to the Economic and Social Council its overall military budget for the fiscal year ending in 1983. For the purposes of this Chapter, such disclosed 1983 military budget shall be called Peak Military Expenditure (PME).

2. Each Member shall establish a special account at its central bank, under the title '2000' FUND.

3. During the fiscal year ending in 1983, each Member shall deposit not less than 10% of its PME to the '2000' account, and shall appropriate for military purposes not more than 90% of its PME.

4. During the next six fiscal years, each Member shall reduce its military budget to not more than 30% of its PME, and shall increase its annual deposit to its '2000' Fund to not less than 70% of its PME, in accordance with the following schedule:

Fiscal Year	Deposit to '2000' Fund	Military Budget
1984	20%	80%
1985	30%	70%
1986	40%	60%
1987	50%	50%
1988	60%	40%
1989	70%	30%

5. Commencing in fiscal year of 1989, each Member shall deposit every year not less than 70% of its PME to its '2000' Fund, and shall never appropriate for military purposes more than 30% of its PME. The rate of further reductions in military expenditures shall be determined annually by the General Assembly.

ARTICLE 113

1. The International Monetary Fund shall establish the International Monetary Unit (IMU) representing the value of set amount of gold, subdivided into 100 fractional units representing the value of set amount of silver.

2. Commencing in 1983, the International Monetary Fund shall issue annually new paper currency in IMU denominations in the amount equaling the total value of all deposits to all '2000' accounts worldwide, and shall supply the central bank of each member with the appropriate amount of IMUs equaling the value of such

Member's deposit to its '2000' Fund. The course of exchange shall be determined by the Economic and Social Council on the basis of special index reflecting the real purchasing power of each Member's national currency.

3. Upon conversion of its annual deposit to the '2000' Fund into IMUs, each Member shall reduce its national money supply by the amount converted, and shall report promptly to the Economic and Social Council any further change in money supply.

4. The IMUs supplied by the International Monetary Fund in accordance with Paragraph 2 shall be accepted by all Members worldwide as a legal tender for satisfying debts, both public and private, paralelly with their national currencies. All Members shall promulgate the necessary legislation to protect free circulation of IMUs within their respective jurisdictions at the course of exchange set by the Economic and Social Council.

ARTICLE 114

1. All Members shall contribute annually to the World Bank a certain percentage of their '2000' Funds as non refundable deposits, in direct proportion to their Gross National Product per capita, at the rate of 1% of '2000' Fund of each $100 of their GNP per capita. Until the national currency of each particular Member is entirely replaced by the IMUs, such annual contributions shall be deducted by the International Monetary Fund from the amount due to each Member upon conversion of its deposit to the '2000' Fund into the IMUs, and transferred directly to the World Bank.

2. All Members shall be entitled to borrow money from the World Bank with interest set individually for each Member by the Economic and Social Council in direct proportion to such Member's Gross National Product per capita, at the rate of 0.1% interest for each $100 of GNP per capita.

3. All Members shall supply the Economic and Social Council with accurate statistics in accordance with the uniform system of reporting the Council may prescribe from time to time. Members who fail to supply such statistics, shall pay double the highest interest rate on their loans from the World Bank.

4. Aside from annual contributions to the World Bank, the '2000' Fund of each Member shall be subject to taxation by the General Assembly.

5. The remaining part of each '2000' Fund, not contributed to the World Bank and not taxed by the General Assembly, may be used by Member for any non-military purpose, provided, that every project financed from this source shall be commemorated accordingly.

Continued

ARTICLE 115

1. The Economic and Social Council shall present to the General Assembly at the 1983 Session, a comprehensive World Development Program.

2. All Members pledge their full support and co-operation to the United Nations in the effort to accomplish all particular goals set forth in the World Development Program, upon its approval by the General Assembly.

3. Commencing in 1983, the Economic and Social Council shall submit annually to the General Assembly, a budget proposal, including request for taxation of the '2000' Fund, to finance the annual goals of the World Development Program.

***

The author regrets her inability to discover the origin of this excellent proposal, and apologises for lack of proper credit to the originator.

---

self-gain or economic survival. Sooner or later for the health of the global village, this US militarism must be confronted for what it is, and for the danger it poses to the rest of the world. The flow of money and resources into this addictive habit threatens global health, education, shelter and food security. No doubt Canadians think the US addiction is 'under control', or that 'they would never use the bombs'. Whether Canadian co-operation with this addiction is subconscious or conscious is really irrelevant to the destructive effects. The situation can only worsen with time and the general public ignorance become more blameworthy.

The United Nations' '2000' fund proposal appears to offer a viable way out of the financial dilemma for countries such as Canada. Such a radical change in money manipulation in favour of a just and peaceful world order requires a new Breton Woods conference.

Global health cannot be restored without systematically examining the policies underlying local, national, and international operational levels. The fundamental policy questions are: How does this policy depend on, promote or lessen violence on the local, national and international level? How are workers' rights respected or violated? By whom is the financial burden carried? Are decisions being shrouded in secrecy in order to 'manage' those who are exploited? Is international security being enhanced or lessened?

Responsible decision-making can enhance human security and ensure the sustainability of the human system into the future. The nation-state system, established after the Peace of Westphalia in 1648, is less than 350 years old. It is obviously unable to provide either global or national security, well-being and safety in the nuclear age. It has failed to foster an international climate in which the true unfolding of human potential can take place. Its benefits can only be retained with some modification of the more potentially harmful dimensions.

Primary effort and attention needs to be directed to developing a human systems model which allows for maximum national and international organisational participation, for equitable global distribution of goods and services, and for the sustainability of human society into the foreseeable future. This system may take on various organisational forms. One promising way of conceiving the system would be as shown in diagram A.

This 'core' of the human system on the local level can meet with the corresponding core segments on the national and international

Diagram A

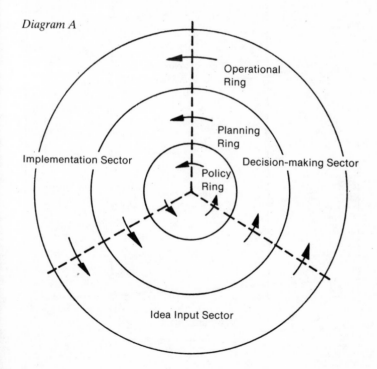

Diagram B

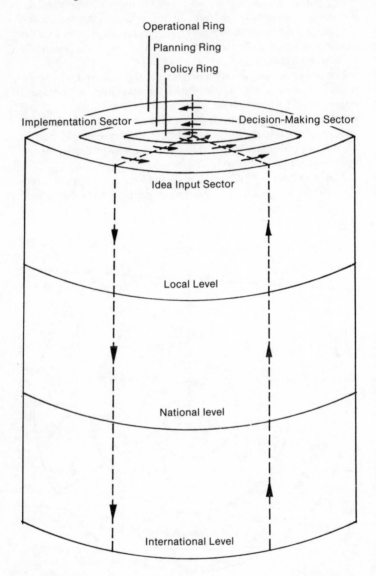

levels. A multilevel dynamic system can be visualised as in diagram B.

The centre core – local, national and international policy – contains the common agreements, treaties, international, national and local laws, and so on. If international law is to have any meaning in the conduct of human affairs, then national and local legislation must be brought into harmony with it. Beyond this and in conjunction with it, much national and local diversity is both possible and desirable.

The action flow around all the rings is from the idea sector to the decision-making sector, to the implementation sector, and then back to the idea sector for modification or redesign. Policy, planning and operations are constantly being refined and modified, although policy tends to be more stable and operations tend to be more changeable.

Within the three sectors the primary directions of flow may differ. The flow in the idea sector would probably be in two directions: from the operational ring to the planning and policy rings, and from the policy ring to the planning and operations rings. An example would be the Council of Churches which responds to the concerns raised by pastors and congregations and which also affects congregations through its local, national and international policy statements.

In the decision-making sector, the primary flow of action is from policy decisions to operational decisions. The same outward flow would also be likely to characterise the implementation sector, although in two cases, namely, physical stress or ideological rejection of planning by the workers, non-co-operation on the operational level should be able to force action back to the policy and planning rings. The familiar strike and boycott are integral and legitimate parts of the human system, forcing a reconsideration of options. If used too frequently or on inconsequential affairs, they can cripple the system. There are many problems common to trade unions and the more recent citizen organisations. These are outlined on p. 383.

The local, national and international levels interface on each ring – policy, planning and operational – and on each sector – the idea input, the decision-making and the implementation. At present, the United Nations formally relates to the national decision-making sector through official representatives of national governments. Informally, the UN implementation sectors interface with international professional organisations, trade unions, transnational corporations and other special interest groups. The balanced system would formalise these latter relationships. In fact, replacing the anachronistic UN Security Council, which merely entrenches the law-of-the-fist mentality, with some sort of representative House of Commons, developed through the present Non-Governmental Organisations affiliated with the UN, would be helpful. An alternative suggestion is a People's Assembly with elected representatives proportional in number to the square root of the population of a nation.

The idea sectors interact through dialogue with international churches, the World Council of Churches, international conferences, international trade unions, university associations, military strategic institutes, the Bilderbergers and the Trilateral Commission, think tanks, and so on. What is proposed here is the formal recognition of these segments and formal channels for balanced input of ideas and experience into the decision-making sector. Let decisions be formal and public, and decision-makers be accountable to the people affected.

The gradual phase-out of the UN Security Council and the introduction of a People's Assembly or International House of Commons would restore faith in the United Nations. Idea input from truly international organisations would strengthen the global village infrastructure. It would play down the law of the fist by which possession of nuclear bombs gives a nation political leverage such as is currently recognised by the permanent membership and veto power of nuclear nations in the Security Council.

Idea input needs to be shifted away from the vested interest organisations such as the International Commission on Radiological Protection (ICRP) and the International Union for the Protection of New Plant Varieties (UPOV), and opened to truly representative scientific advisory groups. One way to accomplish this is to differentiate open from closed (i.e. self-perpetuating) organisations. Which individuals or organisations control membership in international groups is important in deciding whether or not the organisation's interests are vested or general. The record of the group on past international issues is also crucial to their international credibility. Just as national leaders are expected to act on behalf of the good of the nation, and set aside their regional bias and economic self-interests, so too international service requires true internationality of concern. Closed systems need to be labelled lobbying organisations and their informal power reduced.

The proposed systems model for international human interaction builds on present local, national and international structure. It is characterised by formal planning and channels for formal dissent. It depends on voluntary co-operation, not coercion, intimidation, extortion or exploitation. There will still be 'criminals' within this global village, but they are individuals or organisations, not nations. They can be known and restrained locally. It is just as irrational to blow up a country because it harbours criminals as it would be to condemn a family to be killed because one member committed a crime. If the global system is perceived as just, loyalty and co-operation will develop to prevent lawless behaviour and secure the common human goals. Some of the consequences of failure to deal with flaws in present decision-making patterns are given in the insert on p. 385.

We can no doubt think of many transitional steps towards realisation of a global human system which is self-correcting, decentralised and works for the good of people. A few are mentioned here.

The most dramatic attempt to awaken nations to the necessity of curbing national sovereignty, outlawing war and establishing a political will to resolve problems through just laws rather than 'rule of the fist', has come from Japan. In 1946 Article 9 to the Japanese constitution was voted into law as follows:

> Aspiring sincerely to an international peace based on justice and order, the Japanese people forever renounce war as a sovereign right of the nation and the threat or use of force as means of settling international disputes.
>
> In order to accomplish the aim of the preceding paragraph, land, sea and air forces, as well as other war potential will never be maintained. The right of belligerency of the state will not be recognised.

General Douglas MacArthur, head of US occupation forces in Japan, expressed great surprise but whole-hearted support of this Japanese initiative. Unless other nations follow the lead of Japan and develop true alternatives to war this far-sighted and prophetic act will be forever lost to history. Japan has clearly stated that even a so-called defensive war must be renounced.

For general consciousness-raising, we could have international passports, with persons identified by first name, surname and nationality. Changing nationality could become comparable legally to changing one's surname. In this way, nationality reflects origin, culture and language, but does not necessarily confer citizenship rights such as voting or other participation in local or state decision-making. Nationality could be decided at birth and public records of the person kept in the national files for reference in case of marriage, arrest, investigation for approval for international public office or death, and so on. Record-keeping and its confidentiality could be governed by international policy.

Flexibility with respect to citizenship allows for greater international mobility for reasons of work, health, political ideology or other reasons as mutually agreed upon by the host state and individuals. Citizenship can confer rights and duties with respect to the policies of the state in question and could be for a specific time period. While there should be no escape from international policy, state policies and planning may differ, and individuals could be encouraged to choose those which are most compatible with their ideas. In this way, social experiments would become possible without forced oppression of dissidents within the state. This would also allow the oppressed the possibility of non-violent change. It would require redefining 'nationality' and 'citizenship', and rethinking ownership of land, natural resources and resolution of conflicts.

Education of the young is obviously a very important aspect of movement towards a workable global human system. The United Nations University (UNU) was proposed by Secretary U Thant in 1969 and was founded in 1975 to assist in this task and to become an instrument of international scholarship to work on 'pressing global problems of human survival, development and welfare'. The university is located in the Toho Seimi Building in Shibuya, Tokyo, Japan, but its nineteen operating networks, twenty-six associated institutions and more than 100 research and training units located in over sixty nations make it truly global in policy, planning and operation. Within five years of its foundation the UNU has granted 225 fellowships and produced 140 publications, and developed an Endowment Fund of $142 million (US). Besides individual publications, the UNU now publishes two periodicals: *The Food and Nutrition Bulletin* and *Abstracts of Selected Solar Energy Technology*. A quarterly, the *Food and Nutrition Bulletin*, is published under the aegis of the UNU World Hunger Programme. The solar energy monthly, produced under the UN Natural Resources Programme, is now being received by some 550 scientists and engineers in eighty developing countries, whose co-operation and exchange of information forms the basis of a growing network of UNU associates.

One of the greatest frustrations of this new university has been the difficulty in developing international awareness of its existence, its purpose and its work. Existing universities, regional academic associations and research institutions (with their own governing authority) can formally associate with the UNU for specific research and training tasks for designated periods of time, thus allowing students entry into this larger co-operative human venture. It is beyond the scope of this book to describe all the exciting innovative and interdisciplinary projects now in progress under the UNU. Educators concerned about the formation of citizens for the global village can obtain further information directly from the UNU.

Students should be encouraged to consider service in the international sector, to understand UN strengths and weaknesses, to travel and to learn about other nations and cultures. Two years of peace service in a country other than their own is more valuable than many hours of classroom talk.

Schools can be designed as experiments in workable human systems, and student gifts for idea creation, good judgment, organisation or management skills, co-operation, and so on, can be consciously identified and developed.

Within the classroom or school setting, the human talent and energy waste which results from labelling ideas people as 'trouble-makers', 'subversives' or 'traitors' can be learned and experienced. Obviously, in real life, decision-makers would sometimes prefer to hire 'ideas people' and commission them to justify and solidify the status quo. This is one of those anti-survival behaviours which have spawned the uncontrollable arms race. Such a 'comfortable' human system is self-destructive. Ideas people should represent male/female, rural/urban, minority/majority thinking.

The teams of decision-makers from various classes or school organisations need to learn the skills of consensus decision-making. For example, if one group proposes joining a city-wide sports league as a school, there would need to be dialogue within classes and school organisations, and between these groups. After adequate dialogue *about* the proposal (not dialogue *for* and *against* the proposal because that polarises), the proposal could be reworded so that the greatest number of expressed needs are met and objections satisfied. After this rewording, each class and/or school organisation could be asked to take one of four stands towards the proposal:[46]

1. We accept it, and will support and co-operate with implementing it.
2. We accept it, but cannot at this time devote time or energy to implement it.
3. We reject the idea, but will not oppose others doing it.
4. We reject the proposal and we oppose its acceptance by other groups in the school.

The fourth position blocks action by the school. Any group can so block, but it should be understood that this position needs to be explained to others. Blocking may result in setting the proposal aside completely, or in a shift of school populations (comparable to a 'national level' solution) within the region or district so that some schools join the league and some do not.

The third position is known as 'stepping aside'. It is a realistic option in a pluralistic society. The size of the sub-population choosing option (3) signals whether or not the proposal is to be implemented as a 'school' project. If too many step aside the proposal is not really representative of 'school thinking'.

The groups choosing options (1) and (2) are supportive of the proposal. Option (1) people are the present 'workers', who may later be replaced by option (2) people. These two options are the source of planning and implementation teams.

This miniature global human systems model can also function within each class or organisation, as each develops a stance towards the proposal, or develops proposals of its own. The system functions best with honesty, co-operation and respect for individual responses. It breaks down when people are deceived or intimidated.

Consensus decision-making should be reserved for decisions of importance to the group. Decisions related to routine implementation of accepted group policy can be delegated to individuals or small teams. To over-use consensus is to trivialise it or to burden the group unnecessarily.

Once a proposal has been accepted, the implementation begins. The Peruvian example which we explored earlier required forced implementation. Those required to make sacrifices did not enter into the policy or planning phase, nor did the cruelty of their suffering modify the policy goals. The situation was designed to cause revolution within the country, or even war (were it winnable) between the country and those who oppress it from outside. There is nothing 'reasonable' or 'peaceful' in such an oppressive human system even though its policies can be rationalised on the basis of economics. It cannot possibly be sustainable. The financial approach to Peru's problems assumes that the limited wealth of the country would be adequate if the population were smaller. More money could flow out of the country to creditors and the surviving Peruvians could live on the remainder of the money and resources until these were 'needed' by 'outside' policy-makers. There is no limit to such crude waste of life and vitality.

Obviously financial resources are needed to implement and strengthen these suggestions for global reorganisation. That should clearly come from the arms race, which has no redeeming features to recommend it. The United Nations '2000' fund could provide for an orderly transfer of money to worthwhile projects and creative reorganisation within the United Nations.

The Role of Women

In addition to comprising more than half the human race and being the primary cultivators and preparers of food, women traditionally have developed the reconciling arts. While these skills are devalued in a militaristic culture, they become highly prized in a society based on a just peace. The feminist movement does not coincide with the peace movement and anti-nuclear movement by accident.

In spite of the sad history of deceit, manipulation, competition and cruelty which marks the road of so-called civilisation, there is nothing in human nature which appears to make such behaviour inevitable. Rather, there is evidence that humans must be 'psyched up' in order to fight; manipulated and deceived in order to be aroused to wage war; taught and coerced to be secretive; and brutalised to overcome natural fear and compassion.

From an evolutionary point of view, humans are 'flight' animals, not 'fight' animals. When frightened, human reflex behaviour is designed for easier flight: digestion of food stops, bowel and bladder empty, blood sugar level increases, and so on. Humans are 'soft' animals without natural protective armour. The 'fight' animals, on the contrary, have horns, claws or natural armour like a rhinoceros. The fight animals tense and stay quiet when they sense danger. They prepare for attack rather than flight. Human males, having 'acquired' wives, children and property, consciously tried to develop 'fight' behaviour rather than 'flight' behaviour. They generally adopted the rule of the fist. Military trappings, esprit de corps, tanks, planes, bombs, and so on, are all means of counteracting the natural tendency to flight and emulating the strength of the 'fight' animals. A man in a tank has the protection of a rhinoceros. Society severely punishes men who run away from battle.

There exist cultures with no history of all out war and no word for enemy. The people of the Marshall Islands, for example, are considered the gentlest of the Pacific people. They traditionally have a matrilineal society, with land passing from mother to daughter.

Women, who have not become so unnaturally separated from their instincts, need at this moment in history to assume special roles of idea input, facilitating consensus decision-making and seeing to the equitable implementation of plans and substainability of the society's work.

In a special way, women attend to the birthing and dying within society, and they have now turned this concern towards the process of species death or the birthing of a new way of conducting human affairs which might avert such a death. All over the globe women have set aside 'life as usual' to grieve with the dying – the starving and malnourished children, the 'disappeared', victims of urban and government violence, the ravaged peoples subjected to revolutions and wars. Women are also giving birth to and experimenting with new social systems, designed to decentralise decision-making, maximise freedom and diversity, and thrive on co-operation and truth.

The inclusion of women and a feminist perspective in the idea, decision-making and implementation sectors of society is vital for species survival. This implies for males a general reduction of power over other human beings and a playing down of masculine values. Feminist values are needed to reduce the levels of conflict and violence within nations, work-places and families. Present culti-vated competition and violence now 'develops' brutality and ferocity in human relationships.

Although men have always said they go to war for the sake of the women and children, it is very apparent at Greenham Common and other women's peace camps that men are now willing to hurt or even kill women and children in order to go to war. Military and police brutality at Greenham and Cosimo, Italy, has already resulted in women having their bones broken and skulls fractured. Being confronted with their own actions in conflict with their stated objectives must pose a serious crisis in the minds and hearts of these young men. One can only hope that reality will come crashing through fantasy before irresponsible young officers initiate a nuclear holocaust thinking they are saving the world.

Mother's Day Proclamation

Arise, then, women of this day!
Arise all women who have hearts,
whether your baptism be that of water or of fears!
say firmly:
'We will not have great questions decided by irrelevant agencies,
Our husbands shall not come to us, reeking with carnage,
for caresses and applause.
Our sons shall not be taken from us to unlearn
all that we have been able to teach them of charity, mercy and
patience.
We women of one country
will be too tender of those of another country
to allow our sons to be trained to injure theirs.'
From the bosom of the devastated earth a voice goes up with
our own. It says 'Disarm, Disarm!
The sword of murder is not the balance of justice!'
Blood does not wipe out dishonour
nor violence indicate possession.
As men have often forsaken the plow and the anvil at the summons
of war,
Let women now leave all that may be left of home
for a great and earnest day of counsel.
Let them meet first, as women, to bewail and commemorate the
dead.
Let them then solemnly take counsel with each other as to the means
whereby the great human family can live in peace,
each bearing after his own time the sacred impress, not of Caesar,
but of God –
In the name of womanhood and of humanity, I earnestly ask
that a general congress of women without limit of nationality
may be appointed and held at some place deemed most convenient
and at the earliest period consistent with its objects,
to promote the alliance of the different nationalities,
the amicable settlement of international questions,
the great and general interests of peace –

Julia Ward Howe
Boston 1870

It is the armed services, not the peace groups, which have the preponderance of naive, manipulated and trusting youth. These military youths are not exempt from the drug culture. Mike D'Arcy, a drug counsellor with the elite Berlin Brigade of the US European forces, estimates (based on his experience and military personnel urine analysis tests) that about 65 percent of the men engage in 'recreational use' of heroin, about 85 percent use 'non addictive soft drugs like hashish or mad dog', and 7 percent to 10 percent are hard-core heroin abusers. General George Blanchard, senior field commander in Europe, estimates that 8 percent of personnel are 'hard drug users'. When this is applied to the overall active US forces, the number of hard drug users is not less than 8,400 and may be as high as 167,000. The US Department of Defence operates a personnel reliability programme to 'prevent assignment of unreliable or potentially unreliable persons to nuclear duties . . . and then to remove from nuclear duties those persons whose reliability, trustworthiness and dependability become inconsistent with standards' (Assistant Secretary of Defence for Atomic Energy, D. R. Cotter). Roughly 5,000 military personnel were removed from the nuclear weapons programme annually from 1975 to 1977. Yet fear of 'peace-niks' from the Haight-Ashbury area of San Francisco so destabilised an Air Force major that he allowed three men described as having 'dangerous psychiatric problems' to guard the nuclear weapons at the base. One of the men went berserk with a loaded carbine and had to be subdued with force.[47] There are many other well-documented stories of the fragility of the human control of nuclear weaponry. There is no reason to believe that it will be better controlled in the Soviet Union or on an orbiting space platform.

It is women who contemplate with sadness this soft underbelly of the war monster, perhaps because they are not unfamiliar with the judgments made by the youths of nineteen to twenty-three years of age. These women who watched over knives and household poisons, keeping them out of reach of their children, now see their hardly matured sons carrying machine guns or manning a Trident submarine capable of blowing more than 400 cities into oblivion.

Preparations for warlike behaviour begin early in life both in Eastern and Western bloc nations. In East Germany militarism is primarily attributed to the West, and in particular to the 1979 NATO 'two track' decision to modernise its nuclear weaponry while simultaneously negotiating for disarmament. Yet East Germany has long waged 'battles of agriculture and harvesting' and 'battles of production'. In the West, there has been the familiar 'war on poverty' and the 'war on cancer'. Recently, state-purchased war toys were introduced into East German kindergartens. They have been commonplace in Western homes, schools and textbooks for many years. Warlike behaviour has been and is being cultivated and taught in most modern societies. Yet no baby born comes to us with hatred or prejudice or thought of war. Women will undoubtedly need to play key roles in uprooting war from home and school if a transition to peace is ever to take place. We must love and nourish peacemaking in the young.

Internationalisation

The process of becoming a global village involves many complex questions not yet addressed here. One of the most important is the co-operative use of land. It is my opinion that the common heritage of humankind includes land, as well as water, air and outer space. A solution to this might be that whilst people could own their homes, offices or factories they would 'reserve' the use of land. 'Owning' land seems at present to entail people having the right to produce cash crops at the expense of locally needed food, or to pollute the land with toxic and/or radio-chemical waste. This is contrary to a policy of human survival and enjoyment of life.

Transnational corporations also pose international problems. They frequently make transnational decisions affecting the lives of many people who are unable to participate in or even know about these decisions. At present the United Nations is engaged in developing a Code of Conduct for transnational corporations, making them subject to international and national control. While they form an important part of the global job and communication infrastructure, their economic orientation limits their social and human usefulness in building a peaceful global human system. Because of the profitability of global military sales, the transnational corporations sell to both sides in an armed conflict. Conversion to non-military socially useful production will require much creativity of thought and probably outside incentives, to assist transnational corporations to survive in modified form the transition to a world of peace. Transnational corporations belong in the input and implementation sectors of the human system, not the decision-making sector. Their ability to relocate factories while workers are confined within present national boundaries constitutes an intolerable oppression.

Because the recognition of human rights is integrally related to a just and peaceful society, efforts to ratify and implement the UN Charter of Human Rights within each nation are of prime importance. This Charter should be taught in all schools, but especially law and medical schools. Violations of human rights frequently result in medical problems, although this is seldom discussed in medical courses on disease etiology. Amnesty International should be recognised as an investigative arm of the International Court. The court itself should be strengthened and should develop a criminal law branch. International Churches, trade unions, professional organisations and other comparable groups should be recognised as international 'persons' able to bring suit and be brought to suit before the court. This provides an alternative from always having to deal with national governments. Alliances of people across national boundaries reduce the likelihood of war.

The myriad of embassies and consulates throughout the world could be replaced by single UN embassies and consulates. These could attend to passports, assist travellers and serve as a local liaison to the UN agencies. Besides reducing spying and hostile nationalistic behaviour, such a UN service would increase the visibility and relevance of the UN and convey the international concern for the well-being of non-citizens temporarily staying within a nation. The UN embassy or consulate would become a focus for dealing with human rights violations or for observing the first signs of a recurrence of militarism.

Nations have already assumed certain 'personalities' (or 'nationalities') in the global village. These special services could be utilised and enhanced. For example, Switzerland might become the global medical research centre, Canada might develop a global arbitration service, Australia might be the centre for handling labour concerns, India might become spokes-nation for economically developing countries, and so on. These services would, of course, involve all other nations in idea input, decision-making and implementation, but action initiation would always emanate from one place and would involve skilled persons who would need to be in geographical proximity with one another to be effective. An already developed national profile on a particular issue facilitates growth in international trust in the service emanating from that nation.

The Energy Crisis

Viewed from a militaristic perspective, energy is a 'strategic resource'. It can replace manpower; it can be the source of incredibly destructive forces; it can be utilised to kill people and preserve buildings or tanks; and it can provide a research and development façade for military wants.

Stripped of all these artificial needs, energy requirements are reduced and energy sources are multiplied. The United Nations Conference on New and Renewable Sources of Energy identified many promising and available energy sources. As was pointed out earlier, bio-gas research in India has attracted global attention.

The most promising energy resource for Less Developed Countries (LDCs) appears to be decentralised hydro and bio-mass. P. B. Baxendell, Chairperson of the Shell Oil Transport and Trading Company and a Group Manager of the Royal Dutch Shell Group, estimated:

> The LDC hydro resources have, as yet, hardly been tapped. The countries of Latin America have done most, but they have only tapped between 10 per cent and 15 per cent of their estimated potential. The countries of Asia have tapped only 2 per cent, and Africa, 6 per cent.[48]

Hydro development could be carefully combined with irrigation to improve food yield, or with metal refining for export to provide capital inflow.

Small-scale solar water heaters, stills, crop driers and cookers are now commercially available. Photoelectric panels can be used as a local source of electricity for light, irrigation pumps and other mechanical projects. All these energy sources meet the requirements of being labour intensive rather than capital intensive, being understandable and controllable locally, being unlimited if properly managed, and favouring decentralisation and rural development rather than further urbanisation.

In 1970 the LDCs estimated that rural to urban migration was taking place at an average rate of 45,000 people per day. This aggravates the energy situation because urban slum dwellers no longer have access to wood or dung for fuel as they did in rural areas.

First world plans to sell locally unwanted nuclear technology to LDCs further complicates development problems by encouraging

---

Australia Installs Wind Generators

About 3,000 wind-driven electrical generators are being installed near the southern and western coasts of Australia. They are expected to provide 20 MWe power annually, filling two-thirds of the energy needs of the states of Western Australia, South Australia and Victoria.

---

urbanisation. Of the 100 countries usually listed as LDCs only nine have urban centres large enough to support a 600 MWe generator. Until recently, developed countries have been building and promoting 1,000 MWe nuclear plants as a solution to LDC problems. The US Westinghouse Corporation, for example, has built one nuclear plant in South Korea, and is currently constructing five others, with an additional two plants in planning stages. Both Westinghouse and General Electric hope that this nuclearisation of Korea will allow them to open labour-intensive nuclear component manufacturing plants in Korea. They would then avail themselves of unorganised cheap Korean labour, while at the same time closing down their US plants. It is thought that they hope eventually to achieve 'assembly line' production of small prefabricated reactors of 50–200 MWe size.

Korean citizens have no say in their government's nuclear plans or in the US transnationals' plans to regain the international nuclear reactor market at the cost of exploiting Korean workers. It is alleged that the US ambassador to Korea threatened to put the US/Korean alliance in jeopardy if Korea did not purchase the US reactors.[49] The reactor sales were financed by Export-Import Bank guaranteed loans.

In developed countries where free discussion and debate has been allowed, serious strategies for a nuclear energy-free future have been proposed. The Commission of Inquiry on 'Future Nuclear Energy Policy' of the West German Parliament issued a report on 25 June 1980 calling for energy conservation and 'that a greater use of renewable energy resources [be] vigorously undertaken and pursued'. Although the Commission has a majority of nuclear proponents, it agreed to defer decisions on future nuclear technology dependence until 1990, with extensive experiments in other strategies during the interim period. The Commission specified that the public's accustomed level of consumption, and the present structure of the economy, must be maintained by any energy strategy proposed. The fact that alternatives to nuclear policies, within these restraints, can be described was clearly demonstrated by the West German Commission.[50]

Divorced from its military backing and from financial promotion by lending institutes and governmental privileged status, nuclear power projects would collapse naturally. In a global human system appropriate energy technologies could be directed, tested and expanded as needed. Human intelligence now engaged in the defence and promotion of this nuclear industry could be directed to more socially useful and creative ways to improve the human condition and to allow more people to share in technological benefits.

Human Health

We have tried to analyse the interdependence of the triad: national security, energy and health. It is time to say one last word about deteriorating health, the strongest indicator of self-destructive human behaviour. As we examine global health problems already spawned by the preparations for global war – the Third World War – certain victimised peoples claim our immediate serious attention. By ignoring their plight and feigning helplessness we are being brutalised and prepared for still greater hostilities and destruction. Global healing must involve global attention to the past military addiction, admission of complicity or passivity and involvement in future policy development to maximise the survival probability of the victimised peoples. These victims are our brothers and sisters. They are unique jewels, adding irreplaceable value to our global family and home.

The list of close to 13 million pre-Third World War victims fails to include the hundreds of thousands of people killed in recent wars supported by superpowers desirous of maintaining buffer nations or spheres of influence. Hundreds of thousands of other victims die of hunger and malnutrition because of the inordinate global spending on weapons. Millions are homeless or in need of medical care, education and jobs because of the distorted national preoccupations with war-making.

There are other areas of acute human suffering directly caused by the nuclear age which have not yet been mentioned: some are geographically isolated, like the uranium mining area of Namibia, some are nations within nations like the Aboriginal and Indian peoples exposed to weapon-testing and uranium mining in Australia and North America. Some victims quietly suffer unnoticed among us, like the US, Canadian, British and Australian military personnel who participated in the weapon-testing programmes of the 1950s. Some victims are becoming more vocal, like the nuclear workers who are becoming more acutely aware of the nuclear-related health problems.

These early victims demand our attention. On their healing rests the hope of healing our human global system. Ignoring these early warnings, failing to heed the negative feedback from these victims before the process is irreversible, spells doom for the human social experiment on earth.

That such tragedies may not happen again, a human health monitoring system needs to be developed. We cannot turn back the clock and unlearn the science and technology of the late twentieth century. It is not the scientific understanding of the global system in all its basic chemical and physical properties which is disordered; it is rather the policy, planning and operational judgments which have gone awry. Not every possible technology needs to be implemented. Human control of choices is basic to human habitation of the global village. Therefore, idea input needs to be broader, including for example, cultural, health and ethical values, and feedback must be more systematic and explicit. The present vague statements to the effect that 'one cannot bother about radiation because everything causes cancer', 'there are too many variables affecting health so detecting causes of disease is impossible' or 'the health problems are all caused by irrational fears or irresponsible journalism' are counterproductive, paralysing and suicidal. In societies able to deal with the financial and physical variables required for creating technological mega-projects of cosmic dimensions, sorting out health problems is minor. What is lacking is the will to do so. Put more bluntly, the negative feedback into idea input is not wanted, therefore not sought or not welcomed. Human health is wrongly considered a 'renewable resource' and the true costs of mega-projects are externalised, paid by the individual and/or the family. Awareness of impending species extinction is screened out of the collective consciousness. The slow erosion of health and genetic integrity is judged 'acceptable' to policy-makers who rely on 'experts' who rely on funding from military strategists or big business.

Even if a de-escalation of the arms race and demilitarisation of the planet were to begin immediately, a new era of public health research would be needed. Just as the industrial age generated worker health problems and the discipline of occupational health, so the nuclear age is generating new public health problems and the discipline of environmental health.

As industrialisation progressed, the worker soon learned that his or her health was placed in jeopardy by hazards in the workplace. The nature of the toxic material was usually unknown to the worker and sometimes the substance was undetectable to the senses. Information about the chemicals handled by workers was frequently labelled a 'trade secret', carefully guarded by management to preserve the company's exclusive rights to a profit-creating product. Workers had to learn by reflecting on their own experiences what health effects resulted from exposures to these materials. In self-defence they organised into trade unions, lobbied for 'right to know' legislation, hired their own legal and medical advisers, developed collective bargaining and arbitration skills, and invented the strike, a forceful but usually non-violent protest method. The historical development of trade unions has been tortuous and sometimes violent; disunified within or cruelly crushed from without. Workers had to learn survival skills for the industrial era, such as financially supporting one another during strike action, developing transnational levels of co-operation to deal with transnational corporations, forming their own political parties and developing pro-worker legislation. One very important labour development was the establishment of the pre-employment physical examination which provided a base line for later worker compensation claims of health damage due to employment-related exposures. This simple precaution was a protection for worker and management alike, since pre-employment physical problems were clearly identified.

Although workers are still struggling for recognition of their basic human right to dignified work and to reasonable care for health and life, their past struggles and gains provide basic guidelines on which the structures of citizen organisations and environmental health disciplines can be modelled. With respect to labour, the protagonists are clearly identified as workers and management, i.e. the implementation and planning levels of the corporate human system. Both these levels are subservient to the policy level, usually controlled by corporate owners and financiers. The political activity of unions and their lobbying of government is an attempt to affect this corporate policy level, while arbitration and negotiation with

Analogy Between Trade Unions and Citizen Organisations

Basic problems or need	Organisational response to problems	
	Trade unions	Citizen organisations
Toxic by-products of industrial and technological activities	Developed health and safety sections to deal with toxics in the work place.	Sporadic attempts by individuals to identify toxics in the living space: air, water, food.
Military and industrial secrets to protect information	Attempts to develop legislation protecting a worker's right to know.	Attempts to develop freedom of information legislation.
Legal, medical and scientific advisers monopolised by rich	Unions hire their own lawyers, doctors and scientific advisers.	Unorganised and unfinanced groups understand their need for legal, medical and scientific advisers.
Decision-making processes are inaccessible	Unions have developed collective bargaining, binding arbitration and strikes.	Public hearings are usually without power; citizen boycotts, demonstrations or stockholder action attempted.
Polluting agency or company usually accountable to no one	Unions stand behind the individual who is harmed or wronged.	Citizens are beginning to explore the strength of concerted action.

management take place on the planning level.

Early attempts to organise labour met with myriad problems related to the workers' diverse philosophies, ideologies, cultures, languages and national biases. There were wide differences in the problems posed by heavy industries such as steelmaking and light industries such as garment-making; temperate zone work and work in the tropics; construction workers and teachers or hospital employees.

With a stroke of genius, some person whose name is now lost to history (but who probably lived in western Canada) proposed the eight-hour day as a simple rallying issue on which all unions could unite. The effect was astonishing. The issue cut across all differences of sex, race, occupation and ideology, relating very directly to life, health, dignity and the quality of human relationships. Primarily through this clear and simple eight-hour work

Consequences of the Structural Weakness which Favours Industry and Government over Workers and Citizens		
Human sufferings	*Organisational responses*	
Causes	*Trade unions*	*Citizen organisations*
Immediate, identifiable damage	Workers' Compensation Board established; some national legislation in place.	Problems unaddressed unless they reach emergency evacuation level.
Slowly developing long-term health damage	Workers' pre-hiring physical examination can serve as a basis of claim.	No base line health data collected; statistical efforts usually inappropriate to need.
Problems are blamed on the victims	Workers are organised to compare problems and aid in identification of causes.	Individuals made to feel their smoking or food choices are the main health problems.
Toxics can damage sperm or ovum, causing defective offspring	Not recognised or compensated for workers.	Not recognised or compensated for the general public.
Toxics can damage developing embryo or foetus	Individual worker might obtain compensation; usually unrecognised.	Not recognised or compensated for the general public; foetus has no right to protection

day demand, unions attained not only the eight-hour day but also recognition as legitimate participants in decision-making. Once legitimised, it became possible for labour to bargain for other more focussed local needs.

A further important characteristic of the struggle of workers for survival in the industrial age should be noted, and this is the opposition which they experienced. The main tactic used against labour was an attempt to cause disunity within the movement. Other tactics included economic penalties against union organisers, dishonest dealing over health issues and appeals to the general public to blame shortages or the high cost of living on unreasonable labour demands.

This somewhat brief sketch lays the groundwork for understanding the role of citizen organisations and the task of public health workers in the nuclear age. Citizen groups are here viewed as a normal extension of the trade union movement.

Under the pressure of militarism, the quantity and toxicity of materials being produced and tested in the developed world has increased steadily since 1945. These new materials include pesticides, herbicides, plastics, uranium derivatives, fission fragments and other radioactive chemicals, and chemical and biological warfare agents. They are for the most part non-biodegradable, i.e. they cannot be broken down into re-usable simple chemical parts through the earth's natural recycling means. Some of these products, like styrofoam, were developed as components for atomic or hydrogen bombs, others were used to further US military strategy in Vietnam, where the 'enemy' was to be defeated through deliberate altering of the environment. The vegetation, living plants and trees, which provided cover for the 'enemy' was destroyed. Food crops belonging to the 'enemy' were sprayed with poison. The people were forced to move from their farms and villages and relocate in 'strategic hamlets' which could be better controlled by the military 'pacification' plan. Human life and habitat were organised to serve military needs.

Great areas of farmland and tropical forests were completely destroyed by what the military called 'trigger factors'. These are elements injected into the environment that touch off widespread ecological changes. The best-known 'trigger factor' is Agent Orange, a defoliant chemical which kills trees by first stripping them of leaves. Gradually the rich tropical rain forests, precious source of earth's atmospheric oxygen, were converted into the relatively life-poor tropical savannas. These military trigger factors also directly affect other plant life, animal life and human life. In humans they cause damage to the central nervous system and to the embryo or foetus unfolding in the womb. It is estimated that the number of Vietnamese poisoned between 1966 and 1969 was of the order of 300,000 per year.

Because of the interdependence between animals and plants, the Vietnamese animals were not only directly killed by the toxic defoliants and pesticides but they were also starved to death because of the loss of those crops which were their food. The animals and plants were, of course, the basic food of the people, who also starved to death with their disappearance. One spraying of Agent Orange was sufficient to kill great areas of mangroves, which in turn resulted in local loss of many animal species and even of species of fish which rely on nutrients flowing from the mangrove vegetation into the surrounding waterways.

The military attitude towards the spraying betrays a profound lack of understanding of the interconnecting web of life. The prestigious Midwest Research Institute (MRI) report reassures the reader:

> The presence of toxic residues on rice from sprayed fields appears not to be a problem since no rice develops in these fields.

The MRI report further states:

> The trees died but so did 84 to 99 percent of the bark beetles; the author suggests this procedure as a possible method of controlling this beetle.

The unprecedented use of herbicides and pesticides in Vietnam during the 1960s means that war against the Vietnamese people and all living things in the region will continue into foreseeable future time. Much of the land has been rendered infertile and the animal population decimated. Information on the human health effects of these chemicals has been withheld as 'classified and confidential', even from American military personnel who poisoned themselves when using them in Vietnam.

In the 1980s, we are faced with deliberate and widespread introduction of hazardous materials into the human living space and the general earth biosphere. The public is now at risk from a variety of toxic agents which were once restricted to the workplace. These are introduced as warfare agents, or commercial spin-offs of warfare agents (pesticides or herbicides). They are deliberately 'tested' in inhabited areas, released as factory effluence to reduce workplace contamination and incinerated or dumped as toxic waste.

The USA alone has some 74,000 toxic waste dumps, with more than 30,000 classified as potentially severe threats to human health. Like the industrial workers, the public is beginning to realise that many human illnesses are being caused by this cynical carelessness. The problem of military toxic waste is most acute in the United States, the USSR, Vietnam, Afghanistan and elsewhere where the chemical warfare agents are produced or used. However, the air/water/earth circulating system will slowly spread these pollutants throughout the world. They will cause deterioration of the earth's ability to support life, climate changes, and species extinctions, further reducing biological vigour and diversity. An arms agreement or cessation of war-making cannot halt this earth process which is already initiated. It could, however, begin a new era of vigilance and earth-care, a common effort to restore, as much as possible, the integrity of the damaged biosphere.

Spontaneous responses to pollution in the living space, with its corresponding threat to survival and health, are assuming forms analogous to those seen in trade union unfolding. Citizen demonstrations and boycotts are the natural analogues to worker strikes. As yet, citizens are not recognised partners in decision-making with respect to military/industrial planning and toxic pollution, even though they have a right to vote for government leaders. Direct citizen input is frustrated with show-case-type hearings, secrecy, pacification or delay with unkept political promises, misinformation or intimidation. Citizens are made to feel that decisions should be left to the experts. Their protests are described as anarchy and they are labelled 'naive' or 'duped'.

Interestingly, citizen protests focus directly on the major obstacles to security and human growth in the global community. In Western bloc countries they focus on the growing militarism and escalation of the arms race. In Eastern bloc countries and LDCs they focus on human rights violations, food costs, unemployment and development needs. Violence within and between countries is both the sickness and the focus of human efforts to heal. Security priorities for individuals are being sacrificed for what are perceived as security needs for nation-states. The basic struggle is for a just world in which human beings can live in security and freedom. The global village is longing for a time to bloom.

Nation-states are political organisations developed to serve the human need for orderly social living, just as transnational corporations are organised for the efficient production of goods and division of labour. Neither are ends in themselves. Human survival is more important than survival of the power structures of the nation-state or corporate world as we now know them. Just as production expediency and profit-making had to be made subservient to worker survival, so military and political strategy for national growth must not be allowed to undermine survival of the population. This is true whether national survival is defined in terms of suppressing legitimate dissent, borrowing development capital, or turning a nation into one big weapon factory or battlefield.

We should realise, however, that it is also counterproductive to destroy the industrial sector by excessive worker demands, since we are dependent upon machines and technology for modern life. So also, some national structure is helpful for human development and global exchange relationships. We cannot function in a totally unstructured society. Cultural diversity enhances human survival chances and contributes to richness of thought. In fact, this very cultural diversity now speaks against war. In the past, war brought about a mixing of thought and culture, as victor learned from vanquished. Now thought exchange can be achieved through international communication channels and travel, and war merely destroys such fruitful exchange.

If the global movement of people were freer, and the security of individuals increased, social experiments might be undertaken voluntarily and new 'nations' formed in some reasonable and humanly enriching way. Since natural resources are geographically immovable, their use and preservation would need to be regulated internationally as a common heritage of humankind. This proposal undoubtedly poses problems of land and resource ownership, human dislocation and equitable sharing, but the problems are hardly more formidable than those posed by the present violent self-destructive relationships.

Basic to the survival of the global village is the security, health and well-being of the local community. Local decision-making and local fulfilment of needs should take primacy over national desires for power and/or prestige. The national government should serve to facilitate equitable distribution of goods, services and jobs within the nation, while the international public sector should endeavour to do this between nations. These functions require honest human relationships, openness to creative ideas and attention to human feedback from the implementation sector. The system could become self-regulating and evolving if left open and responsive to co-operative human input and feedback. It is a deliberate choice of life and growth. It would usher in a time to bloom.

Citizen protest movements provide incipient international infra-structures for the global village. The simple survival issues which unite citizens internationally should include human rights, development, non-violence and environmental concerns. Perhaps what will prove to be the one directly unifying issue (like the eight-hour day) will be the right to live without fear of arbitrary killing, whether it be by 'disappearance', imprisonment, drugs, pollution or war. Along with the right to not be killed, there must be a right to not have to kill others – both are encompassed by the phrase 'arbitrary killing'.

In the expanded labour analogy, the protagonists are recognised as citizens versus governments, rather than labour versus corporate management. The citizens are the ones who implement government planning: they make the bombs, carry out the midnight searches and seizures, dump the toxic waste and conceal the information which might arouse the public to action. The government officials are planners who carry out the policies developed by international financiers, corporate owners, national leaders and military strate-gists. They try to 'manage' the public and 'please' the policy-makers. Governments directly threaten the lives of citizens through their failure to control pollution, their policies of funding military production which causes inflation and unemployment and their lack of protection for legitimate dissent. By taking to the streets in protest, citizens are sending a strong message to the inaccessible policy-makers who actually control both national government leaders and transnational corporation policy. Their lack of accoun-tability is no longer tolerable, their dreams can no longer be unquestioningly fulfilled.

Further human growth in the nuclear and post-nuclear eras requires an open participatory international structure for policy-making, planning and implementation. The United Nations, suitably strengthened and supported, offers a basis for such a system. Ironically its present weak structure may better suit development of a consensus decision-making style than would have been possible had it evolved as a strong centralised force in the nation-state model. Just as the labour versus management disputes were often solved by national intervention or legislation, so citizen versus national government disputes must systematically be refer-red to the international level for resolution.

As workers had to fund their own representatives and hire their own legal and medical advisers, so too citizens are discovering that they need legal, scientific and medical advisers. It is easier for legal and medical professionals to contribute their services and speak out publicly in the crisis because they usually depend directly on the citizens for economic support. On the other hand, scientists who attempt to aid citizen groups are black-listed by both government and industry, their institutional affiliations are cut off if possible and their research funding discontinued. These scientists are maligned and their work 'critiqued' by other scientists who are frequently unable or unwilling to understand it. Citizens, unfamiliar with scientific jargon or unable to assess the value of the critiques, are fearful of trusting the outspoken scientists. Thus cut off from all support and funding, the 'dissident' and outspoken scientist is sometimes unable to document thoroughly his or her position because of lack of financial resources. Many courageous scientists need and deserve citizen support in the same way that strikers and labour organisers need to be supported by unions. Cut off from scientists, the citizen movement will remain ineffective, unable to read government jargon or understand proposed regulations, unaware of the pollution slowly undermining health. Citizens are sometimes unaware of the fact that scientists who try to help prevent irreparable damage to health and to the earth system are often blamed for causing the health effects by encouraging the public to panic. This tactic is reminiscent of the blaming of economic problems on the demands of trade unions.

Other government attempts to undermine citizen protests include provocateur infiltrators, harsh imprisonment or fines and oppressive crowd control tactics. As infertility and birth defect incidence rates increase because of pollution, women are encouraged to practise birth control and amniocentesis with abortion, to reduce the negative impact on society. Workers are made to think that environmentalists are their enemies, and that pollution is a necessary by-product of their work. Young men are taught that they must kill compatriots and threaten the world with nuclear holocaust in order to protect their nation-state.

In the last analysis, the whole humanly designed structure depends on individual co-operation with the leaders of the nation-state, a co-operation rapidly eroding as people begin to comprehend the suicidal nature of their collective assigned tasks.

If the world does not work for the good of the people, then the people must make it do so. They must empower themselves for the sake of preserving life, just as the worker had to do in the new industrial era.

After all, peace comes down to respect for human inviolable rights, while war springs from the violation of these rights and brings with it still graver violations of them. [51]

Pope John Paul II, one of the most conservative international voices, called for public opinion to set stringent limits on the kinds of action national governments can legitimately take in his 1982 World Peace Day message:

Peace cannot be built by the power of rulers alone. Peace can be firmly constructed only if it corresponds to the resolute determination of all people of good will. Rulers must be supported and enlightened by a public opinion that encourages them, or, where necessary, expresses disapproval.[52]

In moving towards a citizen-action structure able to enter into check-and-balance dynamism with leaders of nation-states, it will be necessary to ensure citizen access to accurate information and objective proof of claims. The incipient structures for such a grounding are medical associations, other professional organisations and non-governmental institutions. Among their many tasks will be the design and execution of data-banks providing relevant information on sensitive health parameters. Among the monitored parameters should be: fertility rate, infant death and birth defect incidence rate, numbers of severe asthmatic or allergic reactions per day per 100,000 people, scholastic ability of children, and the average age at diagnosis of chronic diseases such as diabetes or hypertension, and life-threatening diseases such as cancer. An honest and complete audit of health is as important to human species survival as a financial audit is to economic health.

Such base-line health data can serve as a direct measurement of immediate health loss due to an industrial accident or of slow loss due to routine pollution. It could also measure health gains from the clean-up of an environmentally threatening toxic waste dump. Skill in sleuthing environmentally caused diseases must be deliberately developed, since at present public health efforts largely focus on infectious disease control.

The focus on health in this post-nuclear era is in keeping with the declaration drawn up at the International Conference on Primary

Health Care at Alma-Ata in the Soviet Union, on 12 September 1978. The Conference at Alma-Ata was jointly sponsored by the World Health Organisation and the United Nation's Children's Fund. Out of this meeting a programme, Health for All by the Year 2000, has developed. Unless the excellent groundwork of international policy developed through this programme becomes operative at national and local levels, it will be futile. The knowledge and skill needed to handle global human problems is present in the world's people. It can be channelled into use, ushering in a time to bloom, or it can be stifled by neglect or blasted out of existence with a nuclear bomb.

This has been the story of the birthing of the global village as well as the story of the death of old ways of relating within the world community. The story is not yet over – the birthing is not yet accomplished. The new life is fragile and weak, while the outmoded system is in its death throes. Seeing to the safe birthing of the global village is the task at hand. Its nurturing and evolution belong to future generations. If our generation fails to turn away from self-destruction and towards a new period of growth and human well-being, the human species will decline and die out. We can buy days, but they will not be 'good ones'. A potential for greatness will die, and there will be no time of blooming for the human species on the planet earth.

Declaration of Alma-Ata

The International Conference on Primary Health Care, meeting in Alma-Ata this twelfth day of September in the year Nineteen hundred and seventy-eight, expressing the need for urgent action by all governments, all health and development workers, and the world community to protect and promote the health of all the people of the world, hereby makes the following Declaration:

I

The Conference strongly reaffirms that health, which is a state of complete physical, mental and social wellbeing, and not merely the absence of disease or infirmity, is a fundamental human right and that the attainment of the highest possible level of health is a most important world-wide social goal whose realisation requires the action of many other social and economic sectors in addition to the health sector.

II

The existing gross inequality in the health status of the people particularly between developed and developing countries as well as within countries is politically, socially and economically unacceptable and is, therefore, of common concern to all countries.

III

Economic and social development, based on a New International Economic Order, is of basic importance to the fullest attainment of health for all and to the reduction of the gap between the health status of the developing and developed countries. The promotion and protection of the health of the people is essential to sustained economic and social development and contributes to a better quality of life and to world peace.

IV

The people have the right and duty to participate individually and collectively in the planning and implementation of their health care.

V

Governments have a responsibility for the health of their people which can be fulfilled only by the provision of adequate health and social measures. A main social target of governments, international organisations and the whole world community in the coming decades should be the attainment by all peoples of the world by the year 2000 of a level of health that will permit them to lead a socially and economically productive life. Primary health care is the key to attaining this target as part of development in the spirit of social justice.

VI
Primary health care is essential health care based on practical, scientifically sound and socially acceptable methods and technology made universally accessible to individuals and families in the community through their full participation and at a cost that the community and country can afford to maintain at every stage of their development in the spirit of self-reliance and self-determination. It forms an integral part both of the country's health system, of which it is the central function and main focus, and of the overall social and economic development of the community. It is the first level of contact of individuals, the family and community with the national health system bringing health care as close as possible to where people live and work, and constitutes the first element of a continuing health care process.

VII
Primary health care:
1. reflects and evolves from the economic conditions and socio-cultural and political characteristics of the country and its communities and is based on the application of the relevant results of social, biomedical and health services research and public health experience;
2. addresses the main health problems in the community, providing promotive, preventive, curative and rehabilitative services accordingly;
3. includes at least: education concerning prevailing health problems and the methods of preventing and controlling them; promotion of food supply and proper nutrition; an adequate supply of safe water and basic sanitation; maternal and child health care, including family planning; immunisation against the major infectious diseases; prevention and control of locally endemic diseases; appropriate treatment of common diseases and injuries; and provision of essential drugs;
4. involves, in addition to the health sector, all related sectors and aspects of national and community development, in particular agriculture, animal husbandry, food, industry, education, housing, public works, communications and other sectors; and demands the coordinated efforts of all those sectors;
5. requires and promotes maximum community and individual self-reliance and participation in the planning, organisation,

operation and control of primary health care, making fullest use of local, national and other available resources; and to this end develops through appropriate education the ability of communities to participate;

6. should be sustained by integrated, functional and mutually-supportive referral systems, leading to the progressive improvement of comprehensive health care for all, and giving priority to those most in need;

7. relies, at local and referral levels, on health workers, including physicians, nurses, midwives, auxiliaries and community workers as applicable, as well as traditional practitioners as needed, suitably trained socially and technically to work as a health team and to respond to the expressed health needs of the community.

VIII

All governments should formulate national policies, strategies and plans of action to launch and sustain primary health care as part of a comprehensive national health system and in co-ordination with other sectors. To this end, it will be necessary to exercise political will, to mobilise the country's resources and to use available external resources rationally.

IX

All countries should cooperate in a spirit of partnership and service to ensure primary health care for all people since the attainment of health by people in any one country directly concerns and benefits every other country. In this context the joint WHO/UNICEF report on primary health care constitutes a solid basis for the further development and operation of primary health care throughout the world.

X

An acceptable level of health for all the people of the world by the year 2000 can be attained through a fuller and better use of the world's resources, a considerable part of which is now spent on armaments and military conflicts. A genuine policy of independence, peace, détente and disarmament could and should release additional resources that could well be devoted to peaceful aims and in particular to the acceleration of social and economic development of which primary health care, as an essential part, should be allotted its proper share.

* * *

The International Conference on Primary Health Care calls for urgent and effective national and international action to develop and implement primary health care throughout the world and particularly in developing countries in a spirit of technical cooperation and in keeping with a New International Economic Order. It urges governments, WHO and UNICEF, and other international organisations, as well as multilateral and bilateral agencies, non-governmental organisations, funding agencies, all health workers and the whole world community to support national and international commitment to primary health care and to channel increased technical and financial support to it, particularly in developing countries. The Conference calls on all the aforementioned to collaborate in introducing, developing and maintaining primary health care in accordance with the spirit and content of this Declaration.

References

Introduction: pp. 1–14

1. This is a typical leukaemia death, with tissue breakdown and massive internal haemorrhage. Dr Elizabeth Kübler-Ross – well known for her counselling of the dying – often uses a similar story to stress the importance of home care for the dying. Jimmy is a fictitious name, but this is the story of a real child treated at Roswell Park Memorial Institute in Buffalo, New York, USA.

2. U. Saffiotti and J. K. Wagoner (eds), *Occupational Carcinogenesis*, Annals of the New York Academy of Science, ANYAA9-271-1-516, New York, 1976.

3. 'Biological Effects and Measurement of Radio Frequency/Microwaves', Symposium Proceedings, US Department of Health, Education and Welfare, HEW Publication (FDA) 77-8026, 1977. Available from the World Health Organisation, United Nations.

4. 'The Effects on Populations of Exposure to Low Levels of Ionising Radiation', Report of the Committee on the Biological Effects of Ionizing Radiation (BEIR), 1972. Revised in 1979; revised again in 1980. These will be referred to as BEIR I, II and III.

5. E. Apfelbaum, *Hunger and Disease*, Interpress, Warsaw, 1946, pp. 10–11. See also Czeslaw Pilichowski, *No Time-limit for These Crimes*, Interpress, Warsaw, 1980.

6. International Military Tribunal at Nuremberg, PS 1014; W. Baumgart, *Vierteljahreshefte für Zeitgeschichte*, vol 2, Stuttgart, 1968, pp. 120–49.

7. Emma Rothschild, 'Banks: The Coming Crisis', *New York Review of Books*, 27 May 1976.

8. Continental of Illinois, the Bank of America, Bankers Trust, Manufacturers Hanover, Morgan Guaranty, Wells Fargo, (First National) City Bank and Chase Manhattan. Documented by GATT-FLY, 600 Jarvis Street, Toronto, Ontario, Canada.

9. Patricia Mische (ed), 'Earthscape. Transitions Toward World Order', Global Education Associates, A Whole Earth Paper, 1979.

10. Gerald and Patricia Mische, *Toward a Human World Order: Beyond the National Security Straitjacket*, Paulist Press, New York, 1977.

Part One: pp. 15–63

1. 'Fission Products' in *The Effects of Nuclear Weapons* (3rd edn), US Department of Defence and US Department of Energy, 1977, Section 1.60 to 1.66.

2. Karl Z. Morgan, 'Suggested Reduction of Permissible Exposure to Plutonium and Other Transuranium Elements', *American Industrial Hygiene Association Journal*, August 1975, pp. 567–75.

3. Karl Z. Morgan, 'Hazards of Low-Level Radiation', *Yearbook of Science and the Future*, Supplement of the *Encyclopedia Britannica*, 1980.

4. Karl Z. Morgan, 'Reducing Medical Exposure to Ionizing Radiation', *American Industrial Hygiene Association Journal*, May 1975, pp. 361–2.

5. H. Müller, 'Radiation and Heredity', *American Journal of Public Health*, vol 54, no. 1, 1964, pp. 42–50.

6. R. Bertell, 'X-ray Exposure and Premature Aging', *Journal of Surgical Oncology*, 9, 1977, pp. 379–91.

7. N. Kochupillai, I. C. Verma, et al., 'Down's Syndrome and Related Abnormalities in the Area of High Background Radiation in Coastal Kerala', *Nature*, 262: 60, 1976.

8. P. M. E. Sheehan and I. B. Hillary, 'An Unusual Cluster of Down's Syndrome, Born to Past Students of an Irish Boarding School', *British Medical Journal*, vol 287, 12 November 1983. Also Letters and Author's reply, *British Medical Journal*, vol 288, 14 January 1984.

9. L. S. Penrose and G. F. Smith, *Down's Anomaly*, Churchill, London, 1966.

10. Carl J. Johnson, 'Cancer Incidence in an Area of Radioactive Fallout Downwind from the Nevada Test Site', *Journal of the American Medical Association*, vol 251, no. 2, 13 January 1984.

11. H. B. Jones, 'Estimation of Effect of Radiation Upon Human Health and Life Span', Proceedings of the Health Physics Society, June 1956.

12. L. H. Gray, 'Biological Damage by Different Types of Ionizing Radiation' in *Biological Hazards of Atomic Energy*, Clarendon Press, Oxford, 1950.

13. Karl Z. Morgan, 'Risk of Cancer from Low Level Exposure to Ionizing Radiation', American Association for the Advancement of Science, Washington, DC, 17 February 1978.

14. R. M. Sievert, 'Tolerance Levels and Swedish Radiation-protection Work', Proceedings of the Health Physics Society, June 1956, p. 181.

15. M. Eisenbud, 'Radioactivity in the Environment – Radioactive Concentration in the Fetal Thyroid', *Pediatrics* (Supplement), 41, Part II, 1968, p. 174.

16. ICRP Publication 2, Pergamon Press, Oxford, 1959.

17. H. Müller, 'Radiation and Heredity', *American Journal of Public Health*, vol 54, no. 1, 1964, pp. 42–50.

18. 'Biological Effects of Ionizing Radiation', BEIR III, US National Academy of Science, National Academy Press, 1980, pp. 74–5.

19. R. Bertell, 'Ionizing Radiation Exposure and Human Species Survival', *Canadian Environmental Health Review*, vol 25, no. 2, June 1981.

20. 'Basic Radiation Protection Criteria', US National Council on Radiation Protection Report no. 39, pp. 58–60.

21. A. M. Stewart, 'Delayed Effects of A-Bomb Radiation: A Review of Recent Mortality Rates and Risk Estimates for Five-year Survivors', *Journal of Epidemiology and Community Health*, vol 36, no. 2, June 1982, pp. 80–6.

22. R. Bertell, Letter to the Interagency Task Force on Low-Level Ionizing Radiation (director F. Peter Libassi); published in *Public Comments on the Work Group Reports*, US Department of Health, Education and Welfare, June 1979.

23. S. Macht and P. Lawrence, 'National Survey of Congenital Malformations Resulting from Exposure to Roentgen Radiation', *American Journal*

of Roentgenology, 76, 1955, pp. 442–66.

24. 'Biological Effects of Ionizing Radiation', BEIR III, US National Academy of Science, National Academy Press, 1980, pp. 74–5.

25. ICRP Publication 9, Pergamon Press, Oxford, 1965.

26. Recommended by pioneer researchers A. Mutscheller and R. M. Sievert in 1925. Recommended for international use by the forerunner of the International Commission on Radiological Protection (ICRP) in 1934. Used in most countries until 1950.

27. Recommended by the US National Commission on Radiological Protection (NCRP), 17 March 1934.

28. Recommended by the US NCRP, 7 March 1949 and by ICRP, July 1950, for total body exposure.

29. Recommended by ICRP, April 1956 and US NCRP, 8 January 1957, for total body exposure. This allows for 5 rem per year combined dose from sources external to the body, ingested or inhaled sources. This standard is used in most countries of the world today.

30. ICRP 26, Pergamon Press, Oxford, 1978.

31. See *op.cit.*, note 12, pp. 96 and 123. British, Canadian and American nuclear physicists met in Chalk River, Canada in September 1949 and at Buckland House, UK in August 1950 to agree on radiation-dose levels for workers. Their recommendations were accepted by ICRP.

32. Karl Z. Morgan, 'Hazards of Low-Level Radiation', *Yearbook of Science and the Future*, Supplement of the *Encyclopedia Britannica*, 1980.

33. ICRP Publication 2, Pergamon Press, Oxford, 1959. See also *Statistics Needed for Determining the Effects of Environment on Health*, US DHEW Publication no. (HRA) 77-1459 (1977).

34. R. Mole, 'Radiation Effects on Pre-natal Development and their Radiological Significance', *British Journal of Radiology*, 52:614, 1979, pp. 89–101.

35. 'Ike Sought Confusion Over Nuclear Testing', Associated Press Report, 20 April 1979.

36. For information on past and present nuclear tests in Nevada contact: Citizen's Call, 1321 East 400 South, Salt Lake City, Utah 84102, USA. Official reports on nuclear tests in Nevada are available from the US Department of Energy which operates the test site: *Announced United States Nuclear Tests*, July 1945 to December 1983, NVO-209 (REV.4), January 1984.

37. 'Sources and Effects of Ionizing Radiation', UN Scientific Committee on the Effects of Atomic Radiation, Report to the General Assembly, nos. 90-91, 1977.

38. C. E. Land, 'The Hazards of Fallout or of Epidemiologic Research', Editorial in *New England Journal of Medicine*, 300 (8): 431-2, 1979.

39. For ongoing information on atomic veterans contact: International Association of Atomic Veterans, 236 Massachusetts Avenue, N.E., Suite 306, Washington, DC 20002, USA (tel: 202-543-7711).

40. Karl Z. Morgan and J. E. Turner (eds), *Principles of Radiation Protection, A Textbook of Health Physics*, John Wiley, New York, 1967. See Preface.

41. Dade Moeller, 'The President's Message', *Health Physics Journal*, vol 21, 1971, p. 1.

42. See dialogue in Letters, *The Health Physics Society Newsletter*, vol

XII, May and August 1984.

43. Karl Z. Morgan, 'Hazards of Low-Level Radiation', *Yearbook of Science and the Future*, Supplement of the *Encyclopedia Britannica*, 1980.

44. Personal correspondence to the author from Karl Z. Morgan, first Chairperson of the Committee on Internal Exposures, ICRP and director of the health physics programme at Oak Ridge National Laboratory and Neely Professor, School of Nuclear Engineering, Georgia Institute of Technology.

Part Two: pp. 65–134

1. Personal correspondence. Test results included with permission of Ted Lombard.

2. For articles detailing the story of Ted Lombard see: *Maine Times*, Vol 12, No. 34, 23 May 1980, and *Church World*, Maine's Catholic Weekly, 14 April 1980.

3. Kathleen L. Stanton, 'Four Corners Findings Flash Radiation Danger Signal', report of the work of Donald Calloway, Research Director, Navajo Health Authority, The Arizona Republic, 5–12 December 1982. The entire series of articles may be obtained as a reprint: *Uranium: In the Atom's Shadow*, from The Arizona Republic, 120 East Van Buren Street, Phoenix, Arizona 85004, USA.

4. S. Poet and E. Martell, 'Plutonium 239 and Americium 241 Contamination in the Denver Area', *Health Physics*, vol 23, 1972, pp. 537–48.

5. Carl Johnson, 'Cancer Incidence in an Area Contaminated with Radionuclides near a Nuclear Installation', Royal Swedish Academy of Sciences, AMBIO, vol 10, no. 4, 1981, pp. 176–82.

6. For a complete report of the accident and its medical consequences see Robert A. Conard, 'A Twenty-Year Review of Medical Findings in a Marshallese Population Accidentally Exposed to Radioactive Fallout', Brookhaven National Laboratories, BNL 50424 (Biology and Medicine TID 4500).

7. Glen H. Alcalay, 'The Aftermath of Bikini', *The Ecologist*, vol 10, no. 10, December 1980 (UK).

8. 'Ailin in Enewetok Rainin: The Enewetok Atoll Today', US Department of Energy, Washington, DC, September 1979. Note: The United States tested 43 nuclear devices in Enewetok between 1948 and 1958.

9. W. L. Robinson et al., 'The Northern Marshall Islands Radiological Survey: Sampling and Analysis Summary', Lawrence Livermore Laboratory, UCRL-52853, Pt. 1, 23 July 1981. There is dispute over the findings of the radiochemical analysis of food in the Marshall Islands by this nuclear weapons laboratory. For example, they de-boned the fish before testing, whereas the Marshallese eat the bones.

10. Giff Johnson, 'A Strategic Trust: The Politics of Underdevelopment in Micronesia', *Oceans*, vol 11, no. 1, January–February 1978 (a publication of the Oceanic Society, PO Box 65, Uxbridge, Massachusetts, 01569, USA).

11. Giff Johnson, 'The Pentagon Stalks Micronesia: Strategic Interests vs. Self-Determination', AMPO (Japan–Asia Quarterly Review), vol 14, no. 4, 1982.

12. Robert A. Conard et al., 'March 1957 Medical Survey of Rongelap and Utirik People Three Years After Exposure to Radioactive Fallout',

Brookhaven National Laboratory, Upton, New York, USA, June 1958.

13. See Reference 6, Appendix 3. The Japanese Peace Society for the Lucky Dragon has helped to recover the ship and restore it for public exhibition. A new museum, Lucky Dragon Exhibition Hall, is scheduled for completion in Tokyo in 1985.

14. Ruth Leger Sivard, 'World Military and Social Expenditures 1979', World Priorities (Box 1003, Leesburg, Virginia 22075, USA), p. 13.

15. Suliana Siwatibau and B. David Williams, 'A Call to a New Exodus: an Anti-Nuclear Primer for Pacific People', Pacific Conference of Churches, Lotu Pasifika Productions (PO Box 208, Suva, Fiji), 1982, p. 53.

16. Z. A. Medvedev, 'Two Decades of Dissidence', *New Scientist*, 4 November 1976.

17. Z. A. Medvedev, *Nuclear Disaster in the Urals*, translated by George Saunders, W. W. Norton, 1979.

18. G. Caldwell et al., 'Leukemia Among Participants in Military Maneuvers at a Nuclear Bomb Test', *Journal of the American Medical Association*, vol 244, no. 14, 3 October 1980.

19. This is known as the 'healthy worker effect'. The general public contains persons too ill or disabled to have been hired for a government position or inducted into military service. The post-hiring medical care is generally better than average. This results in better than average rates for most serious illnesses among such health-screened individuals.

20. G. G. Caldwell et al., 'Mortality and Cancer Frequency Among Military Nuclear Test (Smoky) Participants, 1957 through 1979', *Journal of the American Medical Association*, vol 250, no. 5, 5 August 1983.

21. 'The Forgotten Guinea Pigs: A Report on Health Effects of Low Level Radiation Sustained as a Result of the Nuclear Weapons Testing Program Conducted by the US Government', US House of Representatives, August 1980, US Government Printing Office Document no. 65-703 0.

22. J. Lyons, 'Childhood Leukemias Associated with Fallout from Nuclear Testing', *New England Journal of Medicine*, vol 300, no. 8, 22 February 1979.

23. Carl J. Johnson, 'Cancer Incidence in an Area of Radioactive Fallout Downwind from the Nevada Test Site', *Journal of the American Medical Association*, vol 251, no. 2, 13 January 1984.

24. R. Bertell, 'The Nuclear Worker and Ionizing Radiation', *American Industrial Hygiene Association Journal*, 40, no. 5, 1979.

25. NIOSH Report, 'The Risk of Lung Cancer Among Underground Miners of Uranium-bearing Ores', US National Institute of Occupational Safety and Health, 30 June 1980.

26. J. Pearson, 'A Sociological Analysis of the Reduction of Hazardous Radiation in Uranium Mines', US Department of Health, Education and Welfare, National Institute for Occupational Safety and Health (Salt Lake City, Utah), HEW Publication no. (NIOSH) 75-171, 1975.

27. J. Wagoner, 'Uranium: The United States Experience, a Lesson in History', Environmental Defence Fund, 1525 18th St., N.W., Washington, DC 20036, USA.

28. 'Australia – From Massacre to Mining', Colonialism and Indigenous Minorities Research and Action, 5 Caledonian Road, London, N.1.

29. 'Report of the Panel for Hearings on Namibian Uranium', Chairman: Mr Noel G. Sinclair (Guyana), United Nations General Assembly

A/AC.131/L.163 (Part II), 30 September 1980.

30. T. Barry, 'Bury My Lungs at Red Rock', *The Progressive*, February 1979.

31. Leon S. Gottlieb, MD, FCCP, and Luverne A. Husen, MD, 'Lung Cancer Among Navajo Uranium Miners', *Chest*, vol 81, April 1982, pp. 449–452.

32. D. C. Comey, 'The Legacy of Uranium Tailings', *Bulletin of the Atomic Scientists*, September 1975.

33. Kathleen L. Stanton, 'High Risk? Ills Demand Explanation', The Arizona Republic (120 East Van Buren Street, Phoenix, Arizona 85004, USA). Part of a series on uranium mining, 5–12 December 1982.

34. 'Radiological Quality of the Environment in the United States 1977', US Environmental Protection Agency, EPA 520/1-77-009, 1977. This report gives an overview of US residential pollution.

35. For a discussion of uranium mining and nuclear power in Canada see 'A Race Against Time', Report of the Royal Commission on Electrical Power Planning, Chairperson: Arthur Porter, ISBN 0-7743-2890-8, 1978, and Report of Royal Commission of Inquiry, Health and Environmental Protection, Uranium Mining, Chairperson: David V. Bates, MD, Province of British Columbia, October 1980, vol I.

36. For a discussion of uranium mining in Australia see 'Ranger Uranium Environmental Inquiry: First and Second Reports', Chairperson: R. W. Fox, ISBN 0-642-02260-7, Australian Government Publication Service, Canberra, 1976.

37. House of Commons Debate, vol 127, no. 17, 2nd Session of the 32nd Canadian Parliament, 23 January 1984.

38. See *op. cit.*, note 34, above, for a discussion of natural background radiation and technologically enhanced natural background radiation.

39. An example of an Environmental Impact Statement in the USA would be the one for Black Fox Nuclear Stations 1 and 2, available from the US Nuclear Regulatory Commission, NUREG-0176, 1977.

40. T. Mancuso, A. Stewart and G. Kneale, 'Radiation Exposure of Hanford Workers Dying from Cancer and Other Causes' *Health Physics* 33: 369–85, 1977.

41. 'The Effect of Radiation on Human Health, Volume 1', Hearings before the Subcommittee on Health and the Environment of the Committee on Interstate and Foreign Commerce, US House of Representatives, 95th US Congress, 1978 (serial number 95–179). See pp. 697, 725, 748–49. The hearing brought out the fact that Dr Sidney Marks of ERDA left that agency and accepted a research position at Battelle–Pacific Northwest Laboratory, a private research organisation funded by the US government. About $58 million went to Battelle at the same time. Later Dr Marks presented the re-analysis of the Hanford Workers Study at the IAEA meeting in Vienna on 13–17 March 1978.

42. A. M. Stewart, J. Webb and D. Hewitt, 'A Survey of Childhood Malignancies', *British Medical Journal*, 1958, pp. 1495–508.

43. T. Mancuso, A. Stewart and G. Kneale, ' Radiation Exposures of Hanford Workers Dying from Various Causes', Proceedings of the 10th Midyear Symposium, Health Physics Society, Saratoga Springs, New York, 11–13 October 1976.

44. G. Kneale, A. Stewart and T. Mancuso, 'Re-analysis of Data Relating

to the Hanford Study of the Cancer Risks of Radiation Workers', presented at the International Symposium on Late Biological Effects of Ionizing Radiation, IAEA, Vienna, 13–17 March 1978.

45. G. Kneale and A. M. Stewart, 'Low Dose Radiation', *The Lancet*, 29 July 1978.

46. G. W. Kneale, T.Mancuso and A. M. Stewart, 'Job Related Mortality Risks of Hanford Workers and their Relation to Cancer Effects of Measured Doses of External Radiation', *Biological Effects of Low-Level Radiation*, IAEA, Vienna, 1983, pp. 363–72.

47. R. Bertell, Comments and Response to 'The Nuclear Worker and Ionizing Radiation', *American Industrial Hygiene Journal*, vol 40, October 1979, pp. 916–22.

48. S. Marks and E. Gilbert, 'Cancer Mortality in Hanford Workers', presented at the International Symposium on Late Biological Effects of Ionizing Radiation (IAEA-SM-224), IAEA, Vienna, 13–17 March 1978.

49. John W. Gofman, 'Radiation and Human Health', Sierra Club Books, San Francisco, 1981, pp. 214–19. See also G. W. Kneale, 'Stoll and Chakroborty's Comment on "The Question of Radiation Causation of Cancer in Hanford Workers" by John Gofman', *Health Physics*, vol 40, 1981, pp. 257–58.

50. 'Investigation of the Possible Increased Incidence of Cancer in West Cumbria', Report of the Independent Advisory Group, Chairman: Sir Douglas Black, London, HMSO, 1984, p. 88, 5.10.

51. G. W. Kneale, T. J. Mancuso and A. M. Stewart, 'Hanford Radiation Study III: A Cohort Study of the Cancer Risks from Radiation to Workers at Hanford (1944–77 Deaths) by the Method of Regression Models in Life Tables', *British Journal of Industrial Medicine*, vol 38, 1981, pp. 156–66.

52. Nuclear industry public relations statements have been modified to read: 'no *credible* studies' have ever shown health damage at low levels of radiation exposure. It is the radiation users who determine the 'credibility' of the research. See also BEIR III, 1980, p. 136.

53. A. P. Polednak, 'Long-range Studies of Uranium Workers and the Oak Ridge Worker Population' in K. F. Nubner and S. A. Fry (eds), *The Medical Basis for Radiation Accident Preparedness*, Elsevier North, Holland, 1980, pp. 401–9.

54. H. Checkoway et al., 'Mortality Among Workers at the Oak Ridge National Laboratory', Proceedings of the Fifteenth Midyear Topical Meeting of the Health Physical Society, Albuquerque, New Mexico, 1983, pp. 90–104.

55. This manuscript is still in preparation. The principal author will be N. Abd Elgahny, MD, PhD.

56. Bengt Danielsson and Marie Thérèse Danielsson, *Moruroa Mon Amour*, Penguin, Victoria, Australia, 1976. Bengt Danielsson was Swedish Ambassador to French Polynesia. He and his wife are long-time residents of Tahiti.

57. Report to the regional meeting of the World Health Organisation in Wellington, New Zealand, 23 April 1973.

58. 'France Refuses to Provide Data on Cancer', *The Times* (Papua New Guinea), 7 September 1982. France responded in an interview by *Island's Business* (Fiji) with Ambassador Robert Puissant, published in June 1983.

59. John Ward, 'Australian Doctors and the Movement for a Nuclear Free Pacific Ocean', *Journal of the Medical Association for Prevention of War*, vol 3, Pt. 9, June 1983.

60. The author personally travelled to Tahiti in the summer of 1983 and interviewed Marie Thérèse and Bengt Danielsson and others.

61. World Health Organisation National Workshop on Epidemiological Surveillance, Module Three: Disease Surveillance and Epidemic Investigation, WHO Regional Office, Suva, Fiji islands, 1981.

62. Takeshi Yasumoto et al., 'Ecological and Distributional Studies on a Toxic Dinoflagellate Responsible for Ciguatera', Report to the Ministry of Education, Japan, April 1978–March 1980.

63. Robert Aldridge, 'Wreaking Havoc: The Pursuit of US Military Interests in the Pacific Islands', *Sojourners* (1321 Otis St, N.E., Washington, DC 20017, USA), August 1983.

64. *Handbook for Estimating the Health Effects of Exposure to Ionizing Radiation*, compiled by R. Bertell, published and distributed by the International Institute of Concern for Public Health, 63 Mowat Avenue, Suite 343, Toronto, Ontario M6K 3E3, Canada, and the International Radiation Research and Training Institute, Regional Cancer Registry, Queen Elizabeth Medical Centre, Birmingham B15 2TH, UK.

65. The Three Mile Island nuclear accident occurred on 28 March 1979. For details see E. Sternglass, *Secret Fallout: Low-level Radiation from Hiroshima to Three Mile Island*, McGraw-Hill, New York, 1981.

66. For greater detail on radionuclide releases from power generators see 'Environmental Radiation Protection Requirements for Normal Operations of Activities in the Uranium Fuel Cycle', vols I and II, US Environmental Protection Agency, Office of Radiation Programmes, EPA 520/4-76-016, 1976, or the Black Report, note 50, above.

67. ICRP report 2, 1959; this report was updated in reports nos. 6, 10 and 23, Pergamon Press, Oxford.

68. 'Age-Specific Radiation Dose Commitment Factors for a One Year Chronic Intake', prepared for the US Nuclear Regulatory Commission by Battelle–Pacific Northwest Laboratory, NUREG-0172.

69. *Nucleonics Week*, 16 December 1976 (USA).

70. Ontario Hydro publication, *Nuclear Questions and Answers*, CSD 225–50M, November 1978.

71. Current estimates are 2.3 lbs of uranium concentrate per ton of ore.

72. For information contact: Groupe de Bellerive, Colloque 'Energie et Société', 122 rue de Lausanne, Ch 1202, Geneva, Switzerland.

73. World Council of Churches, Division of Church and Society, Geneva, Switzerland, 21 August 1980.

74. Statement on a Plutonium Economy, US National Council of Churches, 4 March 1976.

75. 'The Ethical Implications of Energy Production and Use', adopted by the Governing Board, US National Council of Churches, 11 May 1979. Voting in favour: 120; voting against: 26; abstention: 1.

76. B. Franke, 'Radiation Exposure and Health Damage Due to Nuclear Power Production: The Question of Standards and the Need for Comparative Health Damage Analysis', presented at the American Association for the Advancement of Science meeting, 3–8 January 1982. For copies: IFEU, Heidelberg e.V, Im Sand 5, 6900 Heidelberg, Federal Republic of

Germany.

77. Nuclear Cargo Transportation Project, American Friends Service Committee, 92 Piedmont N.E., Atlanta, GA 30303, USA. See also John Surrey (ed), 'The Urban Transportation of Irradiated Fuel', Macmillan, London, 1984.

78. The Effects on Populations of Exposure to Low Levels of Ionizing Radiation, BEIR I, 1972, US National Academy of Science.

79. To understand the inadequacy of record-keeping see 'Statistics Needed for Determining the Effects of the Environment on Health: Report of the Technical Consultant Panel to the US National Committee on Vital and Health Statistics', DHEW publication no. (HRA) 77–1457, 1977.

80. Karl Z. Morgan, 'Cancer and Low Level Ionizing Radiation', Bulletin of the Atomic Scientist, September 1978.

81. 'Radioecological Assessment of the Wyhl Nuclear Power Plant', Report of the Department of Environmental Protection, University of Heidelberg, FRG (Im Neuenheimer Feld 360); US Nuclear Regulatory Commission Translation no. 520, 1979.

82. R. Bertell, 'Radiation Exposure and Human Species Survival', *Environmental Health Review*, vol 25, no. 2, June 1981.

83. Peter C. Nowell, 'The Clonal Evolution of Tumor Cell Populations', *Science*, 1 October 1976 (USA).

84. V. Archer et al., 'Uranium Mining and Cigarette Smoking Effects on Man', *Journal of Occupational Medicine*, vol 15, 1973, pp. 204–11.

85. R. Bertell, I. Bross et al., 'Proceedings of a Congressional Seminar on Low Level Ionizing Radiation', Subcommittee on Energy and the Environment, Committee on Interior and Insular Affairs, US House of Representatives (76-767 0), 1976.

86. *Nuclear News*, February 1984.

87. BEIR III, 1980, p. 135, US National Academy of Science, National Academy Press.

88. I. V. Petryanov, 'A Story of Man and His Environment' in *Things to Come*, Collected Articles, Mir Publishers, Moscow; English translation 1977.

Part Three: I: pp. 135–198

1. Francis F. Miller, *History of World War II*, Universal Book and Bible Company, Pennsylvania, 1945.

2. Takesi Araki, Mayor of the City of Hiroshima, and Yoshitake Morotani, Mayor of the City of Nagasaki, 'Appeal to the Secretary General of the United Nations', 1977.

3. Kenzaburō Ōe, *Hiroshima Notes*, translated by Toshi Yonezawa, edited by David L. Swain, YMCA Press, Tokyo, 1981. Excerpt reprinted with permission.

4. 95th Congress, United States of America, Public Law 95–91, 4 August 1977.

5. Rachel Carson, *Silent Spring*, Houghton Mifflin, Boston, 1962.

6. Philip M. Boffey, *The Brain Bank of America: An Inquiry Into the Politics of Science*, McGraw-Hill, New York, 1975.

7. See BEIR III, 1980, Appendix B. Review and analysis of selected studies on record, pp. 455–71.

8. US Department of State, 'The International Control of Atomic Energy: Growth of a Policy', Publication 2702, 1946, p. 120.

9. John G. Stogssinger, 'Financing the UN' in Saul Mendlovitz, *Legal and Political Problems of World Order*, Fund for Education Concerning World Peace Through World Law, New York, 1962.

10. B. Weston et al., *International Law and World Order*, West Publishing Co, St Paul, Minnesota, 55165, June 1980.

11. Permanent Court of International Justice Statute – Declarations under Article 36 (2); registered under no. 3, United Nations *Treaty Series*, vol 1, p. 9.

12. Ibid. Canada's declaration is registered under no. 10415, *Treaty Series*, vol 724. Great Britain's is registered under no. 9370, *Treaty Series*, vol 654. It replaced a declaration of 27 November 1963.

13. International Court of Justice Statute, Declarations under Article 36 (2).

14. Ibid. Registered under no. 8196, UN *Treaty Series*, vol 562, replacing the declaration of 10 July 1959.

15. Virginia W. Brewer, 'The Connally Amendment: A Brief History of the Movement to Repeal', Analysis – Foreign Affairs Division, Legislative Reference Service, Library of Congress, 26 August 1960.

16. Senate Resolution 115: Submission of a Resolution Modifying the Connally Amendment, Jacob Javits, US Congressional Record/Senate, 5 May 1971.

17. Senate Resolution 114: Submission of a Resolution Repealing the Connally Amendment, Hubert Humphrey, US Congressional Record/Senate, 5 May 1971.

18. Bernhard G. Bechhoefer, 'The Disarmament Deadlock' in Saul Mendlovitz, *Legal and Political Problems of World Order*, Fund for Education Concerning Peace Through World Law, 1962, pp. 649 ff.

19. Eisenhower's address is excerpted in E. Mansfield (ed), *Defense, Science and Public Policy*, W. W. Norton, New York, 1968, p. 40.

20. This comment by British historian John Costello was discussed in the *Toronto Sun*, 14 May 1981, p. 11. The disclosure of documents was made in the *New York Times* by defence specialist Drew Middleton.

21. *Documents on Disarmament*, US Department of State, p. 252, vol 1, 1945–56, Publication 7008, August 1960.

22. Jim Garrison and Pyare Shivpuri, *The Russian Threat: Its Myths and Realities*, Gateway, London, 1983.

23. Nevada A-site Picked in a Hurry', *Deseret News*, Salt Lake City, Utah, 9 December 1978, p. D1.

24. *The Peaceful Atom Series*, US Atomic Energy Commission. Some of these booklets are still available from the Atomic Industrial Forum, Inc, 1747 Pennsylvania Avenue, N.W., Washington, DC 20006, USA. This is the public relations and lobbying arm of the nuclear industry.

25. H. Wasserman and N. Solomon, *Killing Our Own*, Delta Publishing Co, New York, 1982.

26. Karl Z. Morgan, 'Health Implications of the Three Mile Island Nuclear Accident', testimony before the US Senate, 4 April 1979. US Government Printing Office, no. 49–543 0, pp. 112–17. When beta particles collide with human skin or a radiation badge some of their energy is converted to heat (causing burns), and some to penetrating X-rays. Only the

X-ray component is recorded on the film badge.

27. Thomas H. Saffer and E. Kelly Orville, 'Countdown Zero', Putnam, New York, 1982.

28. G. Caldwell et al., 'Leukemia Among Participants in Military Maneuvers at a Nuclear Bomb Test', *Journal of the American Medical Association*, vol 244, no. 14, pp. 1575–78, 1980.

29. Press release, National Research Council, 2101 Constitution Avenue, N.W., Washington, DC, 13 July 1983. Contact: Gail Porter.

30. Staff Memorandum, prepared by the OTA Health Programme, 17 December 1983; review of the Report 'Multiple Myeloma among Hiroshima/Nagasaki Veterans'.

31. R. J. Smith, 'Study of Atomic Veterans Fuels Controversy', *Science*, vol 221, pp. 733–34, 1983.

32. *American Foreign Policy*, Current Documents 1959, US Department of State, Document 524, 4 November 1959.

33. 'Leukemia Among Persons Present at an Atmospheric Nuclear Test (Smoky)', *Physicians East*, October 1979.

34. Carl J. Johnson, 'Cancer Incidence in an Area of Radioactive Fallout Downwind from the Nevada Test Site', *Journal of the American Medical Association*, vol 251, 13 January 1984.

35. *Las Vegas Sun*, 3 August 1979.

36. Giff Johnson, 'Nuclear Legacy', *Oceans*, vol 13, no. 1, 1980.

37. J. Lyons et al., 'Childhood Leukemia Associated with Fallout from Nuclear Testing', *New England Journal of Medicine*, vol 300. no. 8. 22 February 1979, pp 394–402.

38. *The Biological Peril to Man of Carbon 14 from Nuclear Weapons*, published by the US Atomic Energy Commission, 1959.

39. E. J. Sternglass and S. Bell, 'Fallout and the Decline of Scholastic Aptitude Tests', meeting of the American Psychological Association, New York, 3 September 1979.

40. 'The Nuclear Radiation/SAT Decline Connection', *The Phi Beta Kappan*, vol 61, no. 3, November 1979, pp. 184–8.

41. E. J. Sternglass and S. Bell, 'Fallout and SAT Scores: Evidence for Cognitive Damage During Early Infancy', available from the University of Pittsburgh, School of Medicine, 21 January 1983.

42. J. O. Mason, Director of the Utah Department of Health, 'Health and the 1950's Nuclear Tests', *The Nation's Health*, published by the American Public Health Association, July 1979, p. 9.

43. '4300 Sheep Near Nevada Nuclear Tests Died in '53', *New York Times*, 15 February 1979. The story was documented by Dr Harold Knapp, formerly with the US Atomic Energy Commission Fallout Studies, after leaving the agency.

44. John G. Fuller, *The Day We Bombed Utah: America's Most Lethal Secret*, New American Library, 1984.

45. 'New Claims of A-test With Cancer', *The Advertiser*, Adelaide, Australia, 17 April 1980.

46. Bengt and Marie Thérèse Danielsson, *Moruroa Mon Amour*, Penguin, Victoria, Australia, 1976.

47. M. Eisenbud, 'Radioactivity in the Environment – Radioactive Concentration in the Fetal Thyroid', *Pediatrics* (Supplement), 41, Part II, 1968, p. 174.

48. 'The Forgotten Guinea Pigs: A Report on Health Effects of Low Level Radiation Sustained as a Result of the Nuclear Weapons Testing Program Conducted by the US Government', US House of Representatives, August 1980, US Government Printing Office Document no. 65–703 0, p. 15.

49. See *op. cit.*, note 48, p. 19.

50. ICRP report 26, Pergamon Press, Oxford, 1978.

51. Jerome H. Kahn was a researcher for the Brookings Institute. He served as senior adviser in the Kissinger State Department.

52. Sidney Lens, *The Day Before Doomsday*, Doubleday, New York, 1977.

53. Proceedings: International Symposium on the Damage and After-Effects of the Atomic Bombing of Hiroshima and Nagasaki, 21 July–9 August 1977, Tokyo, Hiroshima and Nagasaki, Japan National Preparatory Committee.

54. Alice M. Stewart, 'Delayed Effects of A-bomb Radiation: A Review of Recent Mortality Rates and Risk Estimates for Five-Year Survivors', *Journal of Epidemiology and Community Health*, vol 36, no. 2, June 1982, pp. 80–6.

55. J. Rotblat, 'The Puzzle of Absent Effects', *New Scientist*, 25 August, 1977, p. 476.

56. J. Rotblat, 'The Risks for Radiation Workers', *Bulletin of the Atomic Scientist*, September 1978.

57. US Atomic Energy Commission document 43/345, 9 February 1951. See also the Congressional Record, 6 September 1979 (E 4316).

58. Issue no. 6, Birch Bark Alliance, spring 1980.

59. Membership policy of ICRP is given in the Appendix of ICRP 26, 1977, Pergamon Press, Oxford.

60. Apparently ICRP does not consider these stands on public health issues as part of its role. At present there is no officially recognised international body speaking on behalf of public health in the matter of nuclear pollution.

61. Z. Medvedev, 'Two Decades of Dissidence', *New Scientist*, 4 November 1976.

62. Z. Medvedev, *Nuclear Disaster in the Urals*, translated by George Saunders, W. W. Norton, New York, 1979.

63. Press Association report, *The Times*, 8 November 1976.

64. Reuters, *Los Angeles Times*, 11 November 1976.

65. A. Tucher, 'Russian Reveals Nuclear Tragedy', *Manchester Guardian*, 8 November 1976.

66. Letter to the editor, *Jerusalem Post*, December 1976, from Professor L. Tumerman, Weismann Institute of Science Rehovoth. Quoted in full in *Nuclear Disaster in the Urals*, note 62, above, pp. 11–12.

67. Foreign Intelligence Report no. 00E 324/01015–77, 24 January 1977; approved for release to Ralph Nader on 27 September 1977 under the Freedom of Information Act; Exhibit 15, p. 196; *Nuclear Disaster in the Urals*, note 62, above.

68. *Op. cit.*, note 62, above, at p. 165.

69. *Op. cit.*, note 62, above, at p. 166.

70. Granada TV, 7 November 1977, 8.30 p.m.

71. D. Burnham, 'C.I.A. Papers Released to Nader Tell of 2 Soviet

Nuclear Accidents', *New York Times*, 26 November 1977.

72. Foreign Intelligence Report no. 00K 323/20537–76, 20 September 1976, Exhibit 14, p. 195; *Nuclear Disaster in the Urals*, note 62, above.

73. Sanitised copy of Report no. CS–3/389, 785, 4 March 1959, Central Intelligence Agency, Exhibit 6, p. 185; *Nuclear Disaster in the Urals*, note 62, above.

74. Ibid. *Nuclear Disaster in the Urals*, note 62, above, at pp. 27–33.

75. 'Birth Defects Up, Russian Reports', *Milwaukee Journal*, 23 August 1978.

76. N. P. Dubinin, *Vechnoe Doizhenie* (Eternal Motion), Gospolitizdat (state publishers for political literature), Moscow, 1973.

77. Dubinin, *op. cit.*, p. 330.

78. F. A. Tikhomirov et al., 'The Migration of Radionuclides in the Forests and the Effects of Ionizing Radiation on Forest Stands', Proceedings of the 4th International Conference on the Peaceful Uses of Atomic Energy, September 6–16 1971, vol 11, IAEA, Vienna, 1972.

79. *Nuclear Disaster in the Urals*, note 62, above, at p. 115.

80. N. P. Dubinin et al., *Genetic Processes in Populations Subjected to Chronic Effects of Ionizing Radiation*, an annual review edited by Dubinin, vol 4, Nauka Press, 1972, pp. 170–205.

81. L. V. Cherezhanova et al., 'Cytogenetic Adaptation of Plants Affected by Chronic Ionizing Radiation', *Genetika* (Genetics), vol 7, no. 4, 1971, pp. 30–7.

82. Report on Investigation of the no. 106 T Tank Leak at the Hanford Reservation, Richland, Washington, US Atomic Energy Commission, 1973.

83. The US National Security Council was established by the National Security Act of 1947, Statute 496 and amended by the National Security Act of 1949, Statute 579. It was placed in the Executive Office of the President of the USA in 1949. The National Security Council members are: the President, the Vice-President, the Secretary of State and the Secretary of Defence. The statutory advisers are the Director of the CIA, the Military Chairman of the Joint Chiefs of Staff, the Assistant to the President, the Deputy Assistant to the President and the Staff Secretary. See US Government Orgnizational Manual 1978–79.

84. Khrushchev, as Chairperson of the Council of Ministers, USSR, before the Twenty-first Congress of the Communist Party of the Soviet Union, 27 January 1959.

85. Declaration of the Soviet Government handed to Mr Thompson, American Ambassador at Moscow, 25 June 1959.

86. American Foreign Policy 1959, US Department of State, Document no. 1418.

87. American Foreign Policy 1959, US Department of State, Document no. 524.

88. American Foreign Policy 1960, US Department of State, Document no. 371.

89. American Foreign Policy 1960, US Department of State, Document no. 372.

90. Bengt and Marie Thérèse Danielsson, *Moruroa Mon Amour*, Penguin, Australia, 1976.

91. President Nixon is thought to have formalised the policy of first strike in 1972 when he introduced the targeting of Soviet missile bases rather than

Soviet cities. Retaliation against already empty missile bases would be futile. President Carter formalised the counter-force policy by Executive Directive no. 59 in May 1980.

92. 'Council Squares up for Trident Battle', *The Times*, 13 May 1984.

93. Ruth Leger Sivard, *World Military and Social Expenditures*, 7th annual edn, 1981.

94. Ruth Leger Sivard, *World Military and Social Expenditures*, 9th annual edn, 1983.

95. *Critical Mass Journal*, February 1980 (USA).

96. Holly Sklar (ed), *Trilateralism*, Black Rose Books, Montreal, Canada, 1980, pp. 157–60.

97. Alden Hatch, *HRH Prince Bernhard of the Netherlands*, Harrap, London 1962, p. 218.

98. Report of the 1974 Mégève, France, Bilderberg meeting. See also *Trilateralism*, note 96, above, at p. 171.

99. See *Trilateralism*, note 96, above, at pp. 177–8.

100. Z. Cervenka and B. Rodgers, *The Nuclear Axis: Secret Collaboration Between West Germany and South Africa*, Julian Friedmann, London, 1978.

101. John G. Fuller, *We Almost Lost Detroit*, Reader's Digest, New York, 1975.

Part Three: II: pp. 199–290

1. C. D. Darlington, 'The Cell and Heredity Under Ionization', Conference on Biological Hazards of Atomic Energy convened by the Institute of Biology and the Atomic Scientists Association, October 1950, Clarendon Press, Oxford, 1952.

2. 'Announced US Nuclear Tests, July 1945 through December 1983', NVO-209 (Rev. 4), US Department of Energy, January 1984.

3. United Nations Scientific Committee on the Effects of Atomic Radiation (UNSCEAR) Report, 1982, p. 227.

4. *Arms Control and Disarmament Agreements: Texts and Histories of Negotiations*, US Arms Control and Disarmament Agency, Washington, DC 20451, 1980, pp. 34–40.

5. 'Report on Health Effects of Low Level Radiation Sustained as a Result of the Nuclear Weapon Testing Program Conducted by the US Government', US Atomic Energy Commission, 1 January 1955.

6. Testimony of Mahlon E. Gates, Manager, Nevada Operations Office, Department of Energy. Hearings before the Subcommittee on Oversight and Investigations of the Committee on Interstate and Foreign Commerce, 96th US Congress, 1st Session; serial no. 96–129 (348), 1979.

7. 'The Forgotten Guinea Pigs: A Report on Health Effects of Low Level Radiation Sustained as a Result of the Nuclear Weapons Testing Program Conducted by the US Government', US House of Representatives, August 1980, US Government Printing Office Document no. 65–703 0.

8. It is estimated that the nuclear industry in the USA has received about $17 billion, not counting the Price-Anderson Act which provides insurance coverage. The US government has recently offered to pay for nuclear waste disposal. Between 1954 and 1976 the coal industry received $3 billion depletion allowance and another $3.5 billion in government services for exploration, research, development and safety. It is the government not the

mining companies which pays the health compensation costs for coal miners. The oil companies (which own the nuclear industry) received federal incentives of $77.2 billion ($40 billion of which was depletion allowance). National gas companies received about $15.1 billion for depletion allowance and drilling expenses. Subsidy of solar energy has never reached the $1 billion level. In 1979, more than 50% of the US public preferred to develop solar energy.

9. I. F. Stone's Bi-Weekly, vol XI, no. 14, Washington, DC, 24 June 1963.

10. Amending the Atomic Energy Act and Authorization of Stanford Acceleration Project: Hearing by the Joint Committee on Atomic Energy, US Congress, 26 August 1959.

11. American Public Health Meeting, Miami, Florida, 12 December 1963.

12. Report of fallout at Valentia Observatory, Co. Kerry, data collected by the Irish Meteorological Service, Glasnevin.

13. J. Crabtree, 'The Travel and Diffusion of the Radioactive Material Emitted during the Windscale Accident', Quarterly Journal of the Royal Meteorological Society, vol 85, pp. 362–70.

14. K. S. Shrader-Frechette, *Nuclear Power and Public Policy: The Social and Ethical Problems of Fission Technology*, Reidel, London, 1980.

15. Pearce Wright (science writer), 'Inquiry Starts into Cases of Leukaemia near Nuclear Power Stations', *The Times*, London, April 1983 (published the same day as the Yorkshire Television programme was transmitted).

16. P. M. E. Sheehan and I. B. Hillary, 'An Unusual Cluster of Down's Syndrome, Born to Past Students of an Irish Boarding School', *British Medical Journal*, vol 287, 12 November 1983.

17. Author's reply, *British Medical Journal*, vol 288, 14 January 1984 for comments on the Sheehan/Hillary paper and the detailed response of the authors (pp. 146–8).

18. Paul Foot, 'Riddle of the Sands', *Daily Mirror*, 19 April 1984.

19. Letter to Dr R. Bertell from Linda Runyon, Chairperson: Child Health Committee, Tompkins County, New York, Comprehensive Health Planning Council, 1 September 1982.

20. Douglass, Grahn, 'Analysis of Population, Birth and Death Statistics in Counties Surrounding the Big Rock Nuclear Power Station, Charlevoix County, Michigan', ANL-8149, January 1975.

21. B. Franke et al., 'Radiation Exposure to the Public from Radioactive Emissions from Nuclear Power Stations', IFEU, In Sand 5, 6900 Heidelberg, FRG.

22. *How Radioactive is Our Milk: The Urgent Need for Sound Monitoring and Public Disclosure*, Another Mother for Peace, 407 North Maple Drive, Beverly Hills, California 90210, USA.

23. *Handbook for Estimating the Health Effects of Exposure to Ionizing Radiation*, compiled by R. Bertell, Ministry of Concern for Public Health, Buffalo, New York; Institute of Concern for Public Health, Toronto, Canada; International Radiation Research and Training Institute, Birmingham, UK; August 1984.

24. 'Disclosure of Corporate Ownership', 1977 Congressional Study, US Senate, and Anna Gyorgy, *No Nukes*, pp. 163–4, South End Press, Boston.

25. R. Nader and J. Abbotts, *The Menace of Atomic Energy*, W. W.

Norton, New York, 1977.

26. J. Berger, *Nuclear Power: The Unviable Option*, Dell, New York, 1977.

27. Down's Syndrome children have a higher than average risk of leukaemia.

28. Memo of Dr Walter H. Jordon, Atomic Safety and Licensing Board (ASLBP), to James Yore, Chairman: ASLBP, 21 September 1979.

29. Report of the International Solidarity Conference for Survival, 33rd Anniversary Conference Against Atomic and Hydrogen Bombs, 3–4 August 1978, Osaka, Japan, *Gensuikin News*, Tokyo.

30. For a discussion of uranium mining in Australia see 'Ranger Uranium Environmental Inquiry: First and Second Reports', Chairperson: R. W. Fox, Australian Government Publication Service, Canberra, ISBN 0-642-02260-7, 1976.

31. *Financial Post*, 25 April 1981.

32. 'The Cluff Lake Board of Inquiry, Final Report', Chairman: Honourable Mr Justice E. D. Bayda, Department of the Environment, Regina, Saskatchewan, p. 122.

33. Robert Jungk, *The New Tyranny*, Warner, New York, 1979.

34. Robert Jungk, *Brighter Than a Thousand Suns*, Penguin Books, Harmondsworth, 1970.

35. Richard Clevaud, 'Living Off a Polluted Land', *Guardian*, 3 August 1980.

36. Jonathan M. Samet et al., 'Uranium Mining and Lung Cancer in Navajo Men', *New England Journal of Medicine*, vol 310, no. 23, 7 June 1984, p. 1481; Naomi H. Harley, Editorial 'Radon and Lung Cancer in the Mines and Homes', ibid., p. 1525.

37. Zbigniew Brzezinski, *Between Two Ages: America's Role in the Technetronic Era*, New York, Viking Press, 1970.

38. Michel Crozier, Samuel P. Huntington and Joji Watanuki, *The Crisis of Democracy*, New York University Press, 1975.

39. Richard N. Cooper, Karl Kaiser and Masataka Kosaka, 'Toward a Renovated International System', *Trilateral Commission Task Forces Reports 9–14*, New York University Press, 1978.

40. For further information on OPEC contact: Dr Karl Lueger, Vienna 1, Ring 10, Austria. There is also an organisation OAPEC, Arab nations only, which can be reached through: OAPEC, PO Box 20501, Kuwait City, Kuwait.

41. The Centre Report, vol IV, nos. 2 and 3, September 1979, p. 3, by the Environment Liaison Centre, PO Box 72461, Nairobi, Kenya.

42. For an especially poignant story see 'Woman Helps Sink Nuclear Plant in Michigan After $4 Billion Outlay', *Wall Street Journal*, 18 July 1984, p. 1.

43. 'EEC Moves to Boost Nuclear Spending', *Guardian*, 28 September 1979.

44. Jim Browning, 'France Losing A-Contracts', special from Paris to the *Christian Science Monitor*, 30 January 1979.

45. In June 1979 the Government Operations Sub-committee of the US House of Representatives received testimony prepared by Jay Polach and Cathryn Goddard, analysts with the US Treasury, detailing misconduct in oil industry market manipulations. Two notes were attached to the testimony, the first saying that 'leaking' of the report might be 'confusing to

the public since it provides an explanation for high petroleum product prices different from that used to date by administration spokesmen'. The second note concluded with a warning from the Assistant to US Secretary of the Treasury, Michael Blumenthal: 'Mike doesn't want anything to be attributed to the Treasury staff that undercuts the official position. Please make every effort to keep the paper under tight control.' The paper showed that there had never been a shortage of oil. The recession, unemployment, bankruptcy of retail gasoline dealers and small trucking firms, inflation and other economic penalties had been caused by market manipulations. See also a special article in the *Toronto Star*, 9 October 1983, p. B3, by John Kimche, London-based expert on Middle-Eastern affairs.

46. G. Duffy and G. Adams, *Power Politics: The Nuclear Industry and Nuclear Exports*, New York Council on Economic Priorities, 1978, pp. 19 and 28.

47. Duffy, *op. cit.*, p. 45, ref. 81.

48. N. Gall, 'Atoms for Brazil, Dangers for All', *Bulletin of the Atomic Scientist*, June 1976, pp. 5–6.

49. Andrew Whitley (in Rio de Janeiro), 'Why Brazil's Nuclear Plans Are in Disarray', Energy Review, *Financial Times*, London, 19 September 1984.

50. 'Pretoria Said to Obtain Secret Uranium Supply', *New York Times*, 14 November 1981.

51. 'Candu Sales Take a Nose Dive', *Toronto Star*, 15 June 1981, p. A8.

52. 'Cut A-power, Aldermen Tell Hydro', *Toronto Star*, 8 April 1981, p. C20.

53. 'Candu Slump Felt Continuing', *Globe and Mail*, Toronto, 20 November 1981.

54. 'Nuclear Power Seen as Future Energy Source at Oil Sands', *Toronto Star*, 28 September 1981.

55. Royal Commission on Uranium Mining, British Columbia, Chair: Dr David Bates, 1980.

56. Margaret Swanson is a pseudonym to protect the identity of this Amok employee.

57. Murray Dobbin, 'Amok and the NDP: Production for Profit, Not for People', *Briarpatch*, Special Supplement, Saskatchewan's independent monthly newsmagazine, 16 October 1980.

58. 'Amok Connections', *Prairie Messenger*, 30 April 1978.

59. E. R. Sternglass, 'The Death of All Children', *Esquire Magazine*, 1969.

60. Leslie J. Freeman, 'Nuclear Witnesses: Insiders Speak Out', W. W. Norton, New York, 1981.

61. 'Report to the Governor of Pennsylvania on the Shippingport Nuclear Power Station', prepared by the Governor's Fact Finding Committee, 1974.

62. J. Lyons et al., 'Childhood Leukemias Associated with Fallout from Nuclear Testing', *New England Journal of Medicine*, vol 300, no. 8, 22 February 1979.

63. Carl J. Johnson, 'Cancer Incidence in an Area of Radioactive Fallout Downwind from the Nevada Test Site', *Journal of the American Medical Association*, vol 251, no. 2, 13 January 1984.

64. American Foreign Policy, 1960, US Department of State, Document no. 342.

65. *Nuclear News*, February 1984.

66. American Foreign Policy, 1960, US Department of State, Document no. 46.

67. American Foreign Policy, 1960, US Department of State, Document no. 342.

68. Burns H. Weston, Richard A. Falk and Anthony A. D'Amato, *Basic Documents in International Law and World Order*, West Publishing Co, St Paul, Minnesota, 1980.

69. *Op. cit.*, note 68, above. Document no. 372.

70. B. Symons, 'Living With Titan Incidents and Accidents,' *Nuclear Countdown*, winter 1979. B. Symons is Secretary of the Association for International Cooperation and Disarmament (ACID).

71. 'Mishap Tally for A-arms is Challenged,' Reuter, *Buffalo Evening News*, 22 December 1980.

72. 'World Armaments: The Nuclear Threat', SIPRI, June 1977.

73. C. J. Johnson et al., 'Plutonium Hazard in Respirable Dust on the Surface of Soil', *Science*, vol 193, 1976, pp. 488–90.

74. S. E. Poet and E. A. Martell, *Health Physics Journal*, vol 23, 1972, p. 537.

75. Ernest J. Sternglass, 'Health Effects of Low-Level Radiation', in J. F. Carroll and P.K. Kelley, *A Nuclear Ireland*, published by the Irish Transport and General Workers' Union, printed by Brindley Dollard, Dublin, Ireland, 1978, pp. 117–44.

76. R. Gillette, 'Nuclear Safety', a four-part series in *Science* magazine, September 1972. Reprints available from: Environmental Action Foundation, 724 Dupont Circle Building, Washington, DC 20036, USA.

77. The entire transcript of senate deliberations on the 1975 extension of the Price–Anderson Act is reproduced in vol 3, no. 1 of *The Advocate*, 160 Chace Avenue, Providence, Rhode Island 02906. The historical perspective on this nuclear subsidy is given in the same issue, by Doug Wilson, former Washington correspondent of the *Providence Journal*. R. I. Senator John Pastore served on the US Joint Atomic Energy Commission as Vice-Chairman, and later Chairman. It was he who 'managed' the Price–Anderson victory.

78. A. M. Stewart, J. Webb and D. Hewitt, 'A Survey of Childhood Malignancies', *British Medical Journal*, 1958, pp. 1495–508.

79. Jim Garrison and Pyare Shivpuri, *The Russian Threat: Its Myths and Realities*, Gateway, London, 1983.

80. *Report from Iron Mountain on the Possibility and Desirability of Peace*, with an introduction by Leonard C. Lewin, MacDonald, London, 1967. This is presented as a fictional report leaked to the public, but it is remarkably true to life. The Ackley Committee was established by Presidential Order in December 1963. It was called the Committee on the Economic Impact of Defence and Disarmament, and headed by Gardner Ackley. The committee report was issued in July 1965.

81. Kathleen Lonsdale, 'The Scientist's Responsibility as a Citizen', conference convened by the Institute of Biology and Atomic Scientists, London, October 1950, in A. Haddon (ed), *Biological Hazards of Atomic Energy*, Oxford University Press, Oxford, 1952. Reprinted with permission.

82. News and Comments, *Science*, 22 May 1981. Also News and Comments, *Science*, 2 October 1981.

83. Issei Nishimori, MD, Professor of Pathology, Nagasaki University

School of Medicine, 12.4 Sakamoto-cho, Nagasaki, Japan.

84. R. Vital, 'Nuclear Power is a Family Affair', *Resurgence*, vol 9, no. 2, May–June 1978; translated by Petra Kelly.

85. Walter Schütze, 'A World of Many Nuclear Powers' in Franklyn Griffiths and John C. Polanyi (eds), *The Dangers of Nuclear War*, University of Toronto Press, 1979, p. 88.

86. Théo Le Diouron et al., *Plogoff – La Révolte, Textes* Editions Le Signor (BP 23, 29115 Le Guilvinec, France), 1980.

87. Geneva Appeal Association, *Yellow Book on the Plutonium Society*, Editions de la Boconnière, Neuchâtel, Switzerland, 1981, p. 54.

88. Professor Ivo Rens, University of Geneva, 'When the Creys–Malville Breeder Reactor Burns Down', *The Ecologist*, no. 4, April–May 1980. Reprinted with permission.

89. E. F. Schumacher, *Small is Beautiful*, Blond and Briggs, London, 1973.

90. Amory Lovins, 'Energy Strategy: The Road Not Taken?', 55 *Foreign Affairs* 65, 1976.

91. Amory Lovins, *Soft Energy Paths: Toward a Durable Peace*, Penguin, Harmondsworth, 1977.

92. Amory Lovins, 'Energy/War: Breaking the Nuclear Link', Friends of the Earth, San Francisco and Harper and Row, New York, 1981.

93. A. Lovins and H. Lovins, 'What is the Energy Problem?', International Project for Soft Energy Paths, 124 Spear Street, San Francisco, California 94105, USA.

94. World Information Service on Energy publishes information on European energy actions on a regular basis. WISE has offices in Amsterdam, Brussels, Copenhagen, Helsinki, Oxford, Stockholm, Tarragona, Tokyo, Verona and Washington. WISE International, Blasiusstraat 90, 1091 CW Amsterdam, The Netherlands. Information is from a grassroots perspective.

95. 'Swedish Vote Supports Use of Nuclear Power', Associated Press, Stockholm, Sweden, *Buffalo Evening News* (state edition), 24 March 1980.

96. This media distortion was first pointed out by Friends of the Earth, London.

97. Corporate Relations, Ontario Hydro, 700 University Avenue, Toronto, Ontario M5G 1X6, publication dated November 1978 and July 1984 on nuclear waste. Reprinted with permission.

98. 'The Management of Canada's Waste', Canadian National Resources and Public Works Publication, E.P. 776, 1978.

99. Correspondence with N. P. Da Sylva, MD, Director of Medical Services, Canadian Medical Association, 26 September 1984.

100. H. Krugman and F. von Hipple, 'Radioactive Wastes: A Comparison of US Military and Civilian Inventories', *Science*, 197, 1977, p. 883.

101. The average exposure of workers at the West Valley plant increased with each year of plant operation: 1968–2.74 rem; 1969–3.81 rem; 1970–6.76 rem; 1971–7.15 rem. The plant closed in 1972.

102. Draft Generic Environmental Impact Statement on Uranium Milling, US Nuclear Regulatory Commission, NUREG–0511, vol 1, Summary, p. 16.

103. 'Long Term Management of Liquid High-Level Radioactive Wastes, Stored at the Western New York Nuclear Service Center, West Valley', US

Department of Energy, DOE/E15–0081, June 1982.

104. M. Resnikoff, 'Nuclear Wastes – The Myths', Sierra Club, July–August 1980.

105. M. Resnikoff, 'The Next Nuclear Gamble: Transportation and Storage of Nuclear Waste', A Council on Economic Priorities Publication, New York, 1983.

106. Robert Jungk, *The New Tyranny*, Warner, New York, 1979, p. 32.

107. Jungk, p. 23.

108. Jungk, p. 38–9.

109. Jungk, p. 40.

110. Bruce Vandervort, 'Battles Over Nuclear Power Loom', *In These Times*, vol 3, no. 21, 1979.

111. E. W. Titterton, *Facing the Atomic Future*, F.W. Cheshire, Melbourne, 1956.

112. E. W. Titterton, 'An Australian Bomb?' *Bulletin*, 87, 19 June 1965, pp. 20–4.

113. E. W. Titterton, 'The Case For', in *Uranium: Energy Source of the Future?*, Australian Institute of International Affairs, Nelson, Melbourne, 1979.

114. E. W. Titterton, 'Benefits We Cannot Forgo', in Alan Manning (ed), *Uranium, a Fair Trial*, Australian Labour Party, Canberra, 1977, p. 22.

115. D. Smith and D. Snow, 'Our Atomic Cover-Up', *The National Times*, 4–10 May, 1980.

116. J. P. Baxter, 'Environmental Pollution and its Control', *Atomic Energy in Australia*, 14, April 1971, pp. 2–7.

117. J. P. Baxter, 'The Nuclear Way is the Safe Way', *Bulletin*, 98, 24 July 1976, pp. 30–3.

118. J. P. Baxter, 'Why Nuclear Energy is Going Ahead', *Search*, 8, November–December 1977.

119. Sir Philip Baxter, 'Energy in Australia: What of the Future?', *Australian Director*, 8, February 1978, pp. 17–20.

120. A. M. Moyal, 'The Australian Atomic Energy Commission: A Case Study on Australian Science and Government', *Search*, 6, September 1975, pp. 365–84.

121. Sir Philip Baxter, *Search*, 6, November–December 1975, p. 458.

122. Sir Philip Baxter, 'We Must Stand by to Repel Boarders', *Herald*, Melbourne, 9 May 1972, p. 4.

123. J. P. Baxter, 'Atomic Energy and Australia', Royal Australian Chemical Institute Proceedings, Supplement 24, November 1957, pp. 67–7.

124. '5000 million face death from disease and war, says atom expert', *Australian*, 27 October 1972, p. 3.

125. Peter King, 'Nuclear Target Australia', *Nation Review*, vol 8, May 1978.

126. W. O'Connell, 'The Big Lie', *Nation Review*, vol 8, September 1978.

127. *The Secret State: Australia's Spy Industry*, Cassell, Melbourne, 1978.

128. B. Machin, *Accomplices To Armageddon*, B.M. Films, GPO Box 145, Nedlands, Western Australia 6009.

129. 'Top Minds Meet to Solve Global Problems', *The Toronto Star*, 13 May 1983, CP–UPC.

130. Donella H. Meadows, Dennis L. Meadows, Jørgen Randers, and William W. Behrens, III, *The Limits to Growth: A Report for the Club of*

Rome's Project on the Predicament of Mankind, (a Potomac Associates book) Universe Books, New York, 1972.
131. Ibid. Foreword, pp. ix and x.
132. Ibid. p. x.
133. Ibid. p. 179.
134. Speech by Chief Seattle, leader of the Suguamish tribe, marking transferal of the territory of Washington to the US Government in 1854. The entire speech is reproduced in: *Active Nonviolence in the United States, The Power of the People*, Peace Press, Inc., 3828 Willat Avenue, Culver City, California 90230.

Part IV: A Time to Bloom: pp. 291–397

1. Barbara Ward, 'Speech for Stockholm' in Maurice F. Strong (ed), *Who Speaks for Earth?*, W. W. Norton, New York, 1973.
2. *Op. cit.*, note above, at p. 31.
3. M. Amaratunga, 'Bio-Gas in an Integrated Farming System', *Agricultural Engineering*, vol 1, no. 1, Peradenuja, Sri Lanka. Reprinted in entirety with permission in *Anticipation*, no. 28, December 1980. Published in Boston for the World Council of Churches (PO Box 38–206, Cambridge, Massachusetts 02138, USA).
4. For more information on natural recycling see Mike Samuels, MD, and Hal Zima Bennett, *Well Body, Well Earth*, Sierra Club Environmental Health Sourcebook, Sierra Club Books, San Francisco, 1983.
5. Thor Heyerdahl, 'How Vulnerable is the Ocean?' in *op. cit.*, note 1, above, at p. 55.
6. Douglas Foster, 'You are What You Eat', *Mother Jones Magazine*, July 1981.
7. Union of Scientists Against Nuclear Power Plants, Matsubara, PO Box 2, Matsubara-shi, Osaka-fu 580, Japan.
8. Doris W. Jackson, 'New Strategies for Protection of the Pacific: Report on the London Dumping Convention', Department of Biology and Environment, University of California at Santa Cruz, Santa Cruz, CA 95064.
9. *French Polynesia, The Nuclear Tests: A Chronology 1767–1981*, available from Greenpeace, New Zealand, Private Bag, Wellesley Street, Auckland 1, New Zealand.
10. Rachel Carson, *Silent Spring*, Houghton Mifflin, Boston, 1962.
11. Elizabeth Kübler-Ross, *On Death and Dying*, Macmillan, New York, 1969.
12. Leonie Caldecott and Stephanie Leland (eds), *Reclaim the Earth: Women Speak Out for Life on Earth*, The Women's Press, London, 1983.
13. Lynne Jones, *Keeping the Peace*, The Women's Press, London, 1983.
14. Reported in *Information*, German Democratic Republic (East Germany) Peace Council, February 1983.
15. Tom Gervasy, *American Military Power in the 1980s and the Origins of the New Cold War: Arsenal of Democracy II*, Grove Press, New York, 1981.
16. Christopher Chant and Ian Hogg (compilers), *The Nuclear War File: Weaponry, Strategy, Flashpoints, the Balance of Power*, Ebury Press, London, 1983.
17. V. I. Lenin, *Collected Works*, vol 30, p. 453: 'Peace will further our

cause infinitely more than war.'

18. See report of the sinking of the *General Belgrano* in 'Falklands: All Out War', *New Statesman*, 24 August 1984.

19. M. K. Gandhi, *An Autobiography: Experiments in Truth*, copyrighted by Navajivan Trust, Ahmedabad, India, and available in numerous collections of his writings globally.

20. Dorothy Day, *The Long Loneliness: An Autobiography* (1st edn) Harper, New York, 1952; (2nd edn), Harper and Row, 1981.

21. Leonard Mosley, *Dulles: A Biography of Eleanor, Allen and John Foster Dulles and Their Family Network*, Dell, New York, 1979.

22. S. Gladstone and P. J. Dolan (eds), *The Effects of Nuclear Weapons* (3rd edn), US Department of Defence and US Department of Energy, 1977.

23. Alexander Solzhenitsyn, *The Gulag Archipelago* Wm. Collins & Sons, London, 1974.

24. Leonie Caldecott, 'At the Foot of the Mountain: The Shibokusa Women of Kita Fuji' in *Keeping the Peace*, The Women's Press, London, 1983.

25. Helen Caldicott, *Missile Envy: The Arms Race and Nuclear War*, William Morrow, New York, 1984, and *Nuclear Madness, What You Can Do*, Autumn Press, Massachusetts, 1978.

26. Norma Love, 'Seabrook Shaking Public Utility Industry', Associated Press, *The Buffalo News*, 29 April 1984.

27. Joanna Rogers Macy, 'How to Deal with Despair', New Age, 32 Station Street, Brookline Village, Massachusetts 02146, 1979.

28. Natalia Gittelson, 'The Fears That Haunt Our Children', *McCalls Magazine*, May 1982.

29. Robert Jay Lifton, *Death in Life*, Basic Books, New York, 1983.

30. William Beardslee and John Mack, 'The Impact of Nuclear Developments on Children and Adolescents', American Psychiatric Association, Task Force Report no. 20, 1982.

31. *The Friendly Classroom for a Small Planet*, Handbook for the Children's Creative Response to Conflict Program, Fellowship of Reconciliation, Box 271, Nyack, NY 10960, USA.

32. Kathleen and James McGinnis, *Parenting for Peace and Justice*, Orbis, New York, 1981.

33. See *Common Security: A Programme for Disarmament. The Report of the Independent Commission on Disarmament and Security Issues*, Pan Books, London, 1983.

34. Patricia M. Nische, 'Star Wars and the State of Our Souls', Whole Earth Paper no. 20, Global Education Associates, 552 Park Avenue, East Orange, NJ 06017, USA.

35. Robert Aldridge, *The Counterforce Syndrome*, Institute for Policy Study, Washington, DC, 1978.

36. Robert Thaxton, 'Directive Fifty-Nine: Carter's New Deterrence, Doctrine Moves Us Closer to the Holocaust', *Progressive*, October 1980, pp. 36–7.

37. Ira Chernus, 'Mythologies of Nuclear War', *Journal of the American Academy of Religion*, L/2, 255–273, 1982.

38. Robert Jay Lifton, *The Broken Connection*, Simon and Schuster, New York, 1979.

39. Personal correspondence with Petra Kelly.

40. 'The Challenge of Peace: God's Promise and Our Response: A Pastoral Letter on War and Peace, 3 May 1983', National Conference of Catholic Bishops.

41. Rosemary Radford Reuther, 'Feminism and Peace', *The Christian Century*, 31 August–7 September 1983.

42. Kenneth A. Dahlberg, 'Plant Germplasm Conservation: Emerging Problems and Issues', *Mazingira*, vol 7, no. 1, 1983 – United Nations Environment Programme.

43. Held in Iowa City, USA; sponsored by the Stanley Foundation; ISSN 0069 8733, p. 20.

44. UPI Report, *The Toronto Star*, 15 April 1983, p. A3.

45. *New York Times*, 1 November 1981, p. 38.

46. The consensus scheme is based on: *The 14 June Civil Disobedience Campaign Handbook*, New York, 1982, prepared at the time of the United Nations Special Session on Disarmament II.

47. Lloyd J. Dumas, 'Human Fallibility and Weapons', *Bulletin of the Atomic Scientist*, vol 36, no. 9, 1980.

48. Jubilee Lecture, the Imperial College of Science and Technology, London, 10 March 1981.

49. P. Hayes and T. Shorrock, 'Dumping Reactors in Asia: The US Export–Import Bank and Nuclear Power in Korea', AMPO Japan–Asia Quarterly Review, 14:2, 1982, pp. 16–23.

50. Günter Altner, Commissioner, Enquete-Kommission des Deutschen Bunderstag, Zukünstige Energiepolitik, Report of 26 June 1980, Bundestucksache 8–4341.

51. Pastoral Constitution on the Church in the Modern World in W. Abbot (ed), *The Documents of Vatican II*, New York, 1966.

52. Pope John Paul II, World Peace Day message, 1 January 1982.

Index